Machine Learning in Biological Sciences

Shyamasree Ghosh • Rathi Dasgupta

Machine Learning in Biological Sciences

Updates and Future Prospects

Springer

Shyamasree Ghosh
School of Biological Sciences
National Institute of Science Education and Research (NISER)
Bhubaneswar, Odisha, India

Rathi Dasgupta
Heathcare
Data Core Systems
Philadelphia, PA, USA

ISBN 978-981-16-8880-5 ISBN 978-981-16-8881-2 (eBook)
https://doi.org/10.1007/978-981-16-8881-2

This Springer imprint is published by the registered company Springer Nature Singapore Pte Ltd.
The registered company address is: 152 Beach Road, #21-01/04 Gateway East, Singapore 189721, Singapore

Dedicated to God
"Even the impossible becomes possible through devotion" Maa Sarada Devi

Shyamasree Ghosh

Dedicated to my mother Mira Dasgupta who made me what I am today

Rathi Dasgupta

Preface

This book has been designed to introduce the reader to the developing science of machine learning together with an overview of diverse and recent applications of Machine Learning (ML) in diverse domains of biological sciences including animal sciences, agriculture, healthcare, and its applications in the study of complex diseases like epilepsy, diabetes, cardiovascular disorders (CVD), neglected Tropical Diseases, (nTD), cancer biology, epigenetics, autoimmune disorders, infectious diseases, drug resistance, domains with a lack of effective diagnostics, bio-robots and applications in biology, study of basic biology, and plant sciences.

Machine learning is finding major applications in process modeling, computer vision, signal processing, speech recognition and language understanding and processing, and life and health sciences.

This book has been designed to comprise eight major sections, with chapters highlighting the application of ML in a specific domain of biology. The book highlights application studies in the healthcare industry including cloud computing infrastructure, medical knowledge in health, and various applications of Microsoft, Google, Facebook, and PyTorch in the biology and health sector.

With an introduction to ML and the various ML methods, the book covers interesting and timely aspects such as applications of ML in genetics, cell biology, study of cell fate in developmental biology research, and study of plant–pathogen interactions and animal behavior. The book discusses computational methods for toxicity prediction of environmental chemicals and drugs with a major focus on research in the field of biology.

It discusses challenges in precision farming, plant disease biology research, and recent developments in ML in plant disease research, understanding plant–pathogen interactions, and is of importance to animal and plant product-based industrial research.

It highlights ML and its application in recent advances in zoonotic infections, in animal health and veterinary sciences, precision farming, and study of animal behavior.

Lastly, it provides an overview of the ethical issues and challenges in the applications of ML to future developments in the immensely complex biological data in the field of ML and its application to health and life sciences. The first ten chapters are contributed by Dr. Rathi Dasgupta, while the rest are contributed by Dr. Shyamasree Ghosh with inputs on computation and modelling from Dr. Dasgupta.

The book has been written with a didactic approach which we believe will help students, postgraduate students, researchers, and faculty pursuing studies in this domain of interdisciplinary sciences, in understanding and appreciating the recent research-based applications of ML and deep learning in life sciences. We believe it will help and draw the interest of researchers of tomorrow in solving problems that remain hitherto unresolved in the domain of biology.

Bhubaneswar, India Shyamasree Ghosh
Philadelphia, PA, USA Rathi Dasgupta

Acknowledgments

We thank all the researchers, technologists, scientists, and observers working in the domain of Biological Sciences, Healthcare, Agriculture, Animal Biology, and Computational Sciences, Machine Learning, and Deep Learning whose untiring efforts are binging hopes to the many who are affected by the many complex problems in biology.

We also dedicate this book to the researchers across the globe who are tirelessly working towards resolving complex problems in biology through the approaches of Machine Learning.

We thank our teachers and mentors who have introduced us to the study of science. A special mention of thanks to Dr. Dhriti Banerjee, Director and Scientist G, Zoological Survey of India (ZSI), Govt of India, who introduced me first to the ever fascinating world of Biology for her constant encouragement, blessings and guidance all through my life. We thank the Director of NISER for his support.

We are deeply indebted to the entire team of Springer Nature, who have helped us immensely throughout the process of publishing.

Bhubaneswar, India — Shyamasree Ghosh
Philadelphia, PA, USA — Rathi Dasgupta

Contents

About the Authors

Shyamasree Ghosh is working as a Scientific Officer at the NISER Bhubaneswar, India. She has worked and published extensively in the domain of glycobiology, sialic acids, immunology, stem cells, nanotechnology, and computational immunology. She has graduated from the prestigious Presidency College Kolkata in 1998 and was awarded the prestigious National Scholarship from the Government of India. She completed her Master's in Biotechnology from Calcutta University in 2000, ranking second in the university. She did her PhD from the Indian Institute of Chemical Biology (IICB) Kolkata, CSIR, India, in glycobiology, sialic acids, immunology, and cancer biology. She did her Post Doctoral Research in Indian Association for the Cultivation of Sciences (IACS), India, on nanotechnology, stem cells, and cancer biology. She has served as faculty and Chair (2005–2009) of department. Her work has been recognized and accepted globally, and she was awarded by different scientific bodies in India. She is a member of different National Science Bodies and is an Editorial Board member in Scientific Societies.

Rathi Dasgupta has been working in the computer science-based industry since the last 30 years and is currently the SVP, Stealth mode Startup and Founder, Etzchaim Fund, Chicago, USA. He did his Bachelor of Science with a major in Physics, St. Xavier's College, University of Calcutta, Calcutta, India; Integrated MTech in radio physics and electronics, Institute of Radio Physics and Electronics; Master of Science (MS) in Nuclear and Particle Physics, University College of Science and Technology; and doctoral research in theoretical physics, Saha Institute of Nuclear Physics, University of Calcutta. He has been adjunct Faculty, in the domain of Mathematics and Computer Science, teaching MS in CSE and EE Class, Alliance University; visiting faculty, Computer Information Science, MBA, in Indian Institute of Management, Bangalore; and Associate Visiting Professor, Computer Science and Mathematics, Xavier Institute of Management and Entrepreneurship. He also has a few US provisional and full patents in Machine Learning.

A Brief Overview of Applications of Machine Learning in Life Sciences

1

Abstract

Biology or the science of life is immensely diverse and immense in its complexity and processes and interaction with the environment.

Despite diverse applications of conventional methods, tools and resources to decipher the complexity in life forms, encompassing their development, evolution, function, biological processes, ecology, behavior, infection and disease biology their remains, intriguing questions in Biology that remain hitherto unanswered. Thus along with the conventional methods, high throughput technology of sequencing and omics based approaches encompassing genomics, proteomics, transcriptomics, metabolomics, metagenomics, microarray technology are being applied to biological sources to understand the complex processes. This is leading to the generation of high dimensional data. The disciplines of Biostatistics and Computational Biology have become relevant and effective in understanding the processes and discover new interactions. Computational Biology uses Machine learning algorithms to analyze the biological problems. Biological data captured through different measurements form the training database to the machine learning algorithm system to learn and create the knowledge. In this chapter we briefly introduce the learners to the domain of machine learning with its application in Biological Sciences.

Keywords

Machine learning · Regression analysis · Python · Animal behavior · Health · Disease · Toxicity

S. Ghosh, R. Dasgupta, *Machine Learning in Biological Sciences*,
https://doi.org/10.1007/978-981-16-8881-2_1

Abbreviations

AI	Artificial intelligence
AIDS	Acquired immune deficiency syndrome
ANN	Artificial neural networks
B paradoxa	*Bacillaria paradoxa*
BBB	Blood–brain barrier
BO	Bayesian optimization
C elegans	*Caenorhabditis elegans*
CNN	Convolutional neural network
CNS	Central nervous system
Ctaurinus	*Connochaetes taurinus*
Evo-Devo	Evolutionary Developmental Biology
HIV	Human immunodeficiency virus
KO	Knock outs
ML	Machine Learning
MRI	Magnetic resonance imaging
NGS	Next Generation sequencing
NLP	Natural language processing
OSA	Obstructive sleep apnea
QS	Quality score
rdt	Recombinant DNA Technology
SNP	Single nucleotide polymorphism

1.1 Introduction

Biology is a Science of life and immensely complex in its diversity and processes. Both theoretical and experimental studies in Biology have been helpful in deciphering the problems in Biology. While mathematical and theoretical biology focuses on application of theory, mathematical models to understand the principles of biological processes, interactions, development, and behavior experiments in biology deal with the conduction of experiments to validate the scientific theories about properties of biological factors, processes, mechanisms, and the interactions of the organisms with the immediate environment.

Experiments generate a large volume of data. Biostatistics is a branch of science which has been doing major analysis of data to study the process, identify trends of processes, molecular behavior, and also anomaly in terms of structure or behavior. Biostatisticians apply regression analysis as a regular technique to predict behavior and model. One of the examples is the detection of breast cancer through anomaly detection in cell imaging through the approaches of magnetic resonance imaging (MRI).

Recently, machine learning has been able to overcome the inaccuracies of approximate modeling by directly learning the entire set of nonlinear interactions from data. In other words the machine creates a set of mapping and correlation between parameters in the test data. This process is very data intensive. We highlight in this chapter the complexity of biological world and the application of machine learning in solving its complexity.

1.2 Complexity of the Biological World and the Origin of Machine Learning: A Journey Through History

Biology is an immensely complex branch of natural Science, encompassing the study of life. Its domain ranges wide diversity from the viruses, bacteria, prions to unicellular and multicellular life forms in both plant and animal system.

Biology finds importance in the study of all the diverse life forms. While the complexity of molecules in cells, their origin and development is interesting, it is also interesting to understand the underlying mechanisms contributing to generation of study development of gradual complexity in cells, involving their interactions with other molecules, signaling and their entire biomolecular biology and chemistry which is tremendously complex and intriguing and finds importance in the study of the biology of the organism and its relation and interaction with other organisms and the environment. It is also interesting to note and study an organism in context of their behavior, their ecology, responses to external stimuli and their interactions with pathogens and other external agents in the environment.

The application of a myriad of conventional methods of biochemistry, molecular biology, biotechnology, cloning, recombinant DNAtechnology (rdt), generation of knock outs (KO), technologies to generate transgenic and technology boom with the developing techniques in the field of sequencing of the genome or genomics, sequencing of the transcriptome or transcriptomics, sequencing and study of the protein or proteomics, next generation sequencing (NGS) approaches, metagenomics, that enables analysis of DNA from environmental samples, in analysis of microorganisms, metabolomics involving the study of chemical processes involving metabolites, which are products of metabolism, in situ and microarray gene expression data, study of complex nature of molecular interactions in biological pathways, studies to understand biological signals are enabling our understanding of structure and function of genes, proteins and other macromolecules including lipids, carbohydrates and glycol conjugated moieties in all life forms including plants, animals and humans.

These technologies have enabled the study of the function of genes, proteins, lipids, and other bio molecules like the carbohydrate moieties and their function in normal individuals and how their defects and dysregulation lead to diseases.

As the complexity in the life forms gradually started unfolding, due to application of advanced tools and technology like NGS, transcriptomics, metagenomics, sequencing of DNA, RNA, proteins, metabolomics, metagenomics to resolve the complexity of life forms, huge data were being generated. To decipher the desired information from the high through put high data generated, and to screen thousands of molecules as potential drug candidates targeting complex diseases and understand their complex interactions with other molecules in less time, space and cost need for improved effective tools were realized. Gradually computational biology and bioinformatics gained importance in their application to problems in life sciences.

To handle and correctly analyze the complex data from life forms, thus it was the need of the hour to develop advanced algorithms and tools. Bioinformatics tools were thus designed to acquire biological data, data mining, genome annotation, and prediction of gene and protein function, gene expression analyses, and development of databases of biological information. This found tremendous importance in the field of biomedical research, development of tools and devices, and predictive applications in the biology of disease. Few major applications of Bioinformatics tools are highlighted.

The complex RNA Sequencing data analyses using the graphical interface of RNASeq GUI enabled handling and analysis of big data sets from RNASeq experiments in a fast, efficient, reproducible, user friendly manner with statistical compliance.

The Cancer Genome Atlas (TCGA) data includes tumor gene expression patterns, with clinical information, in diverse types of Cancer and may be used to study genomic alterations during the development and metastasis of a tumor and finds importance in designing of tailor made therapeutic applications.

Computational tools have enabled in silico screening of drugs and targets and prediction of their permeation through the blood–brain barrier (BBB) that finds importance in the targeting of disorders related to the central nervous system (CNS).

Certain single nucleotide polymorphism (SNP), i.e., altered gene sequence due to alteration of a single nucleotide from that of the wild type or normal individual can lead to disorders and some could even be life threatening. Targeting the SNPs could be a step towards controlling a disease. Bioinformatics applications of SNP_TATA_Comparator find importance in analyzing and extracting unannotated SNPs from the database 1000 genomes revealing 17 novel candidate SNP markers, associated with diseases. Besides with the application of Insilco tools like PredictSNP, it has been possible to predict the occurrence of SNPs from sequences. With the advent of Insilco tools for structure and function of proteins it has been possible to understand the implications of SNP in the structure and function of a protein.

While bioinformatics tools have contributed immensely in the study of biological sciences and biomedical applications, and have enabled a complete paradigm shift of conventional modes of research in both basic and applied biology, recently Machine leaning with its advanced tools and algorithms have enabled the understanding of intricately complex processes and associations in biological processes contributing immensely to Biological Sciences and is perhaps at its zenith in predicting Animal behavior, designing of Biorobots in disease targeting.

Machine Learning closely associated with statistics and Artificial intelligence (AI) are the boon of the computational Sciences and Technology. Most machine learning approaches bear its origin from statistical methods of regression analysis.

In the last three decades considerable progress has been made in the application of Machine Learning to biological data and have been applied in gene prediction, functional annotation, systems biology, microarray data analysis, pathway analysis, encompassing studies involving pattern matching to identify a similar pattern in an unknown data set from a pattern already studied in a known data set.

Machine learning processes bear similarity to predictive modeling and data mining approaches.

Machine learning broadly includes supervised learning, unsupervised learning, and reinforcement learning. While supervised machine learning algorithms require external assistance with use of classifier (and training set is provided to train the machine and evaluated with test sets), unsupervised learning algorithms can run with it. Unsupervised learning algorithm automatically searches the feature of the pattern from the data and groups them into cluster. Reinforcement learning, however, includes decision making on the basis of taken action that that give more positive outcome. We will discuss each method of machine learning and associated algorithms in details in the subsequent chapters.

The most important application of machine learning has been in identification of gene coding regions in the genome by designing of gene prediction tools, understanding structure–function relationships in proteomics analysis by approaches of text mining, application of training sets to identify new or novel drug targets extracting data from articles and secondary databases. Deep learning with applications to transform the data through a number of layers has enabled study of neural networks. Deep learning also has been applied to decipher information from high throughput biological data, understanding process of regulation of biological processes, effect of mutated DNA sequence, analyzing biology of cells, interaction of cells and factors, regulations of biological processes, and function of cells in their microenvironment.

Artificial intelligence and Machine learning have tremendous applications in healthcare, enabling better patient care and management of disease and design of tailor made personalized therapy, with far more accurate predictions of disease diagnosis and therapeutic outcomes over conventional methods together with enhancement of the quality of life and disease free survival or predicting disease relapse in case of deadly diseases like cancer. It is finding applications in efficient performance of clinical trials, drug discovery, effective targets, and disease specific targeted delivery so that side effects of treatment are minimal.

Some of the Machine Learning Tools in Biology include Cell Profiler designed in 2005 that can enable measurement of quantitatively individual features like fluorescent cell number in microscopy field. Deep Variant Tool finds application to mine genome data. Atomwise is yet another tool used in drug discovery.

Different types of deep learning methods including deep neural network (DNN), recurrent neural network (RNN), convolution neural network (CNN), deep autoencoder (DA), deep Boltzmann machine (DBM), deep belief network (DBN), and deep residual network (DRN) that find application in Biology and biomedical application like prediction of biomarkers specific for disease.

DNN finds application in prediction of potential biomarkers from genome and proteome data. CNN tool has been used in development of computational tool DeepCpG to study and predict DNA methylation states in single cells, that alters DNA structure and bears correlation to alteration of DNA function. TensorFlow is yet another a deep learning tool employing DNN with improved graphical visualization and time kinetics. We discuss in detail in subsequent sections.

1.3 Artificial Intelligence, Machine Learning and Deep Learning

Machine learning (ML) enables inferring of models or relationships by learning from data. ML is finding major applications in process modeling, computer vision, signal processing, speech recognition and language understanding, and processing and life and health sciences. Different technologies, deep learning, artificial intelligence, machine learning relate to each other. While AI refers to the simulation of human intelligence processes by a computer: learning, reasoning, and self-correction, ML refers to computational models that effectively perform a specific task without using explicit instructions, but learning from data to make predictions or decisions, and DL is a subset of ML–DL which is a specific type of ML algorithms which are very good at specific tasks. This relation between AI, ML, and DL (Fig. 1.1) could be best

Fig. 1.1 Relation between AI, ML, and DL. Image adapted with modification from Wikipedia (https://en.wikipedia.org/wiki/Matryoshka_doll#/media/File:First_matryoshka_museum_doll_open.jpg), in public domain

visualized by the Russian Matryoshka dolls or stacking wooden dolls of decreasing size placed one inside another.

In this chapter we will approach Machine Learning algorithm from the view point of its application in biological sciences. The main goal is to understand the system and use machine learning as a tool.

1.4 Languages Used in Machine Learning

Machine learning is an algorithm and typically it can be implemented in several different programming languages. One can use C/C++, Java, Java Script, C#, R, Scala, and Python.

Python is the most popular language with 57% codes adopted and written for machine learning algorithm. There are plenty of python libraries such as Teano, Keras, and scikit learn that are available for machine learning, deep learning, artificial intelligence, Natural Language Processing, etc. Numpy is its library that helps to solve many computations and Pybrain is for using machine learning in python. Its syntaxes are very simple and can be easily learnt which makes algorithms easily implemented. It gives direct access to its users for predictive analytics. SciPy is the python library providing scientific and statistical functions.

Java, R, Java Script, and Scala are also widely used for implementing machine learning depending on the type of application and data set. R is used mostly for statistical manipulation. R is highly preferable in bioengineering, bioinformatics, and biomedical statistics. The packages used for machine learning are: RODBC, Gmodels, Class, and TM.

1.5 A Linear Regression Model and Machine Learning Algorithm

The objective of a linear regression model is to find a relationship between one or more features (independent variables) and a continuous target variable (dependent variable). When there is only feature it is called Uni-variate Linear Regression and if there are multiple features, it is called Multiple Linear Regression.

The linear regression model can be represented by the following equation:

$$Y = \theta_0 + \theta_1 x_1 + \ldots + \theta_n x_n$$

where Y is the dependent variable, θ_0 is the bias term, $\theta_1,\ldots,\theta_n$ are the model parameters, and $x_1, x_2,\ldots,x_n$ are the feature values.

where

- θ is the model's parameter vector including the bias term θ_0.
- x is the feature vector with $x_0 = 1$.

In higher dimensions when we have more than one input (x), the line is called a plane or a hyper-plane. The representation, therefore, is the form of the equation and the specific values used for the coefficients. Here is an example.

Training of the model here means to find the parameters so that the model best fits the data.

The line for which the error between the predicted values and the observed values is minimum is called the best fit line or the regression line. These errors are also called as residuals. The residuals can be visualized by the vertical lines from the observed data value to the regression line. To define and measure the error of our model we define the cost function as the sum of the squares of the residuals. The cost function is denoted by

$$J(\theta) = 1/2\text{m Sigma } \{h(x(i)) - y(i)\}2$$

where the hypothesis function $h(x)$ is denoted by

$$h = \theta_0 + \theta_1 x_1 + \ldots + \theta_n x_n$$

and m is the total number of training examples in our data set.

We can try different preparations of one's data using these heuristics and fit the best for the specific problem. These includes the following:

1.5.1 Linear Assumption

Linear regression assumes that the relationship between one's input and output is linear. It does not support anything else. This may be obvious, but it is good to remember when one has a lot of attributes. One may need to transform data to make the relationship linear (e.g. log transform for an exponential relationship).

Relationship between a continuous dependent variable Y (blood pressure) and one or more independent variables X indicating continuous parameters of age, weight, sex, etc. or categorical like social status. A scatter plot enables revealing if the relationship is linear or nonlinear (Fig. 1.2). A linear regression method is used if the relationship is linear while other methods must be used to study nonlinear relationships.

1.5.2 Remove Noise

Linear regression assumes that one's input and output variables are not noisy. Consider using data cleaning operations that let one better expose and clarify the signal in one's data. This is most important for the output variable and one want to remove outliers in the output variable (y) if possible.

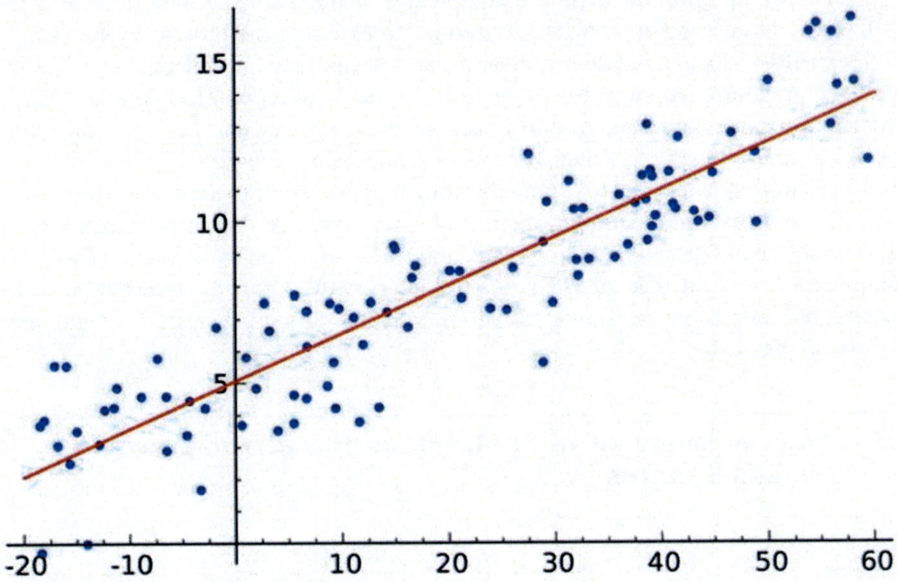

Fig. 1.2 A linear regression model Image adapted from public domain (https://en.wikipedia.org/wiki/Linear_regression#/media/File:Linear_regression.svg)

1.5.3 Remove Collinearity

Linear regression will over-fit one's data when one has highly correlated input variables. Consider calculating pairwise correlations for one's input data and removing the most correlated.

1.5.4 Gaussian Distributions

Linear regression will make more reliable predictions if one's input and output variables have a Gaussian distribution. One may get some benefit using transforms (e.g. log or BoxCox) on variables to make their distribution more Gaussian looking.

1.5.5 Rescale Inputs

Linear regression will often make more reliable predictions if one rescales input variables using standardization or normalization.

In the recent research paper (February, 2020) the authors studied pediatric sleep apnea (Winn et al. 2002) using advanced machine learning algorithm. Pediatric obstructive sleep apnea (OSA) affects an estimated 1–5% of elementary-school aged children and can lead to other detrimental health problems (Laux et al. 2020). Swift diagnosis and treatment are critical to a child's growth and development, but the variability of symptoms and the complexity of the available data make this a challenge. The researchers took a first step in streamlining the process by focusing on inexpensive data from questionnaires and craniofacial measurements. They applied correlation networks, the Mapper algorithm from topological data analysis, and singular value decomposition in a process of exploratory data analysis and then applied a variety of supervised and unsupervised learning techniques from statistics, machine learning, and topology, ranging from support vector machines to Bayesian classifiers and manifold learning. Their results demonstrated that using inexpensive data can still make strong predictions for a binary classification of not having OSA as compared to being at risk for OSA. In particular, random forests, threshold-based density clustering, naive Bayes Bayesian classifiers, and *k*-nearest neighbors performed the best.

1.6 A Brief Overview of Applications of Machine Learning in Life Sciences

In this section we will discuss the recent use of Machine Learning (ML) in different domains of biological sciences including studies in the domain of cell biology, developmental biology, biology of health and disease, detection of disease, plant, and agricultural sciences (Figs. 1.3 and 1.4).

1.6.1 Studies in the Domain of Cell Biology

Image-based cell profiling is a powerful tool to capture the intricacies of cell phenotype. The resolution and the values rapidly stemming from image-based cell profiling has enabled the study of mechanisms of cellular response to disease, drugs, or materials. Together with the explosion of automated and high-throughput microscopy techniques, image-based cell profiling is increasingly relied on as a biological toolkit. Central to image based profiling are software tools devoted to ease the burden of processing a large volume of images by making detection, segmentation, and feature extraction automated. The standard toolbox Cell Profiler already reduces this task by carefully curating the most pertinent and widely-used parameters in cell profiling. Yet selecting an optimum set of cell profiling pipeline parameters (or 'configuration') from the available parameter space is still an onerous task and prone to biases. Optimizing an image processing pipeline is biased against those with limited knowledge in biology, microscopy, or image analysis. The high cognitive load of pipeline optimization can inadvertently lead to decision-making bias that deteriorates the quality of the cell profiling result.

In a recent research in 2020, Lisa Laux and her team in the School of Biomedical Engineering at University of Glasgow (https://en.wikipedia.org/wiki/Linear_regression#/media/File:Linear_regression.svg) presented a new method that integrates user input with machine learning to optimize the configuration of a cell profiling pipeline. A quality score (QS) was obtained from the user and used as a

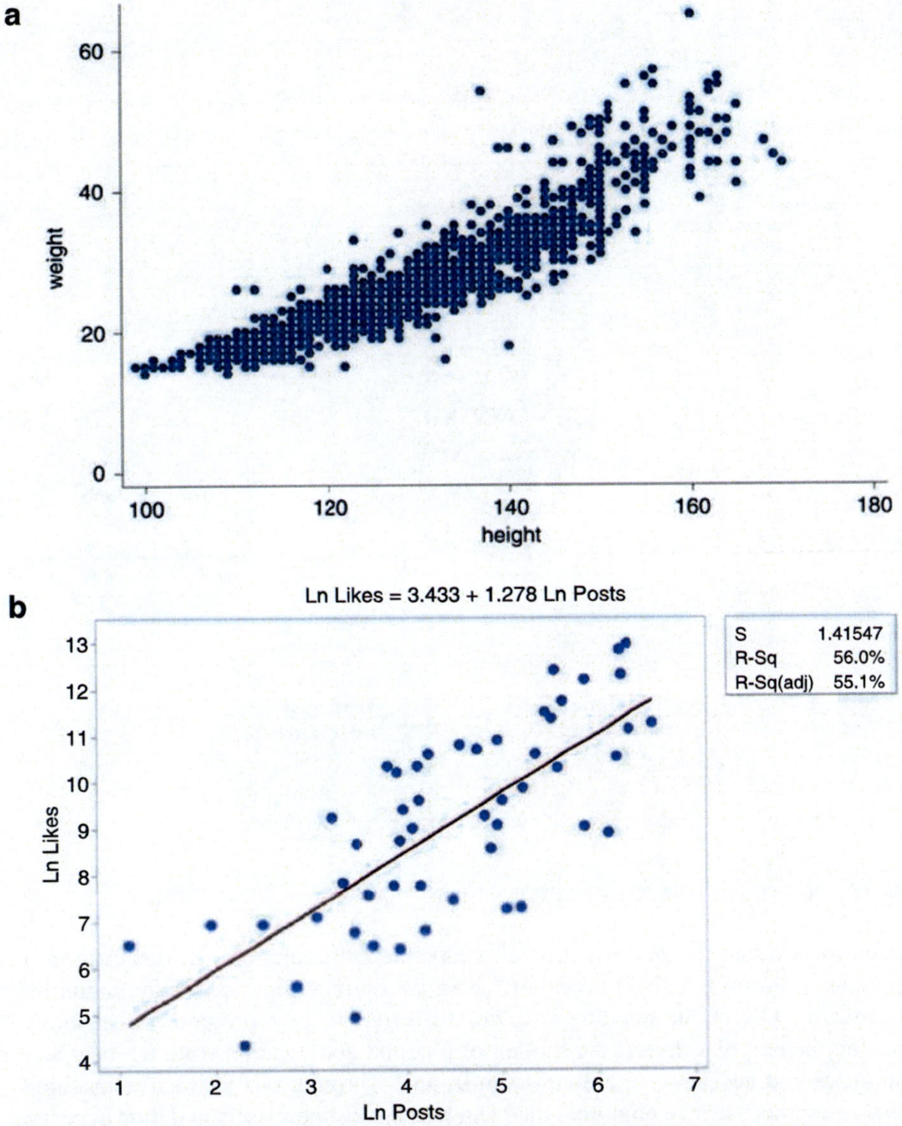

Fig. 1.3 (**a**) A scatter plot of height in cm and weight in kg for 1477 school children with a correlation coefficient (r) of 0.9, showing a linear relationship Image adapted from Nordin et al. (2014) under CCBY license. (**b**) A scatter plot showing an exponential relationship with a regression line. MGE13 Linear Regression (Ln). Scatterplot with Linear Regression Line of Ln Likes (L i) vs Ln Posts (P i) for MGE13 data. Image adapted from Khairuddin and Rao (2017) under CCBY license

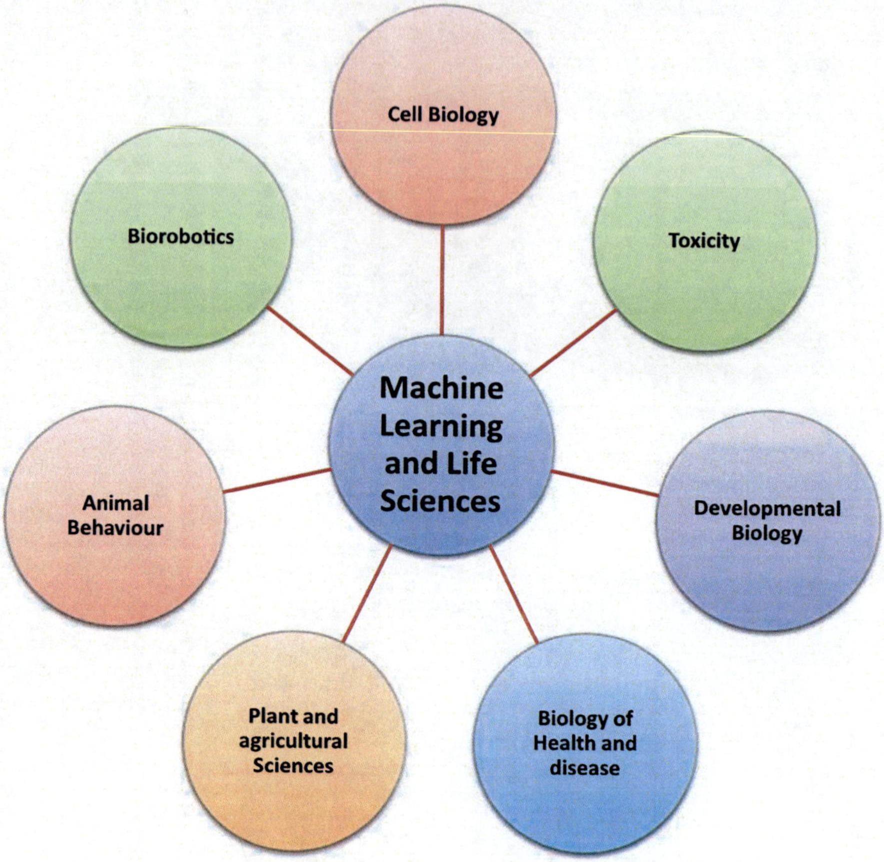

Fig. 1.4 Application of machine learning in the different domains of life sciences

metric to describe the performance of a pipeline configuration. The team used a Bayesian optimization (BO) process to learn the optimal pipeline configuration by maximizing QS in an iterative fashion. Effectively, they presented a machine learning method that diverts the burden of pipeline optimization from the user and automates and accelerates pipeline optimization. Through our interactive machine learning method, reduce cognitive load and bias against new users and thus improve the rapidity and quality of cell profiling.

1.6.2 Application in Study of Toxicity

Computational methods have been designed for toxicity prediction of environmental chemicals and drugs. Machine learning has been profoundly applied on the Tox21 Data. We will discuss the algorithms and architectures and applications of artificial

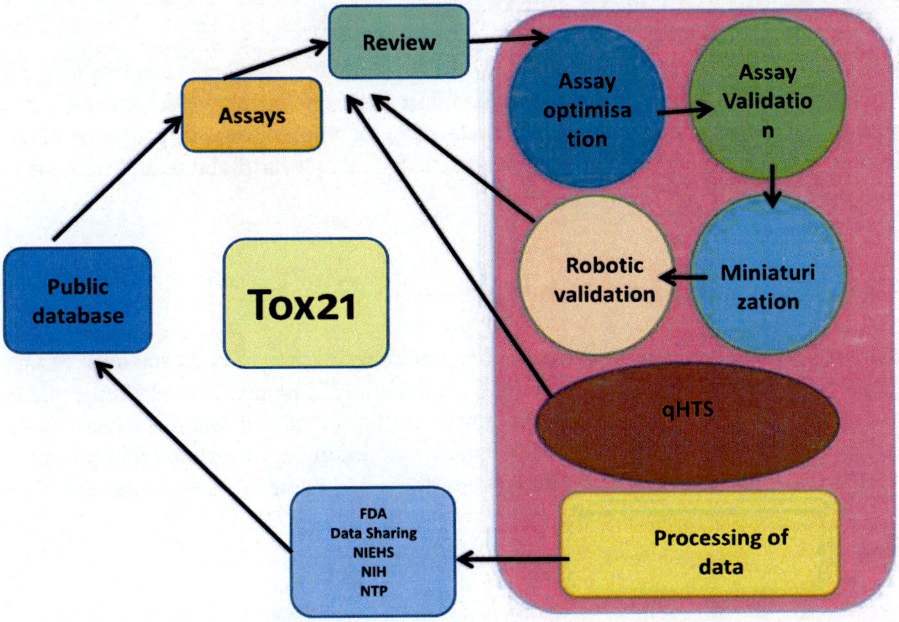

Fig. 1.5 Tox21 screening workflow in drug discovery, qHTS: quantitative high-throughput screening; NCGC: NIH Chemical Genomics Center. Image adapted from (Wu and Wang 2018) under CC BY license

neural networks together with fast computers and massive datasets of higher levels, the DeepTox pipeline applied to the Tox21 Data Challenge is subsequent sections (Fig. 1.5).

1.6.3 Developmental Biology

Evolutionary Developmental Biology (Evo-Devo) entails the science of development of an organism and its modulation by the evolutionary process. Machine learning finds application in the study of Evo-devo biology. Machine Learning has been applied to model the embryo taking into account the cells shape, size, and position. Deep Learning has been applied to the study of embryogenesis of model organism *Caenorhabditis elegans* (*C elegans*), study of colonies of the diatom *Bacillaria paradoxa (B paradoxa), and finds application in the study of* cells, filaments, modeling of movement, currents, or transformational processes. The process of cell division and differentiation are also being modeled as a spatial and discrete temporal process using the application of representational models such as game theory and cellular automata. We discuss in detail in the subsequent chapters.

1.6.4 Biology of Health and Disease

Machine learning algorithm such as K-nearest neighbor has been used in understanding of DNA patterns. Cosine R similarity is used in many identification of similar processes in life sciences. Understanding a complex trend of progression of disease is done multilayer neural networks such as ANN (artificial neural network) (Fig. 1.6).

1.6.5 Plant and Agricultural Sciences

Machine learning has a tremendous application in the domain of plant sciences. Plant diseases are a major threat to the country's economy. Therefore, controlling plant diseases and improving plant production is an important and wanted technology. Convolutional neural network (CNN) models are finding applications in improving plant disease phenotyping. Computer algorithms are being used to determine the

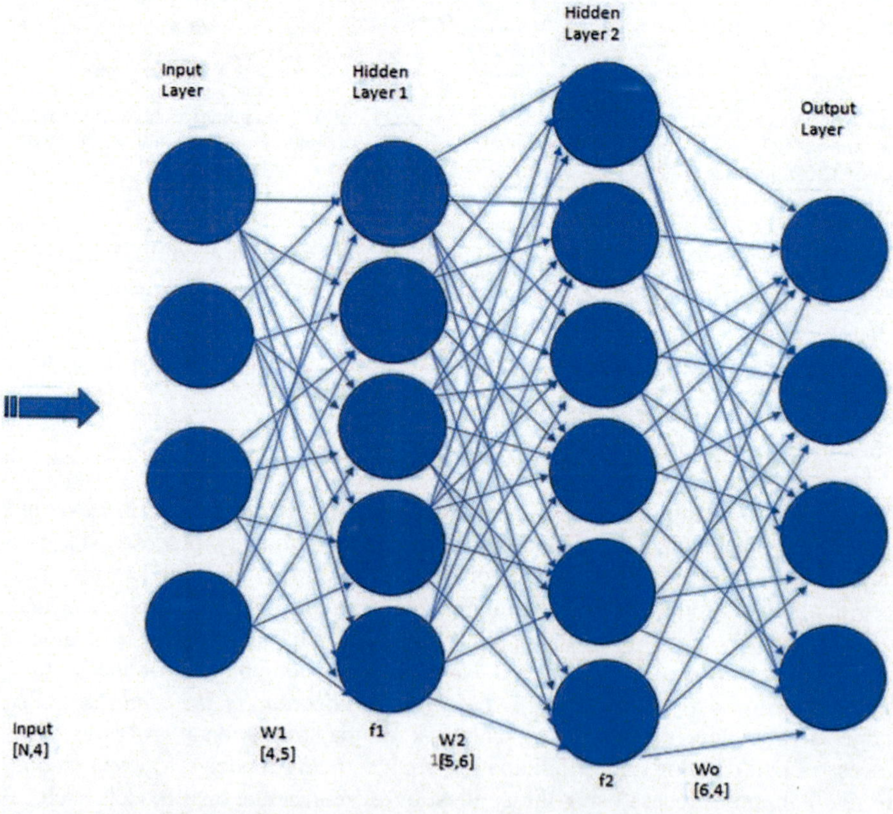

Fig. 1.6 Multi-Layer Neural Network—Input layer, hidden layers (1, 2), and output layer. These layers are mapped using the weight factors W(1), W(2), and W(0)

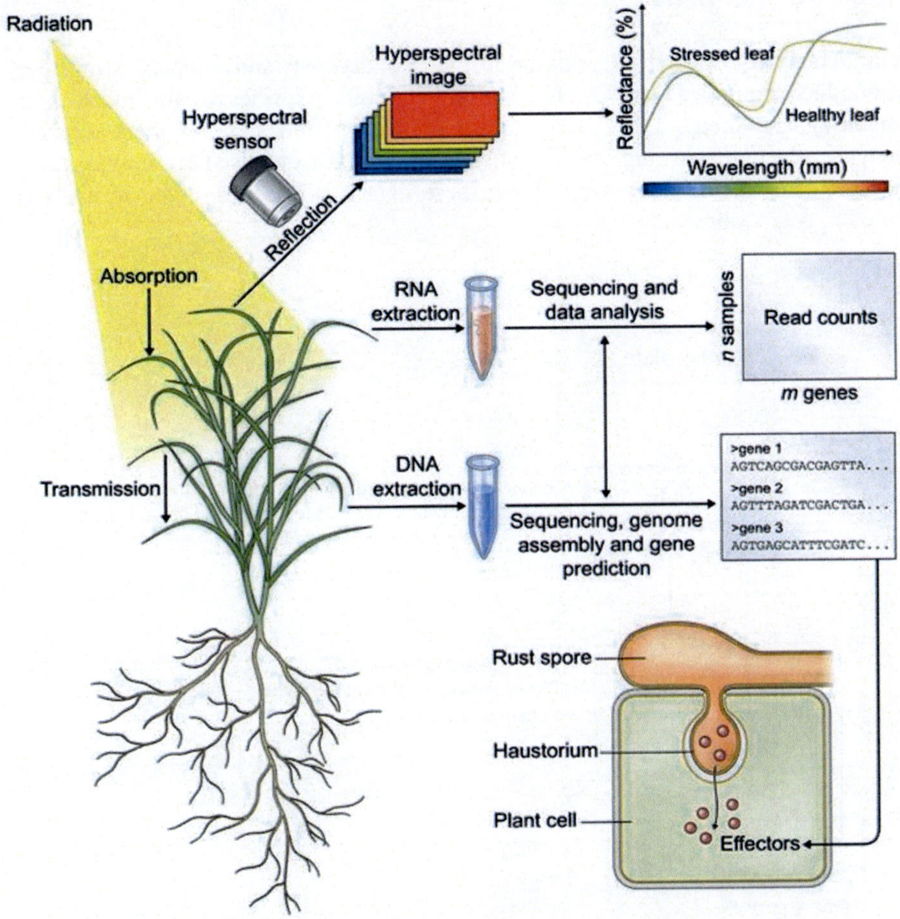

Fig. 1.7 Schematic representation of machine learning (ML) application in plant–pathogen interactions. A rice plant is irradiated with solar rays or artificial light being absorbed, reflected or transmitted, reflection measured by sensors showing differences between stressed and healthy leaves. ML is used to analyze transcriptomics datasets and prediction of gene regulatory networks (grns) and finally genes involved in host–pathogen interactions can be determined. Figure adapted from (Sperschneider 2020) with permission

optimal growing conditions to maximize the concentration of flavored molecules in plants, generation of plant-based biopharmaceuticals, increase or decrease attraction by pollinators, generation of disease resistant varieties. ML has enabled learning to sort through tens of thousands of genes for specialized metabolites in plants (Fig. 1.7).

1.6.6 Animal Behavior

Animal behavior research data has the attributes of diversity and complexity together with added complicacy of its interaction with other species in the niche it is inhabiting and therefore standard statistical methods fell short of their accurate analysis. Machine learning with its unsupervised and supervised learning enables the task of extracting knowledge from these data and has been allied to analyze complex behavioral data (Fig. 1.8).

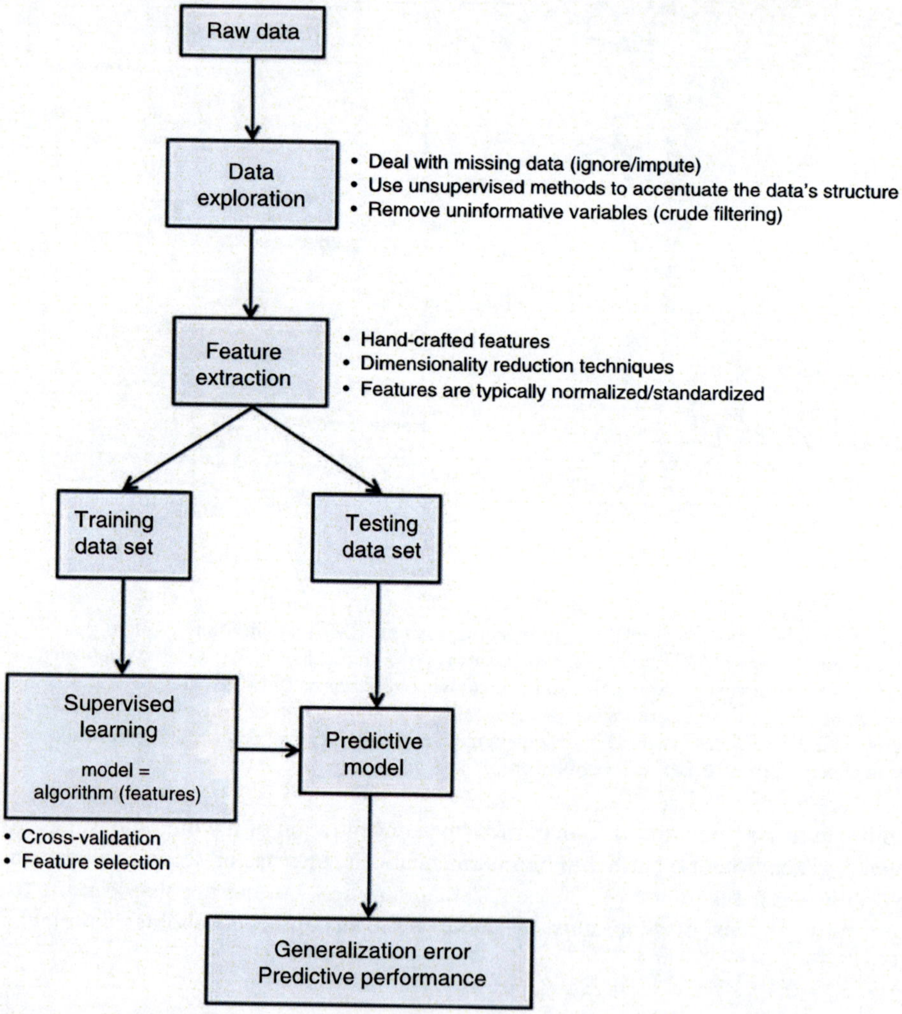

Fig. 1.8 A supervised learning problem workflow. Image and legend adapted with permission from applications of machine learning in animal behavior studies (Valletta et al. 2017)

This is a boon to the study of ecologists. We discuss in subsequent sections the application of machine learning to the study of different animals and their observed behaviors like spectral and morphological features of pheasant. Aerial images are used in training of a classifier to detect the presence and count the number of blue wild beast *Connochaetes taurinus* (*Ctaurinus*) in a population, tracing of movement of animals by attached devices thereby inferring at the ecology of animals, devices. Application of statistical and ML methods by the approaches of supervised and unsupervised learning is enabling our better understanding of animal's ecological and physiology logical behavior and movement. Machine learning and data mining are great tools in the global agriculture to solve pressing problems in animal sciences. Machine learning has been applied with Ebola viral genome to predict its animal reservoirs like bats under suborder Pteropodidforms and could help to prevent future outbreaks in humans. We discuss the viral reservoirs and applications of ML in the detection of such reservoirs.

1.6.7 Nanorobots

Machine learning algorithms have been used in design of microscopically sized nanorobots that are made by the use of bio-sensors, motors, manipulators, power supplies, molecular computers, bio-chips including nanoelectronics. These nanorobots find application in cure of Human immunodeficiency virus (HIV) infection causing AIDS, Cancer, lost tissue restoration, application in disease diagnosis and monitoring disease progression, surgical instrumentation, diagnosis and drug delivery and to monitor neuro-electric signals. We discuss in the subsequent chapters in greater details.

1.7 Discussions

Machine Learning has become an integral part of computational biology. The complexity of biological process and the data made the pattern recognition in such processes difficult and hence machine learning algorithm is playing a big role. Clustering problems are addressed using Support Vector Machine or a K-mean clustering algorithm. Biological image manipulation uses computer vision algorithm such as Convolutional Neural Network. Deep Learning and Reinforcement learning tools are used to understand the problem of a viral infection.

Besides the basic science of Biology, Machine Learning is widely used in image processing (fMRI) in medical (neuroscience) science. Detection of breast cancer uses a combination of Artificial Neural Network and Logistics regression.

IOT based smart agriculture uses Machine Learning algorithm for monitoring the health of farming process and products.

Hence, Machine Learning has become a handy tool for Biologists, Medical Scientists, and Agricultural researchers.

References

Khairuddin MA, Rao A (2017) Significance of likes: analysing passive interactions on Facebook during campaigning. PLoS One 12(6):e0179435. https://doi.org/10.1371/journal.pone.0179435

Laux L, Cutiongco MFA, Gadegaard N, Jensen BS (2020) Interactive machine learning for fast and robust cell profiling. PLoS One 15(9):e0237972. https://doi.org/10.1101/2020.02.20.956268

Nordin P, Poggensee G, Mtweve S, Krantz I (2014) From a weighing scale to a pole: a comparison of two different dosage strategies in mass treatment of schistosomiasis haematobium. Glob Health Action 7:25351. https://doi.org/10.3402/gha.v7.25351

Sperschneider J (2020) Machine learning in plant–pathogen interactions: empowering biological predictions from field scale to genome scale. New Phytol. https://doi.org/10.1111/nph.15771

Valletta JJ, Torney C, Kings M, Thornton A, Madden J (2017) Applications of machine learning in animal behaviour studies. Anim Behav 124:203–220

Winn ET, Vazquez M, Loliencar P, Taipale K, Wang X, Heo G (2002) A Survey of statistical learning techniques as applied to inexpensive pediatric Obstructive Sleep

Wu Y, Wang G (2018) Machine learning based toxicity prediction: from chemical structural description to transcriptome analysis. Int J Mol Sci 19:2358

2 Introduction to Artificial Intelligence (AI) Methods in Biology

Abstract

Artificial intelligence (AI) finds a wide scale application in the science of biology. Modern artificial intelligence will dominate biological data science for its unpreceded learning capabilities to process complex data. Compared to traditional AI techniques (e.g. automated reasoning), machine learning and deep learning are the core to enable machines with intelligence. A deep learning machine has much more complicate learning topologies, which may change dynamically for the sake of learning, besides at least the same complicate-level learning mechanism as traditional machine learning models such as support vector machines. We highlight in this chapter the overview of the applications of AI and machine learning in medicine, infection, and diagnostics.

Keywords

Artificial intelligence · Genomics · Biology

2.1 Introduction

Artificial Intelligence is transforming the way Biological Systems are analysed. Artificial Neural Network (ANN), Convolutional Neural Network (CNN), and Reinforcement Learning (RL) methods are used for the identification and detection of genomic structure and anomaly therein.

Deep learning is good at discovering latent complex relationships among data and handling big data well. More importantly, deep learning merges feature extraction and prediction (e.g. classification) in a single learning procedure and makes feature extraction more adaptive and compatible with prediction. The scRNA-seq data, SNP, interactome, or even clinical data usually need very different but complicated feature extraction procedures before entering downstream learning. Deep learning

S. Ghosh, R. Dasgupta, *Machine Learning in Biological Sciences*,
https://doi.org/10.1007/978-981-16-8881-2_2

prepares itself for a good candidate to process those data and starts to make good progress in handling next generation sequencing data.

Artificial intelligence is expected to dominate biological data science in the near future with the maturity of AI itself. Most state-of-the-art AI techniques are originated from computer vision, image recognition, or natural language processing. It is not easy to migrate the existing AI techniques to the biological data science field though some efforts are being made. The special characteristics of enormous data generated in biological data science calls for building their own AI theory, methods and systems.

- **Artificial Intelligence (AI)** involves the simulation of human intelligence by machines. Here the machines are designed and programmed that think like humans and mimic their actions and learning and solving problem by the human mind.
  The term AI owes its origin as early as in 1956 and involves the application of the science, technology, and engineering in the design of intelligent machines. It has been applied to diverse fields in biology including medicine as robotics, diagnosis, statistics, study of biology of human and omics based approaches of study.
- **Machine Learning (ML)** is a branch of AI where machines are trained to solve problem at their own designed by learning new rules by repeated attempts and feedback. This phenomenon also reveals similarity of AI with that of biological Intelligence. It works on learning of the rules. One could train a system with several images of different objects until the system learn to identify the respective object.
- **Artificial Neural Networks (ANNs)** are directly inspired by neural networks in biological brains. Some similarities have been observed in between artificial and biological neurons in terms of their function in integration of the signal originating from other different neurons, transforming it with a non-linear function and output it to further neurons.
  However, differences are manifested by that biological neurons function by signaling mechanisms, while the artificial ones pass over only numerical values.
- **Deep Learning Algorithms and AI**
  Predictive tools can be created from deep-learning algorithms (DLA), a branch of promising artificial intelligence (AI), based on internal patterns using extraction of raw features from an extremely large, annotated data set, high-throughput imaging like robotic microscopic images for study of brain cells, imagers for study of cells or genomesadvance drug discovery and then the trained algorithms can analyse other data, from other different sources and finds importance in tacklingand visualisation of complex.

  DLA is based on neural networks, since its inception in 1940s, in which neuron-like nodes mimic and analyze human brain (Figs. 2.1 and 2.2).

This enables a trained algorithm to analyze a new image and its correct identification. In 2005, a software package, Cell Profiler was released by the Broad Institute of MIT and Harvard in Cambridge, Massachusetts, to help biologists to

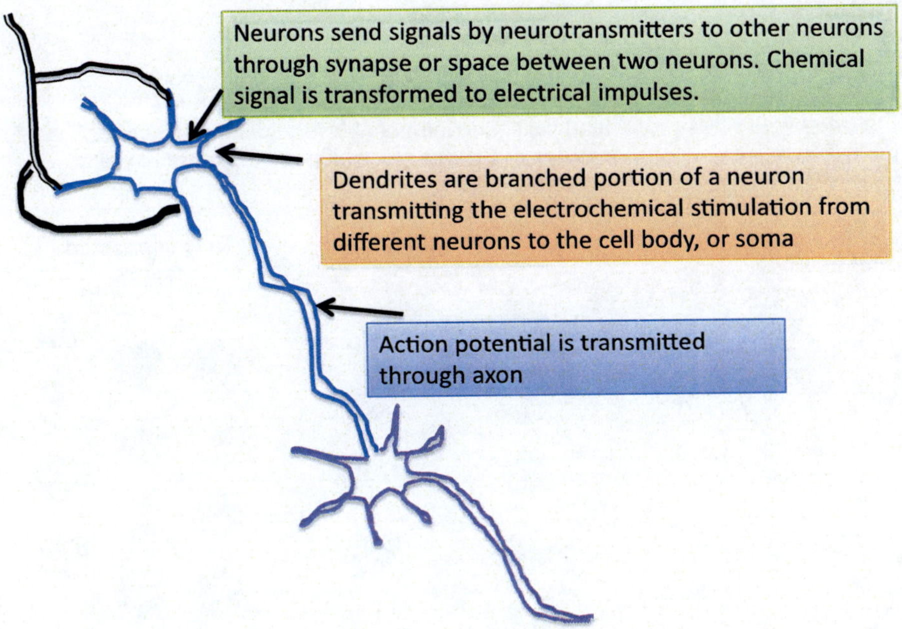

Fig. 2.1 Neuron and transmittance of impulses

quantitatively measure the number cells visualized under fluorescent microscopy. Combining visual features of DNA staining, organelle texture and the quality of empty space in cell, can produce fresh insights.

Verily Life Sciences and Google Life Sciences have developed a deep-learning tool, Deep Variant to identify single-nucleotide polymorphisms (snps) over and above conventional tools that translates genomic information into image-like representations.

We discuss in this chapter the different applications of artificial intelligence in biology briefly.

2.2 Artificial Intelligence in Medicine

Robots sketched initially by Da Vinci are now being designed with the capability of being used in surgery, for different complex surgeries including urologic and gynecologic procedures.

The physical and the virtual are the two branches of the artificial intelligence in terms of its application in medical science.

While the virtual branch comprises of informatics and deep learning based approaches of information management in health management systems, like maintenance of electronic or e-health records, diagnosis by physicians, the physical branch

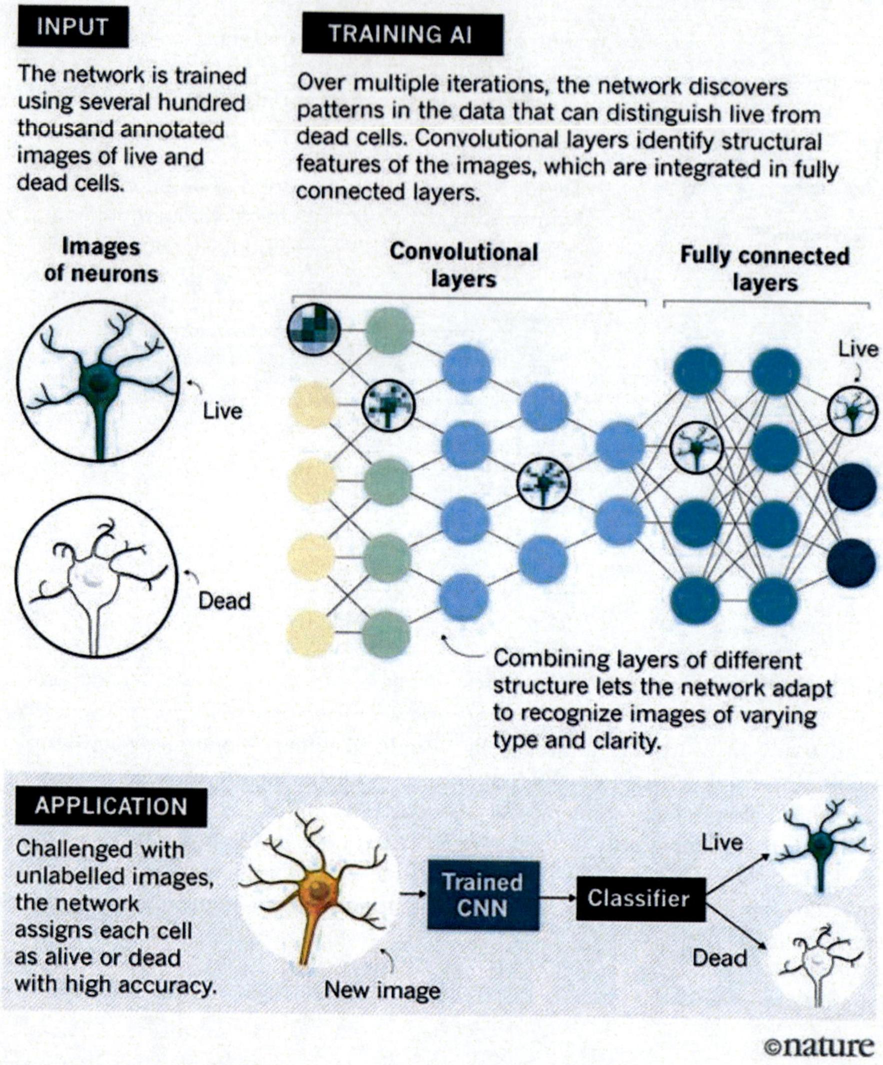

Fig. 2.2 Deep learning algorithm used to understand dead or live neurons. Image adapted with permission from Nature **554**, 555–557 (2018).doi: https://doi.org/10.1038/d41586-018-02174-z

involves the application of robots in surgery. Targeted nanorobots are finding application in drug delivery (Hamet and Tremblay 2017).

AI has enabled the integration of knowledge from diverse domains of classification and diagnosis of disease such as cancer, risk stratification, cancer drugs, factors concerning resistance to drugs, minimal residual disease (MRD), disease prognosis studies from next generation sequencing (NGS) of patients, study of genetic variants, the pathophysiology of cancer biology, and cancer precision drug discovery

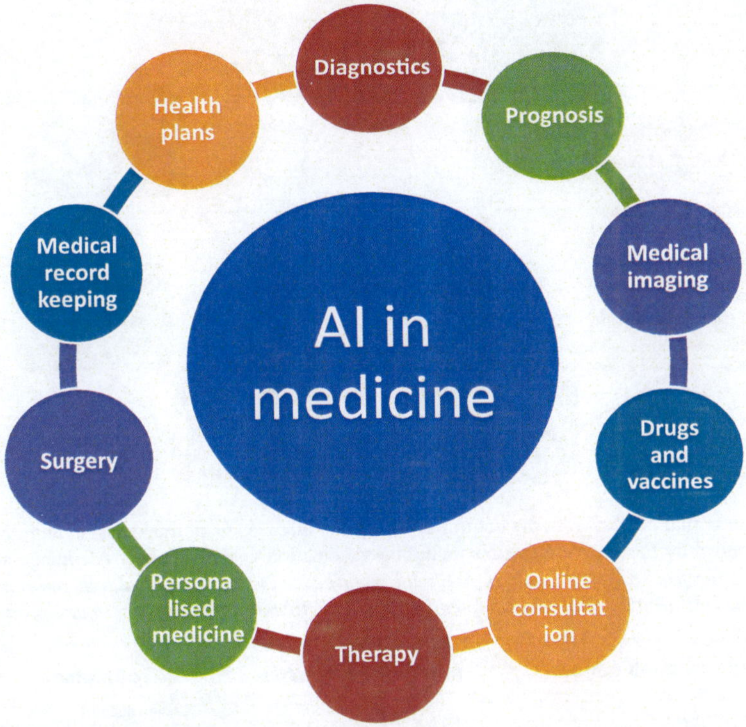

Fig. 2.3 Applications of AI in medicine

(Nagarajan et al. 2019; Aneja et al. 2019; Kantarjian and Yu 2015). It has been applied in the study of neuro-oncology, brain tumors, epigenetics, pathology, with the application of natural language processing (NLP, 3–4), combination therapy in different diseases including HIV, hypertension, infectious diseases and cancer (Tsigelny 2019). DLA enabled feature extraction methods have found application in the study of both basic and clinical cancer research (Shimizu and Nakayama 2020) (Fig. 2.3).

2.3 AI in the Disease Biology of Infection

AI algorithms have found importance in the study of pathophysiology of infection and label free imaging of pathogens (Yakimovich 2019), the complex biology of host–pathogen interaction (Fisch et al. 2019). DLA has been applied to the study and identification of antimicrobial peptides (Veltri et al. 2018) (Fig. 2.4).

Different features of AI including Bayesian network, decision tree, random forest, support vector machine, *k*-nearest joining, unsupervised learning artificial neural network (ANN) have been applied in study of different aspects of infectious diseases

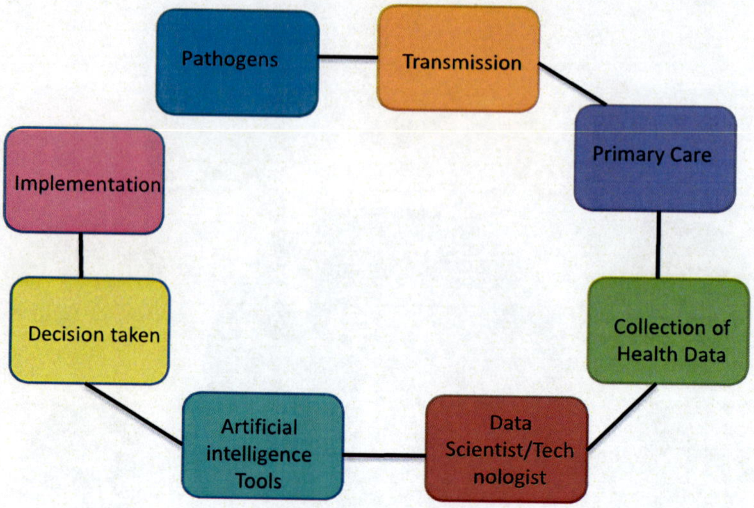

Fig. 2.4 Essential principles in the control of infectious diseases from transmission and improvement of control by preventive measures including vaccination and hygiene by AI. Image adapted *with permission from AgrebiS, Larbi A, Artificial Intelligence in Precision Health, From concept to applications 2020, Pages 415–438 Chapter 18—Use of artificial intelligence in infectious diseases*

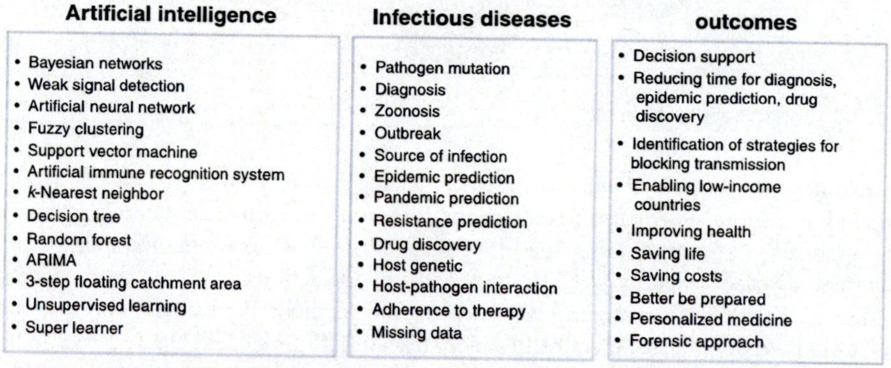

Fig. 2.5 AI and machine learning tools, their use and potential outcomes in the field of infectious diseases and the aspects they target. Image adapted with permission from Agrebi S, Larbi A, Artificial Intelligence in Precision Health, From concept to applications 2020, Pages 415–438 Chapter 18—Use of artificial intelligence in infectious diseases

with the outcomes of support in decision making, early diagnosis, strategies to block transmission, improve health, save life, a step towards personal medicine (Fig. 2.5).

COVID-19 is a highly contagious disease caused by the coronavirus SAR-CoV2. It has affected more than four million people across the globe claiming lives and hence has been declared as pandemic on March 11, 2020 by the World Health Organization (WHO). The immunopathology of the disease reveals affected lungs alveoli with features of lymphocytopenia, cytokine storm, and dis-balanced innate

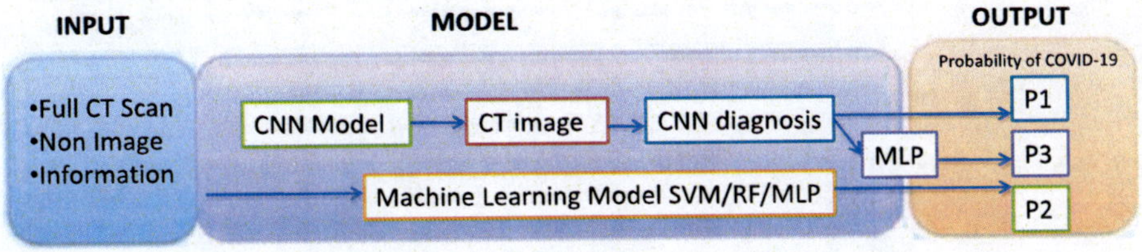

Fig. 2.6 Model of AI used in the detection of COVID-19 using models based on a chest CT scan, clinical information, and combination of them. The top few abnormal CT images per patient were put into the diagnosis CNN to predict the likelihood of COVID-19 positivity (P1). Demographic and clinical data were put into a machine learning model to classify COVID-19 positivity (P2). Features generated by the diagnosis CNN model and the non-imaging clinical information machine learning model were integrated by an MLP network to generate the final output of the joint model (P3). Image and legend adapted with permission from *Mei, X., Lee, H., Diao, K.* et al. *Artificial intelligence–enabled rapid diagnosis of patients with COVID-19. Nat Med (2020).* https://doi.org/10.1038/s41591-020-0931-3

and adaptive immune responses. AI algorithm encompassing CT scan images and clinical information have been designed to identify early diagnose and therefore screen COVID-19 patients, to combat the global disease outbreak accurately similar to that by diagnosis by senior chest radiologist. This method offers advantages as it does not require input by radiologist and physical tests (Fig. 2.6).

2.4 AI in Genomics and Diagnostics

Application of DLA and graphics processing units (GPUs) finds application in training in medical AI. AI based approaches have enabled better and effective image based diagnosis approaches. Algorithms have been instrumental in processing large and complex genomic data involving processes of variant calling, genome annotation, variant classification, and phenotype to genotype relations (Dias and Torkamani 2019). It has found application in the study of cancer genomics (Greatbatch et al. 2019) (Fig. 2.7).

2.5 Discussion

Artificial Intelligence is playing a major role in Computational Biology and Bioinformatics. Understanding complex patterns in biological process is an area where ANN based solutions help. Detection of disease through analysis of fMRI images using computer vision algorithm such as CNN and RNN is a regular practice. Detection of autism through analysis of brain images are also done through AI algorithms.

The Critical Assessment of protein Structure Prediction (CASP) is a nationwide experiment that takes place biannually, which centered around analyzing the best

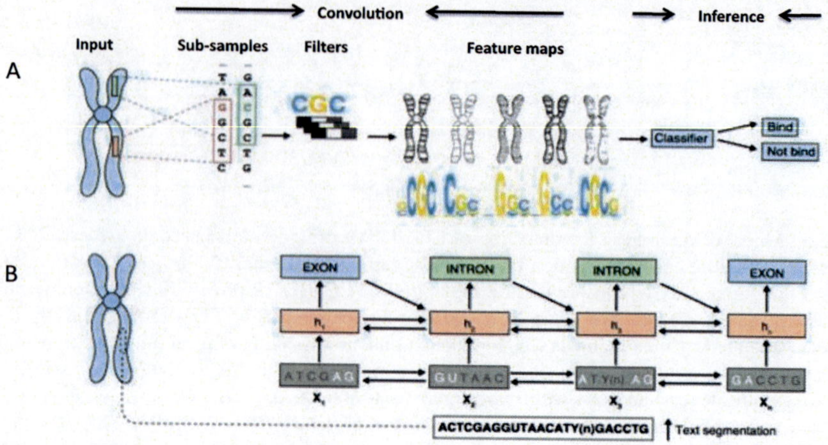

Fig. 2.7 Application of AI in clinical diagnostics (**a**). CNN break the input image or DNA sequence leading to a product with conserved motifs used to map back to the original image used to train a classifier to predict a given label. Filtering of base pairs enables the motif identification. (**b**) Recurrent neural networks (RNNs) to understand connections between input (x) and interconnected hidden states (h). *Image and legend adapted from open access article Dias, R., Torkamani, A. Artificial intelligence in clinical and genomic diagnostics. Genome Med* ***11***, *70 (2019).* https://doi.org/10.1186/s13073-019-0689-8 *under CCBY Licence*

current systems for predicting protein tertiary structures. AI methods are being used to predict exact nature of protein folding phenomena.

Clinical trials and also drug delivery process use clustering algorithm and ANN to find patient groups suitable for a particular strain of a drug.

There are more and more research projects in this area to understand the very complex nature of biological systems.

References

Aneja S, Chang E, Omuro A (2019) Applications of artificial intelligence in neuro-oncology. Curr Opin Neurol 32(6):850–856. https://doi.org/10.1097/WCO.0000000000000761

Dias R, Torkamani A (2019) Artificial intelligence in clinical and genomic diagnostics. Genome Med 11(1):70

Fisch D, Yakimovich A, Clough B et al (2019) Defining host-pathogen interactions employing an artificial intelligence workflow. elife 8:e40560

Greatbatch O, Garrett A, Snape K (2019) The impact of artificial intelligence on the current and future practice of clinical cancer genomics. Genet Res (Camb) 101:e9. https://doi.org/10.1017/S0016672319000089

Hamet P, Tremblay J (2017) Artificial intelligence in medicine. Metabolism 69S:S36–S40

Kantarjian H, Yu PP (2015) Artificial intelligence, big data, and cancer. JAMA Oncol 1(5):573–574. https://doi.org/10.1001/jamaoncol.2015.1203

Nagarajan N, Yapp EKY, Le NQK, Kamaraj B, Al-Subaie AM, Yeh HY (2019) Application of computational biology and artificial intelligence technologies in cancer precision drug discovery. Biomed Res Int 2019:8427042

Shimizu H, Nakayama KI (2020) Artificial intelligence in oncology. Cancer Sci 111(5):1452–1460. https://doi.org/10.1111/cas.14377

Tsigelny IF (2019) Artificial intelligence in drug combination therapy. Brief Bioinform 20(4):1434–1448. https://doi.org/10.1093/bib/bby004

Veltri D, Kamath U, Shehu A (2018) Deep learning improves antimicrobial peptide recognition. Bioinformatics 34(16):2740–2747. https://doi.org/10.1093/bioinformatics/bty179

Yakimovich A (2019) mSphere of influence: the rise of artificial intelligence in infection biology. mSphere 4(3):e00315–e00319. https://doi.org/10.1128/mSphere.00315-19

Machine Learning Methods

3

Abstract

The goal of science is to understand a particular phenomenon, its root case process, analyze, create a model and predict future behavior and outcome. Hence it requires capturing data about a phenomenon, try and find a pattern of behavior of the parameters affecting it and create a model. Statistics is the science that does the data capture and data pre-processing, analysis, and creates a pattern. There is a very fuzzy line between the Machine Learning algorithm and the Statistical model. The goal of the Machine Learning algorithm is to find that pattern. There are various methods of machine learning, starting with the simplistic method of understanding the association and determine the regression pattern, specifically the linear regression. In this chapter, we will address different machine learning algorithms and their use in solving the challenges in Life Sciences.

Keywords

Supervised learning · Unsupervised learning · Association · Classification · Linear regression · Logistics regression · KNN · LDA · SVM · Neural networks · Reinforcement learning

3.1 Introduction

Machine Learning is a statistical learning methodology. In Machine Learning one's objective is to find a pattern of behavior between an input and an output or find the pattern of behavior in a large pool of data. The goal is to find the association between attributes of a system and classify the data based on certain behavior.

In this chapter, we will discuss the methods of Association, Classification for Supervised Learning, and also methods in Unsupervised Learning.

S. Ghosh, R. Dasgupta, *Machine Learning in Biological Sciences*,
https://doi.org/10.1007/978-981-16-8881-2_3

3.2 Associations

Association is a key concept in understanding a set of data. In particular, how X changes with respect to the change of Y or how the change of X, Y affects Z. This can loosely be called finding the correlation between different variables.

An association rule learning method extracts the rules or the mapping between the variables in the data set. In healthcare, association rules can be used to identify a disease-finding relationship between several disparate symptoms. Besides diagnosis, it can also be used to find probable toxicity of drug and effects on a set of population.

One can evaluate the following three metrics to identify the level of associations:

1. Support.
2. Confidence.
3. Lift.

3.2.1 Support

The absolute frequency of occurrence is measured by the support value. An association rule holds with support sup in a data set S if the sup percentage of transactions contain XUY where X, Y Belong to the set S. The support is then measured by:

$$\text{Sup (Support)} = P(\text{XUY}) = \text{count (XUY)}/\text{total transaction count},$$

where is the P is the probability.

3.2.2 Confidence

An association rule holds in dataset S with confidence if the confidence percentage of transactions that contain X also contain Y. This estimates how likely Y is to occur when X occurs.

$$\text{Conf} = P(Y|X) = \text{count (XUY)}/\text{count }(X)$$

3.2.3 Lift

Lift determines how likely item Y occurs, given X occurs, while accommodating Y's popularity.

$$\text{Lift} = \text{supp (XUY)}/\text{supp }(X) * \text{supp }(Y)$$

3.3 Classification

It is a supervised learning algorithm where it learns from input training data and then classifies new observations. An example of classification will be to identify whether it is a black and white photograph or it is a color photograph. The classification of Medical diagnostics through Image classification (such as ultrasonography image of the prostate gland in the categories of BPH or Malignant) is another popular method.

There are many other methods of classification based on several techniques such as linear (logistic regression and naïve Bayes), nearest neighbor, support vector machines, trees (decision, boosted), random forest, and neural network.

3.3.1 Unsupervised Learning

When the data is not labeled, one has to use the method of unsupervised learning. So, it means that we have a set of input data without output. So one needs an algorithm to draw inferences from datasets consisting of input data without labeled responses.

Semi-supervised learning occurs when some output labels are supplied. An example of such a situation in healthcare is a predictive model to detect diabetic retinopathy from a patient's eye scan and partially labeled data.

The most commonly used unsupervised learning methods are clustering and dimensionality reduction. High dimensionality data such as microarray gene expression is analyzed, finding groups or clusters (Fig. 3.1).

3.3.2 Clustering

Clustering refers to the process of determination of relationships within the data. It is done through grouping data in terms of similarities, such as Euclidean distance

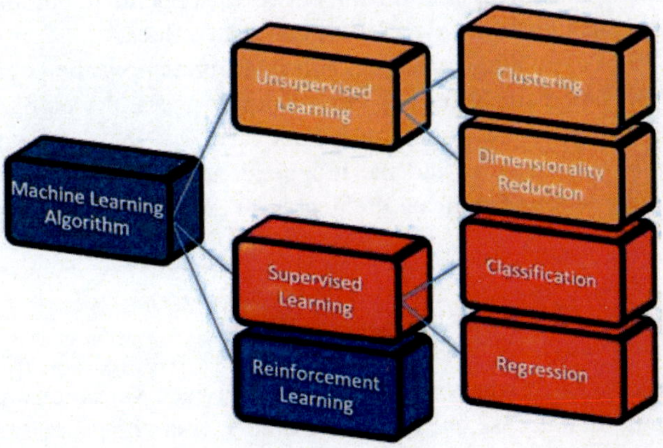

Fig. 3.1 Unsupervised learning

between points or cosine similarities, etc. In Public health studies, one typically uses clustering for cases, such as,

- Grouping patients of similar profiles together for monitoring.
- Defining treatment group bases on medication and associated condition.

In genetic analysis, clustering is done through genetic expression. In a microarray experiment, different tissue samples can be grouped or clustered based on genetic expression or genetic structure.

Clustering can be done using either the method of hierarchical clustering, in a tree like structure or *k*-mean clustering. Any clustering method is iterative and brings the best cluster in the process.

When one analyses biological data, it may be required to do both feature and sample clustering. Let's suppose when analyzes the population in north America in terms of finding the cluster of dermatological disorder, one may create cluster in terms of types of disorder. One may also find cluster of sample or population having similar disorders.

3.3.3 Supervised Learning

It is a scenario where output is defined for a set of input and hence the data is labeled. This is a data set that has been classified to indicate a learning algorithm (Fig. 3.2).

3.3.4 Linear Regression

Linear regression is a supervised learning technique. It is typically used in predicting, forecasting, and finding relationships between quantitative data. An example of the use of multivariate regression technic is finding a relationship between the humidity, water pH, and spread of viral infection. It is essentially finding the linear relationship between a dependent variable and a set of independent variables given the data labels and data values.

For example, if one wants to investigate how the phenomena of patients having high blood pressure, is related to their age and weight. One needs to analyze the mapping between input and output variables. In this example, the dependent variable is blood pressure, and the independent variables are age and weight. Correlation coefficient will indicate the degree of association between the variables.

3.3.5 Spearman's Coefficient

In case of a monotone relationship between variables one uses Spearman's coefficient. In this case, the output variable y, either rises or sinks with the input variable x.

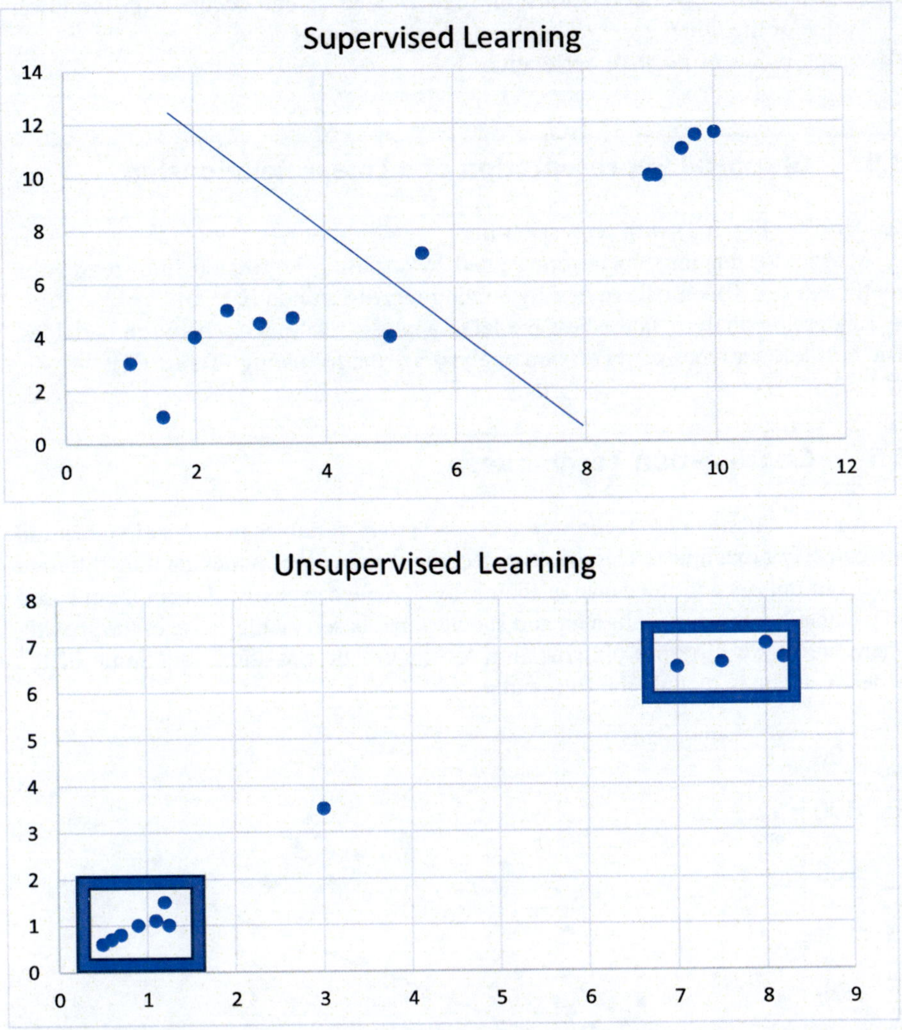

Fig. 3.2 Supervised learning and unsupervised learning

3.4 Pearson's Correlation Coefficient

A linear relationship, between variables, is indicated by the Pearson's coefficient. The value of the correlation coefficient (r) provides the extent of association of the variables. Let's study different values of r:

I. $r = \pm 1$ indicates linear and monotonous.
II. $r = 0$ implies no linear or monotone relationship.

III. $r < 0$ shows inverse relationship, higher value of one occurs with the lower value of the other.
IV. $r > 0$ indicates positive correlation.

3.5 Graphical Representation of a Linear Relationship

Scatter plot (Fig. 3.3) with regression line.

A negative relationship is represented by a falling regression line (regression coefficient $b < 0$), a positive one by a rising regression line ($b > 0$).

Regression analysis is the best model to analysis relationship between variables. The best-known types of regression analysis are the following (Table 3.1):

3.6 Classification Techniques

Typically, a classification technique is used to group variables in terms of values and attributes. For example, a classification technique is used for understanding different groups of tumors and the cause of their growth. One can easily classify them based on whether benign and malignant and subclustered based on the cause of the growth. There are many different classification techniques or classifiers, but some of the widely used ones include the following:

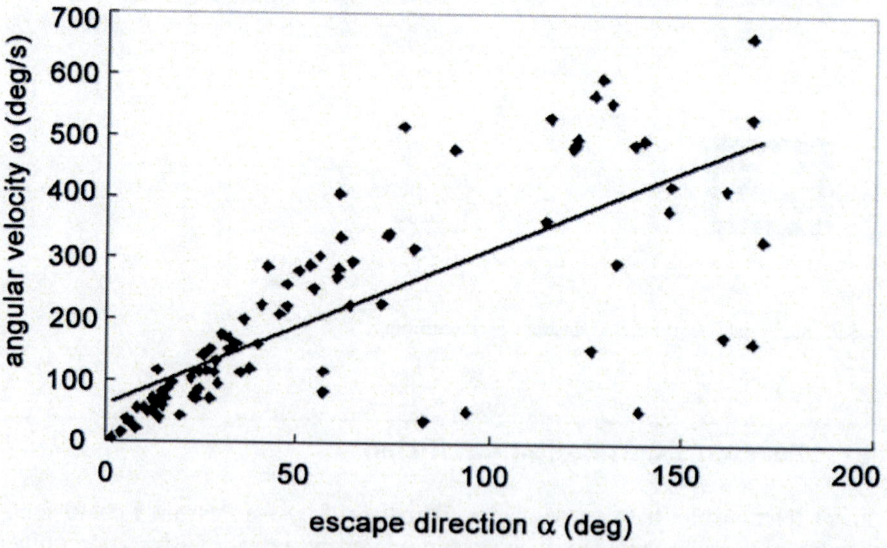

Fig. 3.3 Motor planning modulates sensory-motor control of collision avoidance behavior in the bullfrog, Rana catesbeiana (Image adapted from Nakagawa *H*, Nishida *Y*. Motor planning modulates sensory-motor control of collision avoidance behavior in the bullfrog, Rana catesbeiana. *Biol Open*. 2012;1(11):1094–1101. doi:https://doi.org/10.1242/bio.20121693, under CC license)

Table 3.1 Different regression analysis

Model	Problem	Independent variable	Dependent Variable
Linear regression	Description of a linear relationship	Continuous(weight, blood pressure)	Continuous and/or categorical
Logistic regression	Prediction of the probability of belonging to groups (outcome: yes/no)	Dichotomous(success of treatment: yes/no)	
Proportional hazard regression (cox regression)	Modeling of survival data	Survival time (time from diagnosis to event)	
Poisson regression	Modeling of counting processes	Counting data: Whole numbers representing events in temporal sequence (e.g., the number of times a woman gave birth over a certain period of time)	

(a) Logistic regression.
(b) Linear discriminant analysis.
(c) K-nearest neighbors.
(d) Trees.
(e) Neural Networks.
(f) Support Vector Machines.

3.7 Logistic Regression

When a linear regression fails to provide a feasible output, logistics regression is the technique that is applied—the basic difference between the two lies in the underlying cost function. Logistic Regression provides a binary output.

When there is an analysis of whether smoking affects productivity among men, one uses a functional output (which is a probability function) of 0 or 1. If one uses linear regression, that can generate a negative probability which is meaningless in terms of interpretation.

In machine learning, a cost function is used to represent how far the mathematical model is from the real data. The goal is to minimize the cost function by adjusting parameters within the model (Fig. 3.4).

Here is a linear regression fit example. On the *X* side, we have a parameter that affects the spreading of infection, and the *Y*-axis is the log of the number of death due to that infection. One can use a quadratic cost function.

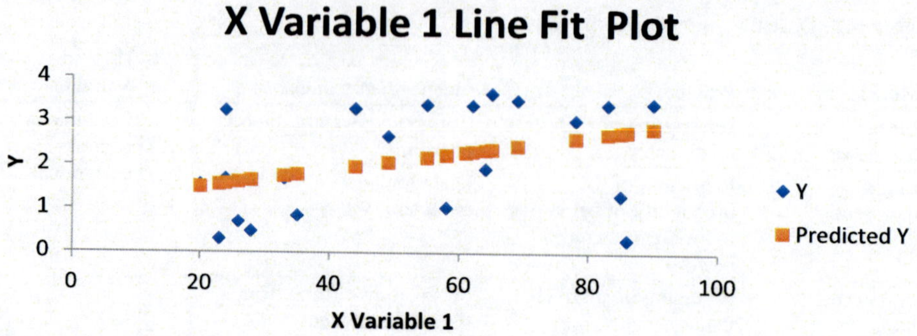

Fig. 3.4 Linear regression fit

$$J(\theta) = \frac{1}{2N}\sum_{n=1}^{N}\left(h_{\theta}\left(x^{n)}\right) - y^{n}\right)^{2}$$

where $J(\theta)$ is the cost function and h_θ is the function linking x and y. This is just the sum of the squares of vertical distances between the points and the straight line.

But the linear regression model can really lead to unfeasible results for situations where there are either binary decisions or probability calculations. Logistic regression works in those circumstances. In such a classification problem, we will fit a function that ranges between 0 and 1 to match the class labels and avoid negative values. Mostly one fits an *S*-shaped sigmoidal function.

$$h_{(\theta)}(x) = \frac{1}{1 + e^{-\theta_0 - \theta_1 x}}$$

The quadratic cost function will not work; instead, the following cost function will work:

$$J(\theta) = -\frac{1}{N}\sum_{n=1}^{N}\left(y^{(n)}\ln\left(h_\theta\left(x^{(n)}\right)\right) + \left(1 - y^{(n)}\right)\ln\left(1 - h_\theta\left(x^{(n)}\right)\right)\right)$$

3.7.1 Linear Discriminant Analysis (LDA)

LDA is a dimensional reduction technique, similar to Principal Component Analysis. The complexity of a learning algorithm lies in the number of input parameters, essentially the dimension d of the input parameter space and also the size of the data sample *N*. So, for optimizing memory and computation, it is required to reduce the dimensionality d. Essentially a complex matrix manipulation consumes so much of memory and becomes analytically challenging that one needs to find an optimal set of parameters to adequately describe a system.

There are two different methods of dimensionality reduction—feature selection and feature extraction.

In feature selection, one is interested in finding k of d dimension that can give us the maximum information and then discard the $(d\text{-}k)$ dimensions.

In the feature extraction method, one finds the k dimension, which is a combination of the original d dimension providing maximum information. The best-known feature extraction methods are LDA and PCA, both based on the linear projection method.

Let us address a two-dimensional LDA situation. Let there be two classes of data C_1 and C_2;we want to find a direction as defined by a vector **A**, such that when the data are projected onto **A**, the data from the two classes are as well separated as possible. Let z be the projection as defined.

$z = \boldsymbol{A}^T\boldsymbol{X}$ is the projection of **X** on **A** and thus is a dimensionality reduction from $d = 1$.

$\boldsymbol{m}_1$ and m_1 are the means of samples from C_1 before and after the projection, respectively. Similarly, we can define $\boldsymbol{m}_2$ and m_2 as means for samples from C_2. We define $\boldsymbol{m}_1 \in \mathbb{R}^d$ and $m_1 \in \mathbb{R}$, $X = \{x^t, r^t\}$ such that $r^t = 1$ if $x^t \in C_1$ and $r^t = 0$ if $x^t \in C_2$. So we find $m_1 = \boldsymbol{A}^T\boldsymbol{m}_1$, $m_2 = \boldsymbol{A}^T\boldsymbol{m}_2$.

The scatter of samples from C_1 and C_2 after projection are

$$s_1^2 = \sum_t \left(\boldsymbol{A}^T\boldsymbol{X}^T - m_1\right)^2 r^t$$

$$s_2^2 = \sum_t \left(\boldsymbol{A}^T\boldsymbol{X}^T - m_2\right)^2 \left(1 - r^t\right)$$

After projection, for the two classes to be well separated, the best result will be that the two means are well separated (as far as possible) and the classes be scattered in as small as a region possible.

Let's take an example of analyzing the protein profile of 1000 proteins for a sample of 100 healthy and 100 cancer patients using an antibody-based microarray. This is a case of high dimension matrix with 1000 variables and 200 samples. Of course we have to use a dimensional reduction technique and find the linear discriminant function that will classify the patients as healthy or diseased (https://raybiotech.com/learning-center/linear-discriminant-analysis-model/).

3.7.2 *K*-Nearest Neighbors (KNN)

In this method, one measures the distances between any new data point and the nearest K of the already classified data. KNN is used for classifying samples according to their numerical features and performing a regression of those values

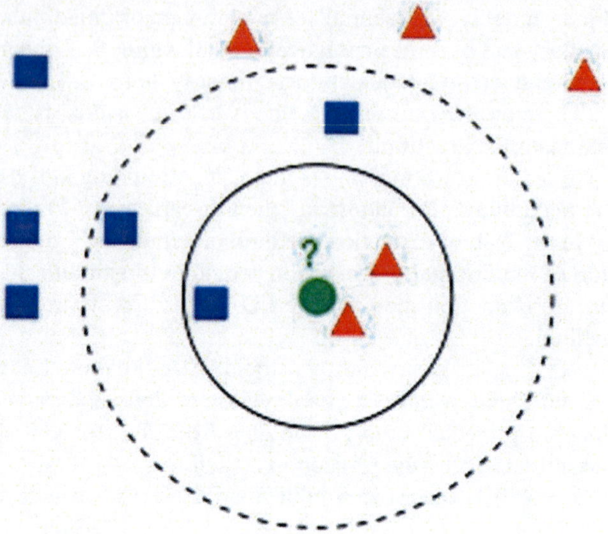

Fig. 3.5 Example of k-NN classification. The test sample (green dot) should be classified either to blue squares or to red triangles. If $k = 3$ (solid line circle), it is assigned to the red triangles because there are 2 triangles and only 1 square inside the inner circle. If $k = 5$ (dashed line circle), it is assigned to the blue squares (3 squares vs. 2 triangles inside the outer circle).Source: https://en.wikipedia.org/wiki/K-nearest_neighbors_algorithmCC BY-SA 3.0

based on the features of the sample. KNN is widely used in DNA/RNA matching (Fig. 3.5).

The nearest neighbor class of estimators adapts the amount of smoothing to the local density of the data. The degree of smoothing is controlled by the number of nearest neighbors $k \ll N$, the sample size. Euclidian distance between two points is the measure that controls the definition of the neighborhood, and it is measured as an absolute value, such as between two points a, b as $\|a - b\|$ and for each x whereas $d_1(x) \leq d_2(x) \leq \ldots \leq d_N(x)$, Whereas d is the distance in ascending order, $d_1(x)$ is the distance to the nearest sample, $d_2(x)$ is the distance to the next nearest sample.

Varying the size of k can change the neighbor radius and hence can change the result of the inference. In the above figure, if we vary k, then the identity of the unknown can change from a red triangle to a blue square and then again probably to a red triangle at a different k value.

3.7.3 Neural Networks (Fig. 3.6)

A neural network is a machine learning algorithm that mimics neuron transmission in brain signal communication. Signals are received by one set of neurons and processed (mathematically manipulated) and then passed on to the next level before they get a decision based on a regression or classification. This method is used for complex modeling relationships between inputs and outputs (Fig. 3.7).

Machine learning using complex biological data. High-throughput data generation techniques for different biological aspects are shown (left) (Fig. 3.8).

Interpretation of machine learning model:

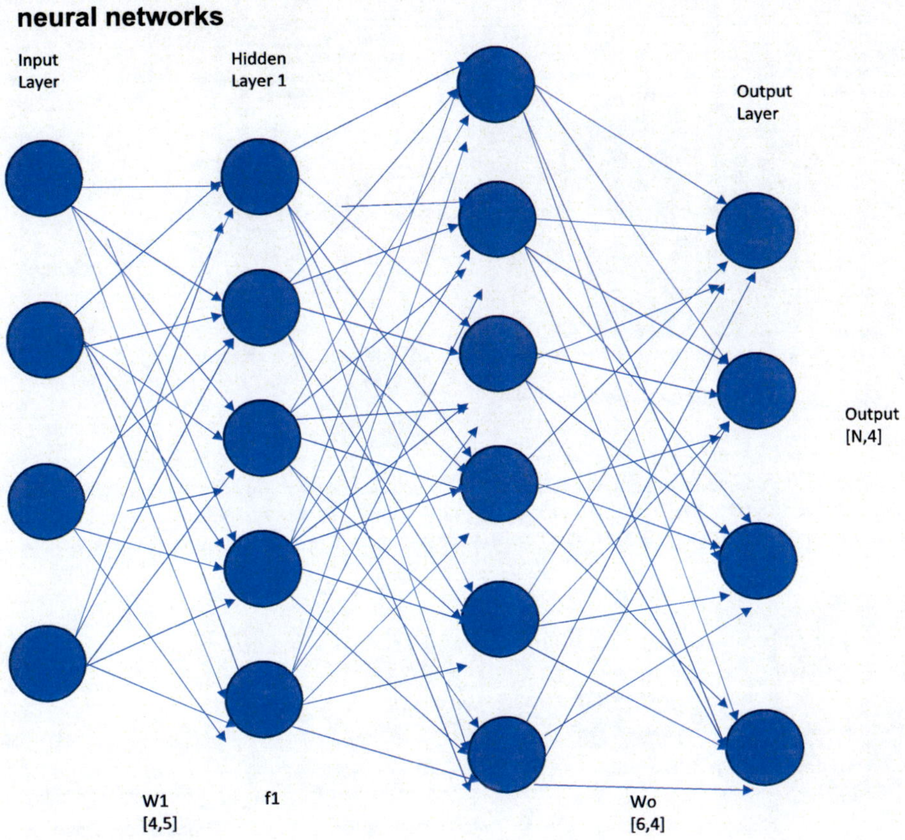

Fig. 3.6 Neural network

Model information may be interpreted directly or be further processed for better understanding. The above model uses a support vector machine or a similar classification model, but an improved method and far more accurate results are expected out of a neural network model either ANN (Artificial Neural Network) or CNN (Convolutional Neural Network) based on the type of problems one encounters.

CNN is also used to address problems such as assessing the risk of breast cancer among women of particular demography and pre-existing health condition.

Mejtabaetal (https://www.ncbi.nlm.nih.gov/pmc/articles/PMC6031801/) studied radiological images and other physical data of 655 women (196 malignant and 459 benign) retrieved from 11,850 screened cases referred to Shahid Motahhari breast clinic affiliated to Shiraz University of Medical Sciences between 2004 and 2012. They used this data on an Artificial Neural Network (ANN) algorithm base model to do the analysis. The nodes in different layers are connected with arcs (connection weights). ANNs "learn" the relationships between input variables and

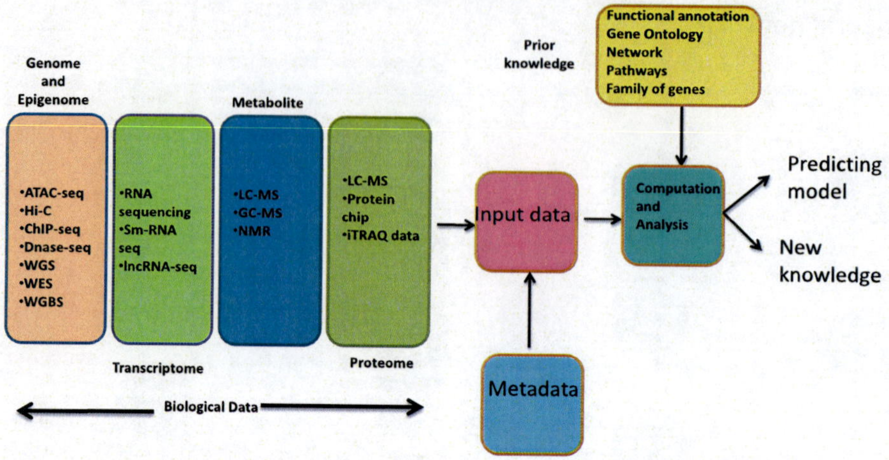

Fig. 3.7 Image adapted from Xu, C., Jackson, S.A. Machine learning and complex biological data. *Genome Biol* 20, 76 (2019). https://doi.org/10.1186/s13059-019-1689-0. https://genomebiology.biomedcentral.com/articles/10.1186/s13059-019-1689-0 under Creative Commons CC BY license

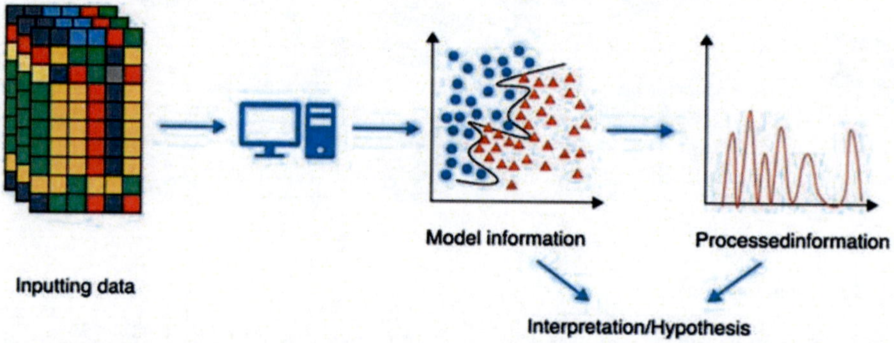

Fig. 3.8 Interpretation of the machine learning model. Model information may be interpreted directly or be further processed for better understanding. Image adapted from Xu, C., Jackson, S.A. Machine learning and complex biological data. *Genome Biol* 20, 76 (2019). https://doi.org/10.1186/s13059-019-1689-0. https://genomebiology.biomedcentral.com/articles/10.1186/s13059-019-1689-0 under Creative Commons CC BY license

the effects they have on the outcome by increasing or decreasing the values of these connection weights on the basis of known cases.

The procedure of estimating the weights is called learning. A back-propagation learning algorithm is used in this study. There are several supervised training algorithms for ANN. The back propagation algorithm is based on the error-correction learning rule. The error back-propagation process contains a forward and a backward pass through the different layers of the network (Chauvin and Rumelhart 1995).

There are the different values and layers used in the ANN model. The group used multi-level networks and used different parameters/conditions to filter the right set of weight factors to determine the probability score to identify the risk of breast cancer.

To train and test the ANN model, tenfold cross-validation was used. In tenfold cross-validation, the data was divided into ten subsets that were approximately equal in size. This process was repeated for ten iterations until all subsets were used once for testing. The process is the following:

- The dataset is firstly divided into ten subsets randomly, nine subsets for training and one for testing.

ANN classification method is evaluated through sensitivity, specificity, and accuracy tests. Sensitivity, specificity, and accuracy are commonly used statistics, using True Positive (TP), True Negative (TN), False Negative (FN), and False Positive (FP) terms.

Variables	Levels
Breast density	Predominantly fatty, scattered fibro glandular, heterogeneously dense, extremely dense
Age groups, *y*	<45, 45–50, 51–54, 55–60, 61–64, ≥ 65
Family history of Brest cancer	Yes, No
Mass shape	Circumscribed, ill-defined, micro-lobulated, spiculate, not present
Mass margins	Oval, round, lobular, irregular, not present
Mass density	Fat, low, equal, high, not present
Mass size	None, small (<3 cm), large (≥ 3 cm)
Lymph node	Present, not present
Asymmetric density	Present, not present
Skin thickening	Present, not present
Skin retraction	Present, not present
Nipple retraction	Present, not present
Skin lesion	Present, not present
Micro-calcifications	Present, not present
History of breast surgery	Yes, No
Menopause	Pre-menopause, Post-menopause
Marital status	Single, Married
History of contraceptive use	Yes, No
Age at first pregnancy	<30y, $> = 30$y
Parity	0,1,2,3, $> = 4$
Age at menarche	<12y, $> = 12$y

3.7.4 Support Vector Machine

Support Vector Machine (SVM) is an algorithm to separate two classes of data. It can typically identify benign breast cancer from a malignant one based on the image and vector interpretations. It is a non-probabilistic binary linear classifier used with both classification and regression problems.

SVM is often used with Natural Language Processing methods to analyze text for topic modeling and sentiment analysis. The algorithm does classification based on the determination of a hyperplane separating the two classes based on support vectors. Support vectors are data points which if altered, change the configuration/ position of the hyperplane.

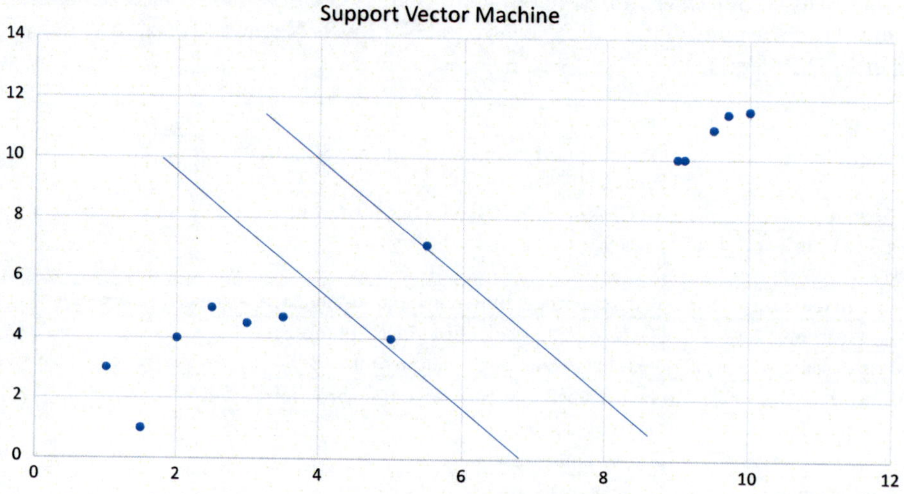

3.7.5 Trees

It is one of the most used supervised learning methods. Trees typically follow a similar structure as that of binary tree, in this case n branches and nodes tree. This ML methods are built by recursively splitting a training sample, using different features from a dataset at each node that splits the data most effectively. The splitting is based on learning simple decision rules inferred from the training data.

This is also called a decision tree method as the branches and nodes provide all possible outcomes, one chooses the most suitable one. Tree methods are normally used for classification and regression problems.

3.7.6 Reinforcement Learning Method

Reinforcement learning method is based on a reward method. The learning algorithm chooses a path where the reward is the most optimized one. The final learning algorithm is decided on trial-and-error basis. Let's try a model situation.

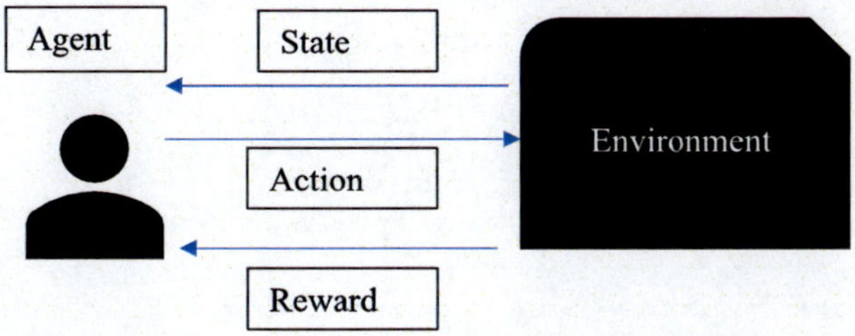

In a Reinforcement learning algorithm, the agent (learning algorithm) reacts with the environment (sample and variables). The environment communicates a state parameter, and the agent reacts with an action and receives a reward. In this repeated process the agent learns the most appropriate algorithm.

So what really happens is that the set of input frames go through a network, neural network and generate an output frame. The challenge here is that the output frame is not known unlike supervised learning situation. The network here is called the policy network and the agent learns the policy through a game, getting the rewards. Reinforcement learning is the basis of modern deep learning algorithm.

3.8 Discussion

In this chapter, we have discussed preliminary concepts on machine learning methods. Among these methods, the most relevant ones such as LDA, Neural Networks, and SVM are discussed in detail. We have also tried to understand the applications of these ideas in Biology and the detection of disease in healthcare. We have also briefly introduced Reinforcement Learning here. Many biological processes can be explained and understood using this award—incentive-based model.

Reference

Chauvin Y, Rumelhart DE (eds) (1995) Backpropagation: theory, architectures, and applications. Lawrence Erlbaum Associates, Inc.

4 Introduction to the Machine Learning Models

Abstract

Understanding the nature of mapping between variables is the goal of machine learning algorithms. The simplest model is to find the linear regression between an independent and dependent variable. It can be far more complex with nonlinear regression. In case of a situation where we try to find the binary answer, we use logistics regression.

Clustering is a typical model in genomic analysis, and one can use many clustering models such as *K*-means clustering or Kernel Methods.

Complex problems involve probabilistic models such as artificial neural networks.

Keywords

The machine learning algorithm · Learning algorithm · Regression algorithm · Regularized regression methods · Decision trees · *K*-means clustering · Naive Bayes Classification (NBC)

4.1 Introduction

Machine learning algorithms assume the structure of data and the nature of the desired output. Hence it is required to choose an appropriate model for the probability of fitment and ensure faster training. There are machine learning models which restrict the existence of outliers. A decision tree-based model classifies a node into two based on a threshold, and hence data outliers are ignored in the model.

In choosing the suitable model, one needs to keep an eye on the following factors:

- Selecting appropriate variables.
- Introduce redundancy.

S. Ghosh, R. Dasgupta, *Machine Learning in Biological Sciences*,
https://doi.org/10.1007/978-981-16-8881-2_4

- Avoid Overfitting.
- Output productivity.

Let us take a simple example of detecting benign and malignant tumors depending on the weight and radius of the tumor. One can capture MRI images of the brain. It is possible to identify the existence of a tumor and measure the radius of the tumor through typical image manipulation tools. From our existing knowledge of the subject, we assume that the threshold radius of the tumor is 5 mm. It is not always true that tumors with that radius $r < 3$ mm will always turn out to be benign and $r > 3$ mm malignant. So one may have to add another feature to the data, such as the thickness of the tumor wall to identify the malignancy, along with the radial measure. The machine learning algorithm we can use may be a support vector machine (SVM) or logistic regression.

4.2 The Classification of Machine Learning Algorithms/Learning Algorithm

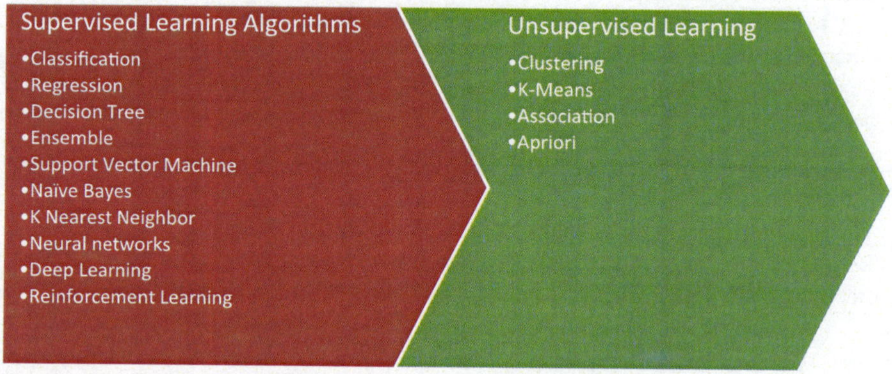

4.3 Machine Learning Models in the Science of Biology

Machine learning has been applied to major biological domains of cell biology, nervous system, genomics, proteomics, microarrays, systems biology, evolution, diagnosis of diseases, and text mining application. We will discuss in subsequent chapters the Model selection and generalization, Multivariate Methods, Dimensional Reduction, Clustering (*K*-means, Adaptive Resonance Theory, Self-Organizing Maps), Kernel Machines, Hidden Markov Model (HMM), Neural nets and Deep Learning, Bayesian Theory for ML and their applications in Biology.

4.4 Regression Algorithm

The regression method is used to predict an output which is a continuous variable. Regression models are evaluated with several statistical parameters such as the coefficient of determination, mean absolute error, relative absolute error, and root mean square error. Linear regression, SVM are examples of regression algorithms.

Depending on the sample in question and the nature of relationship, one uses different regression algorithms such as linear regression, logistic regression, and Poisson regression. One uses regression models to understand genetic epidemiology to detect the association of genetic variants with a trait or disease of interest.

Let us consider a group of k predictors $x_1, \ldots, x_k$, and a predicted outcome y. For ease of use and for a linear regression model, let's consider a linear function of the inputs $x_1, \ldots, x_k$. The equation can be written as: $F[E(y)] = \beta_0 + \beta_1 x_1 + \ldots + \beta_k x_k$, where F is the function and $\beta_0, \ldots \beta_k$ are the regression coefficients to be estimated. Typical form of other regression functions are—$F[E(y)] = E(y)$ (linear regression), $F[E(y)] = \log\{E(y)/[1 - E(y)]\}$ (logistic regression), and $F[E(y)] = \log[E(y)]$ (Poisson regression). Poisson regression may be extended to include the zero-inflated Poisson regression. This model is used in case of a sparse high dimension data, that means if you see 0 values for too many attributes. An example of this scenario is the patient billing for a healthcare establishment doing revenue cycle management. For a particular patient, there are only few ICD10 codes out of 70,000 available codes, so there are many zeros in the matrix.

4.5 Regularized Regression Methods

Overfitting is a problem in machine learning methods. One uses regularization methods to address this problem. Let us understand the overfitting problem.

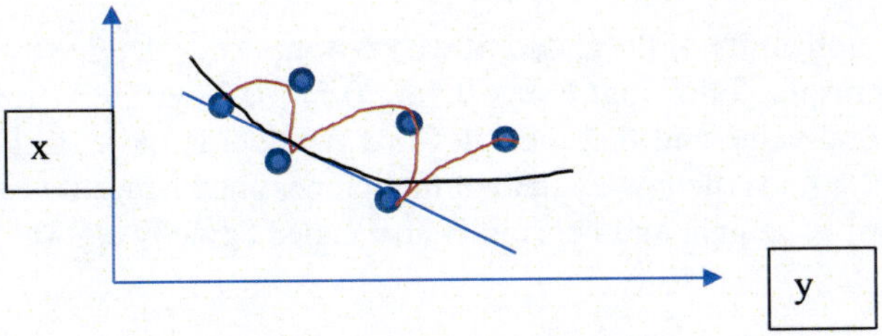

- The blue line is an example of under fit: $y = a$ + mx.
- The red line is an example of over fit: $y = a$ + mx + bx**2 + cx**3 + ...
- The black curve is a called balance fit.

So the overfitting is probably the best fit but it is given by a polynomial in x and analytically as numerically this becomes a challenge. One can reduce the complexity of the polynomial if the weight factors are too small compared to the first few terms.

If x is small and b, c weight factors are small, then the analysis becomes easier. Mathematically this process is called regularization. This is done by calculating the mean square error function and optimizing the same.

4.6 Decision Trees

Decision tree is based on binary tree and the goal is to lead to node leaf. Let's understand how we choose from the root node to branches and how we find the effectiveness of a decision tree.

When we work on a decision tree, we take the most set of data at the root node and then split based on the condition therein. As the root node contains the maximum randomness, it is also measure by the entropy of the state.

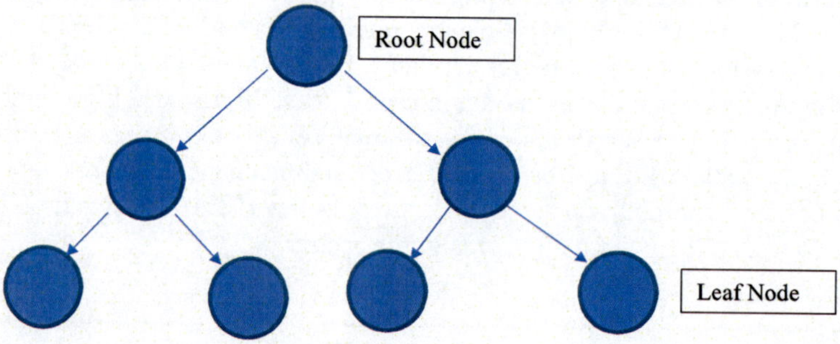

So if one wants to know the entropy of the root node, it is calculated in the following way:

$$E = \sum -p_i \log (p_i)$$

where p_i is the probability of the class i. If we know how to calculate the Entropy for the node, it returns the following: $E = -0.5\log(0.5) - 0.5\log(0.5) = 1$. So that's the maximum entropy value and that indicates that the data is most random. The tree node splits until it finds the lowest state of randomness and hence it aims to find the maximum information gain and hence it is also called "greedy algorithm."

4.7 *K*-Means Clustering

Let's take an example of a data set in two dimensions. The question is how do we cluster this dataset and what is the right number of clusters. We address this problem through k-means clustering. Let's suppose we start with $k = 2$, that means we start with two centroids.

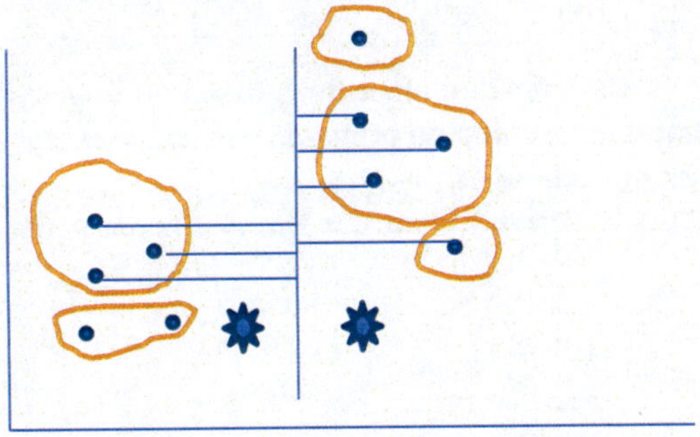

Let's draw a vertical line to create a separation between 2 clusters, based on their centroids, then calculate the distance of those points from the centroids. Distance is calculated drawing perpendicular to the vertical line. If the distance is more, we need reposition centroids.

Now, the question is how we can ideally create more centroids or clusters around the centroids. When do we know that the clusters are optimized? We do that by a method, popularly known as 'elbow method'. The free form brown color clusters are also possible for these data set. So we compute a statistical parameter, sum of squared errors (SSE):

$$\mathrm{SSE} = \sum\nolimits_{i=1}^{2} \mathrm{dist}(x_i - c_i)2$$

In this we have taken two clusters. Now if we repeat the same process of creating centroids based clusters and calculate SSE and draw diagram with SSE versus the no of clusters, we will see the following:

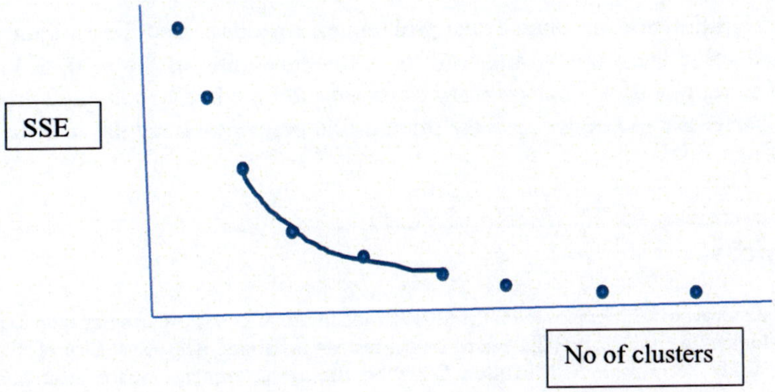

So one chooses the optimal number of clusters as the graph takes an elbow pattern. *K*-means algorithm is widely used in genetic expression analysis (Botía et al. 2017).

4.8 Naïve Bayes Classifier

Naive Bayes classifiers are a collection of classification algorithms based on Bayes' Theorem. It is a family of algorithms that share a common principle and hence every pair of features classified is independent of each other.

The Bayesian classification algorithm is derived from the Bayes Theorem of conditional Probability.

$$P(A|B) = \frac{P(A).P(B|A)}{P(B)}$$

where $P(A)$ is the probability that an event A has occurred, $P(B)$ is also the probability of an event B has occurred, $P(A|B)$ is a probability of occurrence of A provided B has happened and similarly $P(B|A)$ is defined.

Bayesian clustering algorithm has been effectively used in Biological Sequence Analysis (Durbin et al. 1999), Microarray Data Analysis (Speed 2003), and Protein Informatics (Vanucci et al. 2006).

4.9 Discussion

In a later chapter we will discuss a model selection strategy and procedure. It is essential to understand which algorithm or model is appropriate for a particular problem. Feature identification and extraction are essential processes to fit a particular model.

For a high-dimensional sparse data problem, one needs to find the minimal set of features to solve the problem and lead to an understanding of the pattern. In such circumstances, one uses a poison matrix factorization method to solve the problem.

So it is necessary to understand the models first before applying the same to solve the problem.

References

Botía JA, Vandrovcova J, Forabosco P et al (2017) An additional k-means clustering step improves the biological features of WGCNA gene co-expression networks. BMC Syst Biol 11:47

Durbin R, Eddy SR, Krogh A, Mitchison G (1999) Biological sequence analysis: probabilistic models of proteins and nucleic acids. Cambridge University Press, Cambridge

Speed TP (2003) Statistical analysis of gene expression microarray data. Florida Chapman & Hall/CRC, Boca Raton

Vanucci M, Do K-A, Müller P (2006) Bayesian inference for gene expression and proteomics. Cambridge University Press, New York

Model Selection for Machine Learning 5

Abstract

The selection of an appropriate model is a highly critical part of implementing the machine learning algorithm. A particular problem can be addressed with more than one algorithm. However, the choice of that depends on the availability of data type, data set, complexity, use of resources, and statistical cost function. Here in this chapter, we have explored different algorithms for their suitability to solve a particular problem.

Keywords

Model selection · Probabilistic method · Resampling method · AIC · BIC · MDL · Cross validation · Bootstrap

5.1 Introduction

The goal of a learning algorithm is to appropriately understand and reproduce the relationship between different variables in a sample and find the particularity of a sample. The simplest model has always been to find whether is a linear or nonlinear relationship between variable in a set. To achieve that goal, one can easily do a graphical representation after identifying the independent and dependent variable. This exercise becomes difficult when there are more than two variables or features of a dataset. While mathematically it is possible to understand n dimension, it is not easy to visualize the concept, beyond two to three dimensions. Nonlinear models pose bigger challenge and hence there are approximations one can adopt to do an optimal fitment.

Working with flexible models such as neural networks requires careful control of the model complexity and generalization ability of the resulting model. It is widely acknowledged that in most modeling scenarios, there is not sufficient data available

S. Ghosh, R. Dasgupta, *Machine Learning in Biological Sciences*,
https://doi.org/10.1007/978-981-16-8881-2_5

to reliably use these estimators for assessing generalization or select/optimizing models. Therefore, one resorts to resampling techniques like cross validation such as jackknife or bootstrap.

5.2 Model Selection Process and Criteria for Model Selection

Here is an overview of different algorithms and their usage in application areas.

Algorithm	Speed of prediction	Model training	Memory	Tuning	Application areas
Logistics regression	Fast	Fast	Small	Minimalistic	Problems with linear decision boundary—typical examples will be two state results of 1 or 0, yes or no. They are widely used in problems to identify the existence of a disease and the medication's effectiveness
Decision tree	Fast	Fast	Small	Moderate	Decision trees are used for handling nonlinear data sets effectively. It can be divided into two types: Categorical variables and continuous variable decision trees. The stock trading application uses a decision tree. Decision trees are an essential and widely used technique for the classification of data. Recently, they have been adapted for use in computational biology, especially for the analysis of DNA sequence data
Support vector machine	Slow	Slow	Medium	Moderate	The support vector machine (SVM) is a pattern recognition algorithm. SVM has been used to analyze various complex biological data sets, microarray expression profiles, DNA and

(continued)

Algorithm	Speed of prediction	Model training	Memory	Tuning	Application areas
					protein sequences, protein–protein interaction networks, tandem mass spectra
Nearest neighbor	Moderate	Fast	Medium	Minimalistic	Widely used to analyze genetic structures
Naïve bayes	Fast	Fast	Medium	Moderate	It improves classification accuracy in the TMA prostate cancer dataset
Ensembles	Medium	Slow	Depends on the size and complexity of the data structure	Moderate	Most widely used ensemble learning methods and their application in various bioinformatics problems, including the main topics of gene expression, mass spectrometry-based proteomics, gene–gene interaction identification from genome-wide association studies, and prediction of regulatory elements from DNA and protein sequences
Neural network	Medium	Slow	Medium to large	Major tuning required	A wide range of applications in biology and medical science, particularly detection of the tumor through multi-layer perception, is an example of such an application

The model selection process involves the following stages:

I. Choose a sample of sufficient size as to represent the data adequate.
II. Understand whether the data is labeled.
III. Divide the data set into three categories – Training Data, Cross Validation Data, and Test Data.
IV. Choose the algorithm based on the fitment, avoid overfitting or underfitting.
V. Identify the mean square error, probability, and entropy.
VI. Train, cross validate, and test data.

5.3 Model Selection Techniques

Now the question is how we select a model. How do we really understand if the model needs to be linear or quadratic or polynomial? Let's take this set of data.

Blood sugar level (A1C)	Chances of renal problem (%)
2	2
2.5	2.5
3	5
3.5	8
4	10
4.5	15
5	18
5.5	20
6	30
6.5	32
7	47
8	42
9	51
10	52

We can get the following representations:

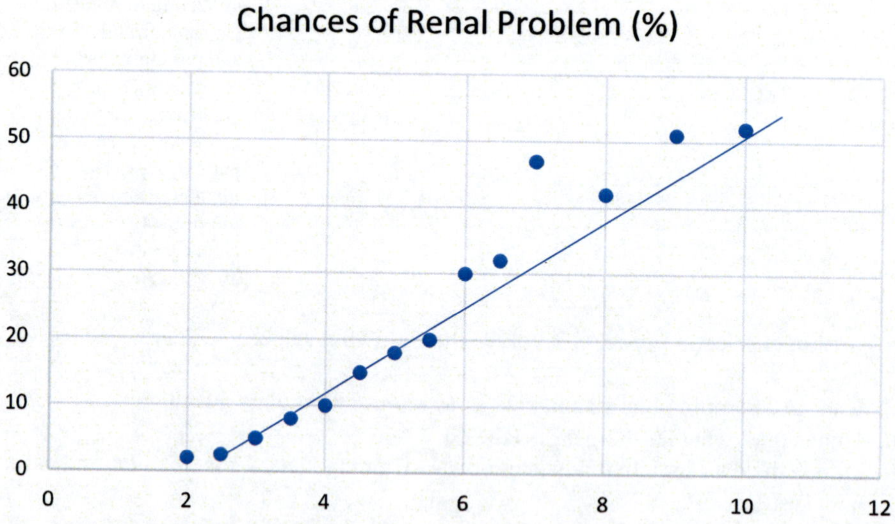

We have tried a linear fitment here. The question is whether it represents the set appropriately and whether I can generate most of the points on this line. This is an example of underfitting.

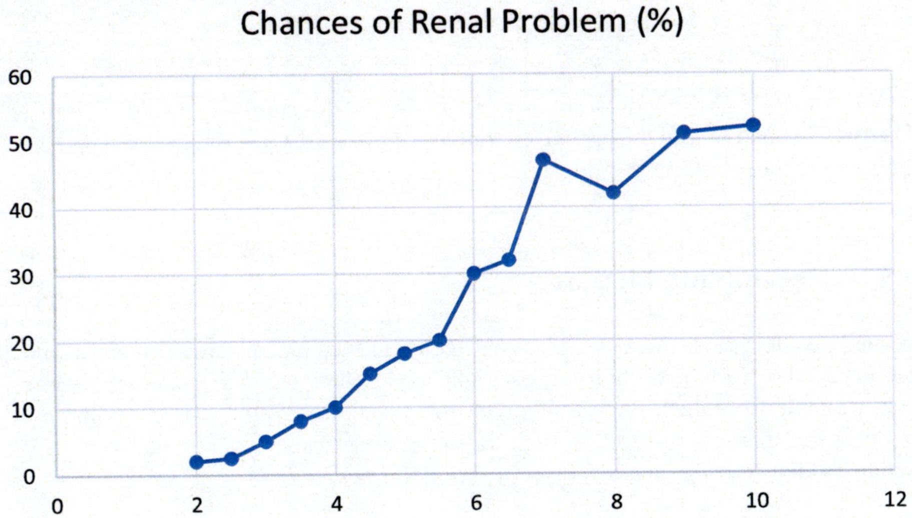

This is an example of overfitting. Can we really represent the curve mathematically to include future points of data? The answer is not affirmative.

So, let's learn how do we approach the model selection. Typically, we can look at a data set and can try right from linear to polynomial curve fitment and let's understand the complexity of the situation.

$$y_\theta(x_i) = \theta_0 + \theta_1 x + \theta_2\, x^2 + \ldots + \theta_i x^i + \ldots n \text{ terms}$$

Now if we try to fit keeping the first two terms, we get a linear model:

$y_\theta(x_i) = \theta_0 + \theta_1 x$, in our example, this generates an under fitment.

On the other hand, if we choose to stay as a polynomial, it generates an overfitment. Hence, we need to choose judiciously the right polynomial, and probably keep up to some m terms where $m < n$.

How do we now practically create and test the model? We divide the whole dataset into training data, cross validation data, and test data. Typical standard for such a classification is that we use 60% of the data for training, 20% for cross validation, and 20% for test.

Training data is denoted as:

$(x^1, y^1), (x^2, y^2), \ldots.$ n terms, n being the number of training data

Cross Validation data as:

$\left(x_{cv}^1, y_{cv}^1\right), \left(x_{cv}^2, y_{cv}^2\right), \ldots n_{cv},$ $n_{cv},$ being the number of cross validation data

Test data as:

$\left(x_{\text{test}}^1, y_{\text{test}}^1\right), \left(x_{\text{test}}^2, y_{\text{test}}^2\right), \ldots n_{\text{test},}$ $n_{\text{test},}$, being the number of test data

Now, we compute the squared error for each category.

Training error:

$$J_{\text{train}}(\theta) = \frac{1}{2m} \sum_{i=1}^{n} (h_\theta(x^i - y^i))2$$

We calculate Cross validation error and test error accordingly using a similar formula. So one can find the optimized training, cross validation and test data and the appropriate model fitting it.

5.4 Resampling Methods

While we discussed Model Selection technique using different models and estimating the errors and choosing the right model, here is another one where one draws sample from the same training data and try and refitting the model with new sample. Resampling method is high in computing resources because of many iterations and drawing data from sample several times.

5.5 K-Fold Cross Validation

This is another method to training data and test data partition method and it works on averaging out the prediction error.

The method involves steps of partitioning the data into K equal sized sample. For each value of *k*, the model fits the data for other *k*-1 components and then estimate the prediction error. All these prediction errors are averaged out to create a mean prediction error.

Further Reading

Akaike H (1969) Fitting autoregressive models for prediction. Ann Inst Stat Math 21(1):243–247

Akaike H (1974) A new look at the statistical model identification. IEEE Trans Autom Control 19(6):716–723

Allen DM (1974) The relationship between variable selection and data augmentation and a method for prediction. Technometrics 16:125–127

Andersen CM, Bro R (2010) Variable selection in regression a tutorial. J Chemometr 24(11-12): 728–737

Arcones MA, Giné E (1992) On the bootstrap of M-estimators and other statistical functionals. In: Exploring the limits of bootstrap (East Lansing, MI, 1990), Wiley series in probability and mathematical statistics. Wiley, New York, pp 13–47

Arlot S (2007) Resampling and model selection. PhD thesis, University Paris-Sud

Burnham KP, Anderson DR (2003) Model selection and multimodel inference: a practical information-theoretic approach. Springer

Greenland S (1989) Modeling and variable selection in epidemiologic analysis. Am J Public Health 79(3):340–349

Guyon I, Elisseeff A (2003) An introduction to variable and feature selection. J Mach Learn Res 3: 1157–1182

Ing C-K, Wei C-Z (2005) Order selection for same-realization predictions in autoregressive processes. Ann Stat 33(5):2423–2474
Johnson JB, Omland KS (2004) Model selection in ecology and evolution. Trends Ecol Evolut 19(2):101–108
Kadane JB, Lazar NA (2004) Methods and criteria for model selection. J Am Stat Assoc 99(465): 279–290
Liu W, Yang Y (2011) Parametric or nonparametric? A parametricness index for model selection. Ann Stat 39:2074–2102
Parry M, Dawid AP, Lauritzen S (2012) Proper local scoring rules. Ann Stat 40:561–592
Shao J (1997) An asymptotic theory for linear model selection. Stat Sin 7(2):221–242
Shibata R (1980) Asymptotically efficient selection of the order of the model for estimating parameters of a linear process. Ann Stat 8(1):147–164
Stoica P, Selen Y (2004) Model-order selection: a review of information criterion rules. IEEE Signal Process Mag 21(4):36–47
Vapnik V, Chervonenkis A (1971) On the uniform convergence of relative frequencies of events to their probabilities. Theory of Probability and Its Applications 16:264–280
Wallace CS, Boulton DM (1968) An information measure for classification. Comput J 11(2): 185–194
Yang Y (2006) Comparing learning methods for classification. Stat Sin 16:635–657

Multivariate Methods in Machine Learning in the Context of Biological Data

6

Abstract

Until now we have discussed the general concepts and models of Machine Learning algorithm, here we will move deeper into the understanding of these methods and models in biological data. Feature extraction and Dimensionality reduction are important methods used in Biology and Healthcare. The relevance of these techniques is more due to the use of sensors and multi-modal data sources. There are number of feature extraction techniques grouped under the Multivariate Analysis (MVA). In this chapter we will discuss Parameter Estimation, Estimation of Missing Values, Multivariate Normal Distribution, Multivariate Classification, and Tuning Complexity in the context of analysis of biological data.

Keywords

MVA · Parameter estimation · Multivariate normal distribution · Multivariate classification

6.1 Introduction

Machine learning techniques need capturing data for a set of events, set of actors and expected or observed output. Several measurements are made on each individual or events generating an observation vector. The sample may be represented in the form of a $(m \times n)$ matrix where n columns correspond to n variables denoting the result of measurements made on an individual or event. These are also called inputs, features, or attributes. The m rows correspond to independent and identically distributed observations, examples or instances on m individuals or events.

Typical example of such a matrix is the following. If one captures details of patients and the corresponding diagnostics codes (ICD 10) for estimating billing for

S. Ghosh, R. Dasgupta, *Machine Learning in Biological Sciences*,
https://doi.org/10.1007/978-981-16-8881-2_6

the payers (health insurance payers), the matrix generated will be a similar $(m \times n)$ matrix where m is the number of rows, one for each patient or one for each patient instance and n is the number of columns for ICD 10 codes. In this case the matrix will be a sparse matrix as there are 72,616 codes (Financial Year 20–21) and each patient instance will contain only handful of those. These codes are related to respective diagnosis, and they are interrelated. For example, if a patient has a fall, he may have a fall due to a stroke, got a head injury, may have a temporary paralysis and a blood loss and the gender of the person is male and age 25, then there will ICD codes (169,354 for left side weakness, unspecified head injury (S09, 90XA), loss of consciousness S06; it will not have an ICD code for O13, which trimester pregnancy induced hypertension ICD 10 code. So, these codes are related to the events and the variables. So, one needs a do a multivariate statistical analysis.

In most cases these variables are correlated and hence to understand the relationship, one needs to do a multivariate analysis. In case the predicted variable is discrete, it is a multivariate classification problem, otherwise it is a multivariate regression problem.

To determine where each variable falls between upper and lower bounds, one needs to compute the mean and standard deviation for each variable and to determine if there are differences between data taken from two groups, p values are computed. However, these methods ignore that the collected data are often correlated and may be due to these measurements describing quantities that are connected by biological networks. Multivariate analysis approaches are more appropriate in these scenarios, as they can detect differences in datasets that the traditional univariate approaches may miss.

6.2 Multivariate Analysis

Multivariate analysis involves the investigation of multiple variables simultaneously. These techniques take on a variety of forms and are used for a few different tasks. Analyses of variance models (Scheffé 1999) are commonly used to test the effects of multiple categorical factors on a measured response variable. The support vector machine (Cortes and Vapnik 1995) is an option for the supervised classification of groups of data consisting of several measurements. Hierarchical clustering (Johnson 1967) can be used for cluster analysis, while partial least squares regression (Wold et al. 2001) offers an approach for parameter estimation.

FDA is a useful method for maximizing separation between two or more groups of data samples (Fisher 1936) and is most appropriate when the input variables are normally distributed (Hastie et al. 2011) continuous variables.

6.3 Multivariate Data

Let us represent a set of data derived out of several measurements for single or multiple events or phenomena by a data matrix.

$$\mathbf{X} = \begin{bmatrix} X_1^1 & \cdots & X_d^1 \\ \vdots & \ddots & \vdots \\ X_1^N & \cdots & X_d^N \end{bmatrix}$$

where d represents the number of variables denoting the results of the measurements made for a particular event and N denotes the number of independent, identically distributed observations or instances. d is the number of inputs, features, or attributes. Typically, these variables are corelated and hence the objective of the Multivariate Analysis (MVA) is to find the minimal set of features or number of variables to describe a system and hence the whole question of dimensionality reduction is addressed.

6.4 Parameter Estimation

Let's define a mean vector $\boldsymbol{\mu}$ such as each of the elements are mean of one of the columns of **X**.

$$E[x] = \boldsymbol{\mu} = [\boldsymbol{\mu_1} \ldots \boldsymbol{\mu_d}]^T$$

where the variance of X_i is denoted as σ_i^2 and the covariance of two variables X_i and X_j is defined as $\sigma_{ij} = \text{Cov}\ (X_i, X_j) = E[(X_i - \mu_i)(X_j - \mu_j)] = E[X_iX_j] - \mu_i\mu_j$.

With $\sigma_{ij} = \sigma_{ji}$ and when $i = j$, then $\sigma_{ii} = \sigma_i^2$.

With d variables there be d variances and there be $d(d–1)/2$ covariances and can be represented by a dXd matrix. This is also called a covariance matrix. The diagonal terms are the $i = j$ terms and hence are the variances and the off-diagonal terms are the covariances.

The correlation between variables X_i and X_j is normalized between -1 and $+1$. And defined as $\text{Corr}\left(X_i, X_j\right) = \rho_{ij} = \frac{\sigma_{ij}}{\sigma_i \sigma_j}$. If two variables are independent, then their correlation is 0.

For a multivariate sample, the maximum likelihood estimator for the mean of the sample mean, **m**, is:

$$\boldsymbol{m} = \frac{\sum_{1}^{N} \boldsymbol{x^t}}{\boldsymbol{N}}$$

The Estimator of the Covariance Matrix is expressed as follows:

$$s_{ij} = \frac{\sum_{t=1}^{N} \left(x_i^t - m_i\right)\left(x_j^t - m_j\right)}{N}$$

These are biased estimates. The Sample correlation coefficients are as follows:

$$r_{ij} = \frac{s_{ij}}{s_i s_j}$$

6.5 Estimation and Handling of Missing Values

There may be several reasons behind the missing values, and it is important to determine the reasons of missing values to handle the other part of the data correctly. Data may be randomly missed and lead to a biased analysis. Hence it is only desired that that studies be designed with minimum missing values in data. Data could be missing completely at random or MCAR, occurring at random, and the events leading to missing value are not dependent either on observable variables or unobservable parameters which may lead to unbiased analysis. In case of data missing at random or MAR, occurs with non-random missing of data. The missing-ness of data is accounted for by variables with complete information. It can induce parameter bias but application of Full Information Maximum Likelihood, MAR can enable unbiased analysis. The other form of missing data is classified as Missing not at random or MNAR in which the value of the missing variable bears relation to the cause of it missing.

There is a way to avoid such situation by discarding such observations all together for a large sample. In case the sample is small, one uses a mean imputation procedure where the values are replaced by the mean value in place of null values. One can also use a imputation by regression method where one replaces the missing values by the most likely values from a known observation. The expectation–maximization algorithm is used to find maximum likelihood (ML) estimates of parameters in a statistical model, depending on unobserved latent variables and it enables computation of statistical values to estimated complete dataset in keeping the pattern of the data that is missing.

Omission is another method to handle such cases of missing data by discarding samples with invalid data from further analysis and or by the application of analysis involving application of methods that are not affected by missing values. Interpolation is another involving incorporation of new data points within the range of known data points. Partial deletion is yet another method to handle missing data to reduce the dataset to a set where no data is missing including methods of case wise deletion or pairwise deletion. Full analysis method of missing data handling involves taking into full account of all available information without the disturbance created by imputation. This is done by using the expectation–maximization algorithm, maximum likelihood estimation, Max-margin classification of data with absent features, Partial identification methods.

Missing values in already complex biological data has a major role in detection and quantification of interactions in biological systems to infer relationships accurately. Mutual information and interaction information have been used to evaluate multivariable dependencies for entire data with missing parts (Uechi et al. 2019).

6.6 Multivariate Normal Distribution

A random vector is said to be d-variate normally distributed if every linear combination of its d components has a univariate normal distribution. Its importance derives mainly from the multivariate central limit theorem. The multivariate normal distribution is often used to describe, any set of correlated real-valued random variables, each of which clusters around a mean value.

The distribution is represented as follows:

$$p(x) = \frac{1}{\frac{(2\pi)^{\frac{d}{2}}}{\Sigma^{\frac{1}{2}}}} \exp\left[-\frac{1}{2}(x-\mu)^T\Sigma^{-1}(x-\mu)\right]$$

where μ is the mean and Σ is the covariance matrix.

$$\frac{(x-\mu)^2}{\sigma^2} = (x-\mu)\left(\sigma^2\right)^{-1}(x-\mu)$$

The squared distance from x to μ, in standard deviation units, normalizes for different variances. In the case multivariate situation, the Mahalanobis distance is used.

$(x-\mu)^T\Sigma^{-1}(x-\mu) = c^2$ is the d-dimensional hyper-ellipsoid centered at μ and its shape and orientation are defined by the covariance matrix Σ. Because of the use of the inverse of Σ, if a variable has a larger variance than other, it receives less weight in the Mahalanobis distance. Similarly, two highly correlated variables do not contribute as much as two less correlated variables. The use of the inverse of the covariance matrix thus has the effect of standardizing all variables to unit variance and eliminating correlations.

6.7 Multivariate Classification

Biological data sets reveal high dimensionality and low effect sizes. Data analysis by multivariate pattern classification (MPC) and also taking into account the systematic similarities between subsets of trials conditions, the nested subclasses within each class characteristic is formed. To detect the differences in multivariate data sets, multivariate pattern analysis (MVPA) has been used in experiments related to neuroimaging data and recording human electroencephalogram (EEG) activity (Jamalabadi et al. 2018).

6.8 Tuning Complexity

When one reduces number of parameters through the covariance matrix, essentially doing a tradeoff of a simple model with generality, one introduces a bias by eliminating other variables. On the other hand if one wants to avoid the bias, the matrices are arbitrary and the quadratic discriminant may have large variances on small datasets.

When one has a small dataset and (even when the covariant matrices are arbitrary), it may be useful to assume a shared covariance matrix. A single covariance matrix has fewer parameters, and it can be estimated using more data. This method corresponds to Linear Discriminant.

There are other methods of using Euclidean Distance to measure similarity or cosine similarity and so on. One can also use Bayesian Classifier, assuming variables are independent. Friedman proposed a method that combines all these special cases and it is called Regularized Discriminant Analysis (RDA, https://citeseerx.ist.psu.edu/viewdoc/download?doi=10.1.1.382.2682&rep=rep1&type=pdf). The regularization corresponds to approaches where one starts with high variance and constraints toward lower variance, at the risk of increasing the bias.

References

Cortes C, Vapnik V (1995) Support-vector networks. Mach Learn 20:273–297

Fisher RA (1936) The use of multiple measurements in taxonomic problems. Ann Eugenics 7:179–188

Hastie T, Tibshirani R, Friedman J (2011) The elements of statistical learning: data mining, inference, and prediction, 2nd edn. Springer, New York, NY

Jamalabadi H, Alizadeh S, Schönauer M, Leibold C, Gais S (2018) Multivariate classification of neuroimaging data with nested subclasses: biased accuracy and implications for hypothesis testing. PLoS Comput Biol 14(9):e1006486

Johnson SC (1967) Hierarchical clustering schemes. Psychometrika 32:241–254

Scheffé H (1999) The analysis of variance. Wiley, New York, NY

Uechi L, Galas DJ, Sakhanenko NA (2019) multivariate analysis of data sets with missing values: an information theory-based reliability function. J Comput Biol 26(2):152–171

Wold S, Sjöström M, Eriksson L (2001) PLS-regression: a basic tool of chemometrics. Chemom Intell Lab Syst 58:109–130

Further Reading

Frye RE, James SJ (2014) Metabolic pathology of autism in relation to redox metabolism. Biomark Med 8:321–330

Lemieux I, Lamarche B, Couillard C, Pascot A, Cantin B, Bergeron J, Dagenais GR, Després J-P (2001) Total cholesterol/HDL cholesterol ratio vs LDL cholesterol/HDL cholesterol ratio as indices of ischemic heart disease risk in men: the Quebec Cardiovascular Study. Arch Intern Med 161:2685–2692

Mika S, Ratsch G, Weston J, Scholkopf B, Mullers KR (1999) Fisher discriminant analysis with kernels. In: Proceedings of the 1999 IEEE neural networks for signal processing IX workshop, Madison, WI, USA, 23–25 August, pp 41–48

Morgan DB, Carver ME, Payne RB (1977) Plasma creatinine and urea: creatinine ratio in patients with raised plasma urea. Br Med J 2:929–932

7 Dimensionality Reduction Methods in Machine Learning

Abstract

The goal of machine learning algorithm is to understand the basic features of a complex system. If the dataset is large and the number of features is large as well, it is possible that one can get features or input variables easily identified. In case the dataset is small, there may be a circumstance that one may miss some of the observations and then eventually ignore some features and find the minimal set of features to define the system adequately. In this chapter we will cover the feature selection methods that choose a subset of important features and skip the rest and the feature extraction methods that form the minimally accepted feature from original set of features.

Keywords

Dimensionality Reduction · Feature Selection · Feature Extraction

7.1 Introduction

When one attempts to solve a problem, one looks into the complexity and the complexity depends on the number of input dimension and size of data sample. The problem is represented in the form of solving a matrix of (m × n) dimension. For finding an optimal solution in terms of machine io and memory, one looks for a reduced dimension adequate to describe the problem. Decreasing sample size also decreases the complexity of the interference algorithm during testing phase.

When an input is flagged as "not important," we save the cost of extracting the same. Simpler models are more robust on simpler datasets. Simpler models have less variance and hence they vary less depending on the particulars of the sample, including noise, outliers, and exceptions.

S. Ghosh, R. Dasgupta, *Machine Learning in Biological Sciences*,
https://doi.org/10.1007/978-981-16-8881-2_7

7.2 Dimensionality Reduction and Biological Data

Dimensional reduction is the process of reducing the number of variables and hence features. Dimensionality reduction can be divided into subcategories of feature selection and feature extraction. Feature selection involves Wrapper, Filters, and Embedded. Feature extraction includes Principal Component Analysis (PCA).

The Critical Assessment of protein Structure Prediction (CASP) is a nationwide experiment that takes place biannually, which centered around analyzing the best current systems for predicting protein tertiary structures. AI methods are being used to predict exact nature of protein folding phenomena.

Clinical trials and also drug delivery process use clustering algorithm and ANN to find patient groups suitable for a particular strain of a drug.

There are more and more research projects in this area to understand the very complex nature of biological systems.

Application of PCA for dimension reduction and its projection has been done in the new coordinate system, through scatter profile plots and has enabled better visualization as compared to the original coordinate system. SpRay technique has been developed by employing linear and nonlinear dimension reduction methods to microarray data in Streptomyces coelicolor, microarray data in normal cells and with trisomies of human 13 and 21 chromosome (Lehrmann et al. 2013).

Proteins are biomolecules and their structure function and interaction with other molecules govern biological functions. Hence modeling protein flexibility is a challenging task. Although X-ray crystallography and computational methods of molecular dynamics are used, since a protein of medium size has few thousands of degrees of freedom, PCA has been used for dimensionality reduction transforming the original high dimensional representation of protein motion into a lower dimensional representation enabling study of dominant protein motion but conformations similar to that of experimental results could be obtained (Teodoro et al. 2002).

While study of secondary structure of protein and access to solvent enables the study of three-dimensional structure of a protein, reducing number of dimensions provides faster model training, prediction and elimination of noise. Feature selection methods, autoencoders, and PCA has been applied to this biological problem of the secondary structure of the protein. Access to solvent enables wherein input space was reduced, while maintaining the prediction accuracy (Aydin et al. 2018).

7.3 Dimensionality Reduction Techniques

So now the question is how dimensional reduction techniques help solving the problem in a less complex way. Essentially it does so by reducing the number of features that are to be considered.

Let's understand in a simpler language how feature extraction and feature selection can happen in an algebraic situation. Let's suppose that there is an equation with five unknowns such as:

$x + y + z + r = p$. Now to determine a set of solutions, it becomes both analytically and numerically, it becomes an unsurmountable task. Let's see we find a constraint equation, typically relationship between x, y where $x + y = xy$. Then we get the equation to transform as:

$xy + z + r =$ p. This is a typical example of feature extraction by imposing a constraint equation, typically, using a known relationship between two variables.

Now, we find that, $z = 0$, which indicates that the variable z is not important or does not influence the result, we can reduce the equation to, $xy + r = p$. This is an example of Feature selection, rather ignoring the feature which is not important. An example of such a situation is whether skin color influences blood pressure.

7.3.1 Feature Selection Methods

Feature selection is the process of selecting subset of relevant features or variables. There are three different methods of feature selections: wrapper method, filter method, and embedded method.

Wrappers use a predictive model that scores feature subsets based on an error rate of the model. While these are computationally io intensive, they usually produce the best selection of features. Mostly used technique is stepwise regression. It is an algorithm that adds the best feature and removes the worst feature iteratively.

Filters use a proxy method that is less intensive in terms of machine io. It generates less accuracy compared to the Wrappers. Filters capture the usability of the data set in terms of error measurements. Feature sets are more general compared to the Wrappers. Filters produce feature sets that are not dependent on the predictive model so that one can adjust the variables good for accuracy or bad for accuracy, to be included or excluded, accordingly.

Embedded Algorithm learns about which features contribute effectively to producing an accurate model during the model building process. Most commonly used one is the regularization model. It is a step by step picking up the features in defining the model.

7.3.2 Feature Extraction Method

Feature Extraction is the process of projecting a space of large dimensions to a smaller dimension. Let's take an example in Geometry. Let's suppose we wish to project a point in three dimension to a two-dimensional situation. In the process we reduce the dimension from x, y, z to x, y or x, z or a combination of any two of them. The most popular technique in Feature Extraction is Principal Component Analysis (PCA). There are many other linear and nonlinear techniques to achieve the feature extraction process.

PCA is the process of reduction of higher vector space to a lower order through projection. It is very similar to the geometric projection system. When the projection from higher dimension happens, the new axis is called the Principal Axis and the new data is the principal components. PCA techniques are used to get the ancestry information of genomic data. It can separate people of Latino and non-Latino origin through dimensionally analyzing them using PCA. In fact in one the experiments in European lab, they could separate out geography from genomic data of thousands of Europeans.

There are other methods of dimensional reduction, such as matrix factorization, manifold learning, and autoencoder methods.

Matrix factorization breaks a complex matrix into a product two metrices, one being the sparse with respect to the other. This process can go until one removes all the sparseness out of the matrix structure.

Manifold learning is also another method of reducing a higher dimensional manifold to a lower and most optimal manifold so the most effective features are identified and extracted in the process.

Autoencoders are simple learning circuits which aim to transform inputs into outputs with the least possible amount of distortion.

7.4 Discussion

Although several techniques are used in dimensionality reduction, there is no ideal method for this process rather for best performance, it may be an integrated approach of the use of systematic controlled experiments, that can fit in a dimensionality reduction technique, paired with the model of choice. While working with input variables of differing scales or units, it is ideal to normalize and standardize the data prior to application of these methods.

References

Aydin Z, Kaynar O, Görmez Y (2018) Dimensionality reduction for protein secondary structure and solvent accessibility prediction. J Bioinforma Comput Biol 16(5):1850020. https://doi.org/10.1142/S0219720018500208

Lehrmann A, Huber M, Polatkan AC et al (2013) Visualizing dimensionality reduction of systems biology data. Data Min Knowl Disc 27:146–165

Teodoro ML, Phillips GN, Kavraki LE (2002) A dimensionality reduction approach to modeling protein flexibility. In RECOMB '02: proceedings of the sixth annual international conference on Computational biology, pp 299–308. https://doi.org/10.1145/565196.565235

Further Reading

Fukunaga K (1990) Introduction to statistical pattern recognition. Academic, San Diego

Jimenez LO, Landgrebe DA (1997) Supervised classification in high-dimensional space: geometrical, statistical, and asymptotical properties of multivariate data. IEEE Trans Syst Man Cybern 28(1):39–54

Naanaa W, Nuzillard J-M (2005) Blind source separation of positive and partially correlated data. Signal Process 85:1711–1722

Pearson K (1901) On lines and planes of closest fit to systems of points in space. Philos Mag 2:559–572

Spearman C (1904) General intelligence objectively determined and measured. Am J Psychol 15: 206–221

Torgerson WS (1952) Multidimensional scaling I: theory and method. Psychometrika 17:401–419

Winter ME (1999) N-FINDR: an algorithm for fast autonomous spectral end-member determination in hyperspectral data. Imaging Spectrom 3753:266–275

Hidden Markov Method

8

Abstract

The mathematical theory of Markov Chain was known to us for 100 years or more. In 80s engineers started using this theory to analyze speech and define speech model. In recent years, bioinformatics applications, in the area of alignment, labeling, and profiling of genetic sequences, protein structure prediction, and pattern recognition, use Hidden Markov Models. We will introduce the method and its application in biological data analysis.

Keywords

Markov chain · HMM · Forward–backward algorithm · Stochastic model

8.1 Introduction: Markov Chain

Hidden Markov Method (HMM) is based on the concept of Markov Chain. It is a model based on predicting the probability of a future state based on the current state probabilities. As an example if we have three probable situations of day's weather, such as rainy day, sunny day and cold day; then the future state may depend on the occurrence of the past state or a combination of past states, depending on the probabilities of occurrence.

Let us consider state variables as $q_1, q_2, \ldots, q_i$. According to Markov Model the probability of the ith state depends on the $(i-1)^{\text{th}}$ state only, such as: $P(q_i = a | q_1 \ldots q_{i-1}) = P(q_i = a | q_{i-1})$.

S. Ghosh, R. Dasgupta, *Machine Learning in Biological Sciences*,
https://doi.org/10.1007/978-981-16-8881-2_8

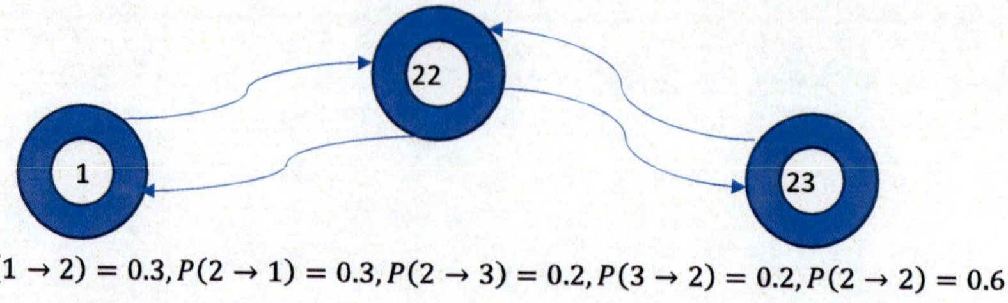

$$P(1 \rightarrow 2) = 0.3, P(2 \rightarrow 1) = 0.3, P(2 \rightarrow 3) = 0.2, P(3 \rightarrow 2) = 0.2, P(2 \rightarrow 2) = 0.6$$

where the states are nodes and the transitions, with their probabilities, edges. The transitions probabilities mandate values of arcs leaving a given state must sum to 1.

A Markov chain is specified by a set of N States ($Q = q_1 \ldots q_N$), transition probabilities between states ($A = a_{11}a_{12} \ldots a_{nn}$) with $\sum_{1,1}^{n,n} a_{ij} = 1$, initial probability distribution ($\pi = \pi_1 \ldots \pi_N$) with $\sum_{1}^{N} \pi_i = 1$.

8.2 The Hidden Markov Model (HMM)

Hidden Markov Model is based on the concept of Markov Chain. Let us take an example to under the HMM. Let us assume that a particular location can have three different kinds of weather—rainy day, sunny day, and cold weather. Depending on the weather, the stock market in the local stock exchange fluctuates.

According to the Markov assumption, whether a particular day be cold or not, depends completely on the state of the previous day. So, if we have the stock market movement as the observables whereas the three weather conditions are hidden variables, therefore, we are trying to find a scenario where the stock market is moving up and we have a situation of a cold day. We have a set of three states (q_i) and set of observables (o_i) and transition probabilities (π_i). Therefore, we will calculate the probability of occurrence of cold day, given the probability of concurrence of a stock market good day.

An HMM is defined by N states, A transition probability matrix, A sequence of observations, A sequence of observable likelihoods (emission probabilities), and π initial probabilities with $\sum_{1}^{N} \pi_i = 1$.

A first-order hidden Markov model has two assumptions:

a. Markov assumption—probability of a particular state depends on the immediate previous state: $P(q_i = a| q_1 \ldots q_{i-1}) = P(q_i = a| q_{i-1})$.

b. Second, the probability of an output observation o_i depends only on the state that produced the observation q_i: $P(o_i| q_1...q_i...q_T, o_1...o_i...o_T) = P(o_i| q_i)$.

Rabiner (1989) introduced the idea that hidden Markov models should be characterized by three fundamental problems:

a. Likelihood: Given a particular HMM ($\lambda = (A, B)$) and an observation O, we have to determine $P(O|\lambda)$.
b. Decoding: Given O and HMM λ, find the best hidden state sequence Q.
c. Learning: Given O, λ, learn the HMM parameters A and B.

8.3 Likelihood Computation and the Forward–Backward Procedure

Let us calculate the probability $P(O|\lambda)$. This can be done considering every possible combination of the number of observations. For every fixed state sequence $I = i_1 i_2 ... i_T$, the probability is

$$P(O|I, \lambda) = b_{i_1}(O_1) \dots b_{i_T}(O_T).$$

The probability of such state sequence I is written as: $P(I|\lambda) = \pi_{i_1} a_{i_1 i_2} \dots a_{i_{T-1} i_T}$, where $\{a_{ij}\}(s)$ are state transition probabilities as given by

$$a_{ij} = P\left(q_j \; at \; t+1 | q_i \; at \; t\right)$$

And $\pi = \{\pi_i\}$ is the initial state distribution as represented by $\pi_i = P(q_i \text{ at } t = 1)$.

The probability that O and I occur simultaneously is given by Bayes' Theorem, the product of the two probability functions as $P(O| I, \lambda)$ and $P(I| \lambda)$ as

$$P(O, I|\lambda) = P(O|I, \lambda) P(I|\lambda)$$

Hence, $$P(O|\lambda) = \sum_i P(O|I, \lambda) P(I|\lambda) =$$
$$\sum_{i_1 i_2 i_T} \pi_{i_1} b_{i_1}(O_1) a_{i_1 i_2} b_{i_2}(O_2) \dots a_{i_{T-1} i_T} b_{i_T}(O_T).$$

The process continues until we reach the transition from state of i_{T-1} to i_T with a transition probability of $a_{i_{T-1} i_T}$ to generate a symbol of O_T with a probability of $b_{i_T}(O_T)$. Computation of this probability requires $(2T - 1)N^T$ multiplication and $(N^T - 1)$ additions. It is computationally unfeasible exercise. Hence, we have more sophisticated method of the forward–backward procedure.

8.4 The Forward–Backward Procedure

Let us consider forward variable $\alpha_t(i)$ as defined:

$$\alpha_t(i) = P(O_1, O_2, \ldots, O_{t,}i_t = q_i/_\lambda)$$

where the probability of the partial observation sequence, until time t and state q_i at time t, given the model λ.

$\alpha_t(i)$ can be solved inductively as,

$$\alpha_1(i) = \pi_i b_i(O_1), 1 \leq i \leq N$$

For $t = 1, 2, \ldots, T - 1$ and $1 \leq j \leq N$

$$\alpha_{t+1}(j) = \left[\sum_{i=1}^{N} \alpha_t(i) a_{ij}\right] b_j(O_{t+1})$$

Then, $P(O|\lambda) = \sum_{1}^{N} \alpha_t(i)$.

Computation of $\alpha_t(j)$ involves N^2T calculations instead of the same in direct computation $2TN^T$ as mentioned in the last section. For $N = 5$, $T = 100$ we need 3000 computations in forward method whereas in direct computation, it needs 10^{72} computations.

Since there are only N states, all the possible state sequences will re-merge into these N states, irrespective of how long the observation sequence is.

At time $t = 1$, we need to calculate values of $\alpha_1(i)$,$1 \leq i \leq N$. At times $t = 2,3,..T$ one needs to calculate values of $\alpha_t(j)$, $1 \leq j \leq N$, where each calculation involves N previous values of $\alpha_{t-1}(j)$, because each of the N grid points is reached by same N grid points in the previous moment.

Similarly, we can define backward variable $\beta_t(i)$ as:

$$\beta_t(i) = P(O_1, O_2, \ldots, O_{t,}i_t = q_i/_\lambda)$$

That is, probability of the partial sequence from t+1 to the end, given state q_i at time t, given the model λ.

We can solve for $\beta_t(i)$ inductively, such as

$\beta_t(i) = 1$ for $1 \leq i \leq N$

For $t = T$-1, T-2, …, 1 and $1 \leq i \leq N$

$$\beta_t(i) = \sum_{j=1}^{N} a_{ij} b_j O_{t+1}(j)$$

Now if we estimate the number of computations to find $\beta_t(i)$, it will turn out to be N^2T which is significantly reduced from direct computation.

8.5 Discussion

HMM has been widely applied for identifying the sequence of events in understanding the biological processes. It is used effectively in understanding the genomic sequences and anomaly therein. Simple applications like identifying the ancestry of an individual based on a population, also uses HMM. Besides, Bioinformatics, HMM has found its use in understanding the speech and speech sequences and in telecom. Moreover, Natural Language Processing also uses the same theoretical assumptions.

References

Rabiner LR (1989) A tutorial on hidden Markov models and selected applications in speech recognition. Proc IEEE 77(2):257–286

Further Reading

Baum LE (1972) An inequality and associated maximization technique in statistical estimation for probabilistic functions of Markov processes. In: Shisha O (ed) Inequalities III: proceedings of the 3rd symposium on in-equalities. Academic, Cambridge, pp 1–8

Baum LE, Eagon JA (1967) An inequality with applications to statistical estimation for probabilistic function of Markov processes and to a model for ecology. Bull Am Math Soc 73(3):360–363

Baum LE, Petrie T (1966) Statistical inference for probabilistic functions of finite-state Markov chains. Ann Math Stat 37(6):1554–1563

Choo KH, Tong JC, Zhang L (2004) Recent applications of Hidden Markov models in computational biology. Genomics Proteomics Bioinformatics 2:84–96

Dempster AP, Laird NM, Rubin DB (1977) Maximum likelihood from incomplete data via the EM algorithm. J R Stat Soc 39(1):1–21

Durbin R, Eddy S, Krogh A, Mitchison G (1998) Biological sequence analysis: probabilistic models of proteins and nucleic acids. Cambridge University Press, Cambridge

Eddy SR (1996) Hidden Markov models. Curr Opin Struct Biol 6:361–365

Eisner J (2002) An interactive spreadsheet for teaching the forward-backward algorithm. In Proceedings of the ACL workshop on effective tools and methodologies for teaching NLP and CL, pp 10–18

Forney GD (1973) The Viterbi algorithm. Proc IEEE 61(3):268–278

Gough J (2002) The superfamily database in structural genomics. Acta Crystallogr D Biol Crystallogr 58:1897–1900

Jelinek F (1997) Statistical methods for speech recognition. MIT Press, Cambridge

Karplus K, Barrett C, Hughey R (1998) Hidden Markov models for detecting remote protein homologies. Bioinformatics 14:846–856

Kruskal JB (1983) An overview of sequence comparison. In: Sankoff D, Kruskal JB (eds) Time warps, string edits, and macromolecules: the theory and practice of sequence comparison. Addison-Wesley, Boston, pp 1–44

Markov AA (1913) Essaid'une recherche statistique sur le texte du roman "Eugene Onegin" illustrant la liaison des epreuveen chain ('Example of a statistical investigation of the text of "Eugene Onegin" illustrating the dependence between samples in chain'). Izvistia Imperatorskoi Akademii Nauk 7:153–162

Markov AA (2006) Classical text in translation: A. A. Markov, an example of statistical investigation of the text Eugene Onegin concerning the connection of samples in chains. Sci Context 19(4):591–600

Needleman SB, Wunsch CD (1970) A general method applicable to the search for similarities in the amino-acid sequence of two proteins. J Mol Biol 48:443–453

Rabiner LR, Juang BH (1993) Fundamentals of speech recognition. Prentice Hall, Upper Saddle River

Reichert TA, Cohen DN, Wong AKC (1973) An application of information theory to genetic mutations and the matching of polypeptide sequences. J Theor Biol 42:245–261

Sakoe H, Chiba S (1971) A dynamic programming approach to continuous speech recognition. In: Proceedings of the seventh international congress on acoustics, vol 3. Akademiai Kiado, Budapest, pp 65–69

Sonnhammer EL, Eddy SR, Birney E, Bateman A, Durbin R (1998) Pfam: multiple sequence alignments and HMM-profiles of protein domains. Nucleic Acids Res 26:320–322

Truong K, Ikura M (2002) Identification and characterization of subfamily-specific signatures in a large protein superfamily by a hidden Markov model approach. BMC Bioinf 3:1

Neural Network and Deep Learning

9

Abstract

In this chapter we will provide an introductory knowledge of Neural Network and Deep Learning and we will use those two words interchangeably meaning the same. Neural network has evolved from a concept of simulating human brain recognition of images through communication involving neurons. We will analyze how handwritten digits can be recognized effectively using neural network techniques. Neural network as a part of statistical learning methods evolved more than 50 years and it got new energy because of application in understanding complex processes in Physics and Life Sciences.

Keywords

Neural network · Artificial neural network · Convolutional neural network · TensorFlow

9.1 Introduction

As the name suggests, neural network is an effort to replicate the communication, manipulation, and output of a neuron based human brain system. One can understand the neural network concept, specially feed forward network, analyzing the computer vision problem of identifying handwritten digits through an algorithm.

S. Ghosh, R. Dasgupta, *Machine Learning in Biological Sciences*,
https://doi.org/10.1007/978-981-16-8881-2_9

The normal eyes can detect these images and our human brain can recognize these digits as 2, 0, 3, 4, 5, 6, 7, 8, 9, and 1. This is possible due to our highly trained neural network present in human brain. So the question is how does the human brain recognizes these digits. Let us try to understand the mechanism.

When one sees the image, one tries to look at the image as a composite of different geometrical structures and puts a weight factor at different levels to try and match the known structure of a digit. The one that gets the maximum score is the winner. This is how a collection of neurons decipher an image and identify the image.

Now the question is how we do the same through a computer vision algorithm. We need to get the image at pixel level of resolution. The light intensity is measured at each pixel. On needs light sensors to turn the light rays into electrical impulses that the computer can detect. The image needs to be discretized and represented by a 2d array. MNIST dataset (https://www.nist.gov/itl/products-and-services/emnist-dataset) provides data for every image and each image is 28×28 array of integers, where white is represented by 0 and black as 255, every other shade is represented by an integer between 0 and 255.

9.2 Perceptron

Perceptron was created as a computational model of neurons. A single neuron typically has many inputs (called dendrites), a cell body and a single output (termed as axon). Similarly, a perceptron has many inputs, hidden layers, and a single output.

In this diagram, x_i (s) are inputs, w_i (s) are weights and $f_\Phi(x)$ is the output. $f_\Phi(x)$ is equal to 1 if $b + \sum_{i=1}^{l} x_i w_i > 0$ otherwise it is 0.

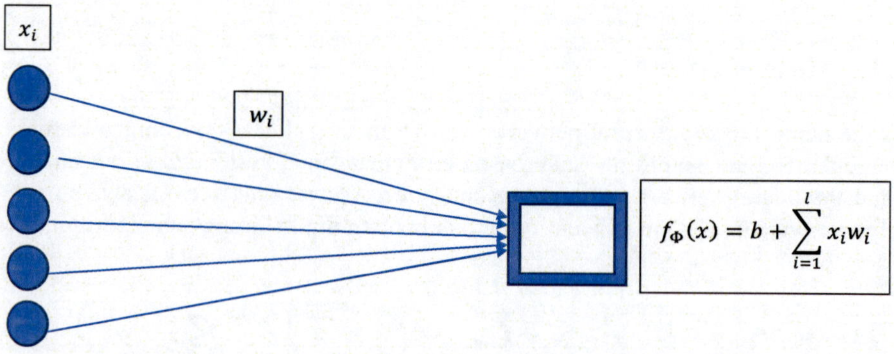

In fact a perceptron consists of a vector of weights $w = [w_1 w_2 \ldots w_m]$, one for each input x_i, a bias b. These are the parameters of perceptron.

In essence, one takes a dot product of the x_i and w_i and add the bias which is a constant weight.

Let us analyze a two input scenario. We assume that there two inputs and single output. We consider a perceptron with two inputs, each having weight factor –2 and a bias 3.

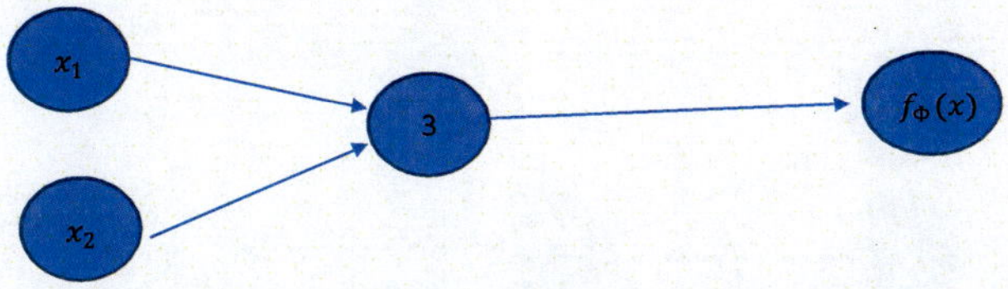

When the input is (0,0) the output is $f_\Phi(x) = 0 * (-2) + 0 * (-2) + 3 = 3$. In case of input (2), the output $f_\Phi(x) = 1 * (-2) + 1 * (-2) + 3 = -1$. So the input can be (0,0), (0,1), (1,0) with a positive value of $f_\Phi(x)$. We can consider cases with multiple perceptrons and hidden layers as well. Since (0,0) is successful and so it is like a logic gate NAND. We can actually implement it slightly differently, introducing a layer of multiple perceptrons.

9.3 Sigmoid Neurons

In the last section, we considered a problem where the output is single and has binary values 0 and 1. For better approximation of results we may create a model where the $f_\Phi(x)$ can assume any values between 0 and 1, any non-integral values. That requires us to create a different neural network with a non-linear $f_\Phi(x)$. The purpose of using sigmoid neuron is to train the learning engine with a greater accuracy. We would like the network to learn weights and biases so that the output from the network correctly classifies the digit. To see how learning might work, suppose we make a small change in some weight (or bias) in the network. What we'd like is for this small change in weight to cause only a small corresponding change in the output from the network.

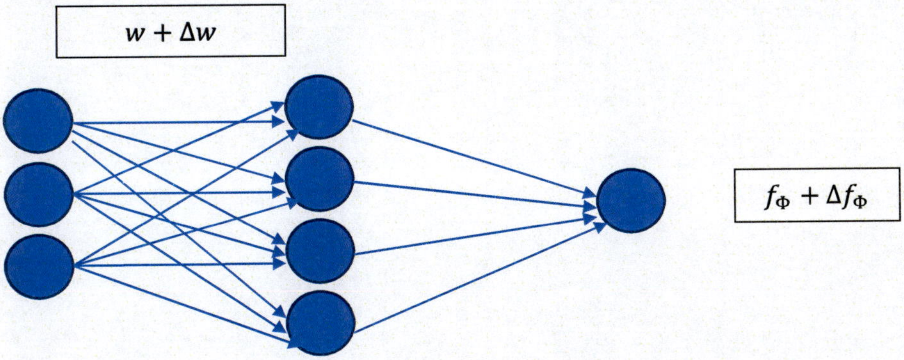

If we could change the output with small change in weight, we would have used the system to learn the way we want. In reality it does not behave accordingly and it is difficult to manipulate all possible parameter values to adjust the output.

So one defines sigmoid neuron in the same way we defined perceptrons.

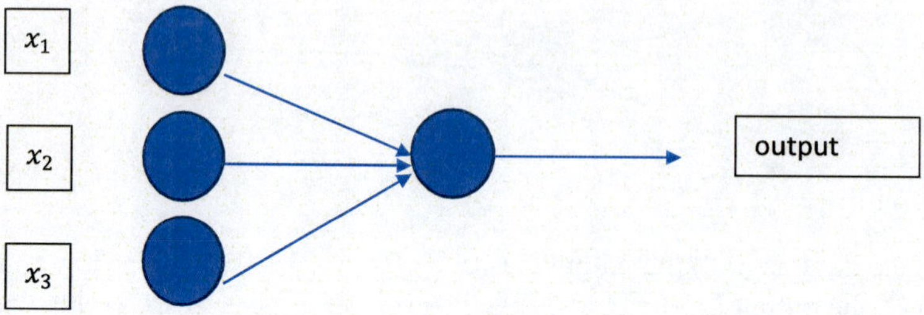

Here $f_\Phi(x) = \sigma(wx + b)$ with σ is called the sigmoid function and is defined as $\sigma(z) = \frac{1}{1+e^{-z}}$ and $z = wx + b$ hence the output is $\frac{1}{1+\exp\left(-\sum_j w_j x_j - b\right)}$. How should we interpret the output from a sigmoid neuron? Obviously, one big difference between perceptrons and sigmoid neurons is that sigmoid neurons don't just output 0 or 1. They can have as output any real number between 0 and 1, so values such as 0.173. . . and 0.689. . . are legitimate outputs.

9.4 Structure of Neural Networks

So a neural network consists of input layer, hidden layers (manipulating input with varying weight factors), and a output layer. The output gets scored based on the output function and that in turn depends on the input parameters, weight functions and bias. So here is a typical structure of a neural network.

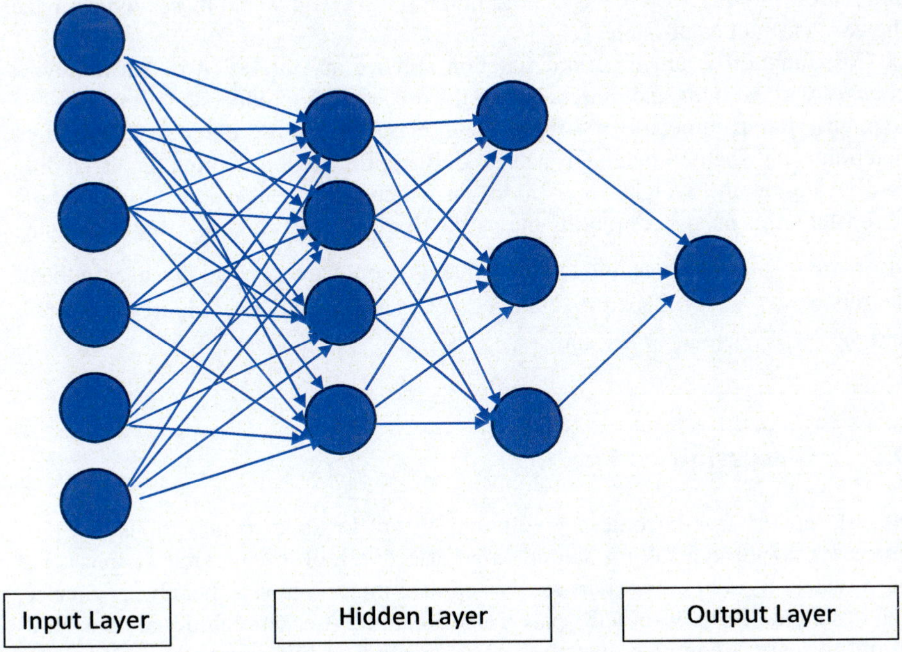

9.5 Cost Function and Back Propagation

When one designs a neural network, one needs to understand the effectiveness of the network model and whether it can be fine-tuned based on some cost function.

$$C(w, b) = \frac{1}{2n}\sum_{x}||y(x) - a||^2$$

Here, w denotes the collection of all weights in the network, b all the biases, n is the total number of training inputs, a is the vector of outputs from the network when x is input, and the sum is over all training inputs, x. Of course, the output a depends on x, w and b.

C is also called the quadratic cost function; it's also sometimes known as the mean squared error or just MSE. Inspecting the form of the quadratic cost function, we see that $C(w, b)$ is non-negative, since every term in the sum is non-negative. Furthermore, the cost $C(w, b)$ becomes small, i.e., $C(w, b) \approx 0$, precisely when $y(x)$ is approximately equal to the output, a, for all training inputs, x. So our training algorithm has done a good job if it can find weights and biases so that $C(w, b) \approx 0$. By contrast, it's not doing so well when $C(w, b)$ is large—that would mean that $y(x)$ is not close to the output a for a large number of inputs. So the aim of our training algorithm will be to minimize the cost $C(w, b)$ as a function of the weights and biases. In other words, we want to find a set of weights and biases which make the cost as small as possible.

Cost function is an important function and we are interested to know how it changes with weights and biases. Hence the objective is to understand how one can fine tune the parameters of weights and biases to adjust the cost function to determine the whether the goal is achieved. Backpropagation is about understanding how changing the weights and biases in a network changes the cost function. Ultimately, this means computing the partial derivatives $\frac{\partial C}{\partial w_{ij}^{l}}$ and $\frac{\partial C}{\partial b_{j}^{l}}$. But to compute those, we first introduce an intermediate quantity, δ_{j}^{l}, which we call the error in the jth neuron in the lth layer. Backpropagation gives us a procedure to compute the error δ_{j}^{l}, and then will relate δ_{j}^{l} to $\frac{\partial C}{\partial w_{ij}^{l}}$ and $\frac{\partial C}{\partial b_{j}^{l}}$.

9.6 Discussion

Neural network has become a revolutionary tool to understand various complex processes. While typically it started as an effective tool in computer vision, it has been widely used in other areas such as clinical image analysis, healthcare finance, and detection of cancer or bell's palsy. Deep learning uses convolutional network as a tool to learn a complex system such as creating artificial languages and doing speech recognition, etc. It is also widely used in cyber security as a tool to understand the essential breaking point in a topological structure of a network and determine the probability of intrusion. This chapter is an introductory one. The purpose was to create interest among the readers and introduce the fundamental concepts.

Further Reading

Abadi M, Agarwal A, Barham P, Brevdo E, Chen Z, Citro C, Corrado GS, Davis A, Dean J, Devin M et al (2016) Tensorflow: large-scale machine learning on heterogeneous distributed systems. arXiv preprint arXiv:1603.04467

Alipanahi B, Delong A, Weirauch MT, Frey BJ (2015) Predicting the sequence specificities of DNA-and RNA-binding proteins by deep learning. Nat Biotechnol 33:831–838

Basdevant C, Deville M, Haldenwang P, Lacroix J, Ouazzani J, Peyret R, Orlandi P, Patera A (1986) Spectral and finite difference solutions of the Burgers equation. Comput Fluids 14:23–41

Baydin AG, Pearlmutter BA, Radul AA, Siskind JM (2015) Automatic differentiation in machine learning: a survey. arXiv preprint arXiv:1502.05767

Hornik K, Stinchcombe M, White H (1989) Multilayer feedforward networks are universal approximators. Neural Netw 2:359–366

Krizhevsky I, Sutskever GE, Hinton (2012) ImageNet classification with deep convolutional neural networks. In: Advances in neural information processing systems. ACM, New York, pp 1097–1105

Lake M, Salakhutdinov R, Tenenbaum JB (2015) Human-level concept learning through probabilistic program induction. Science 350:1332–1338

LeCun Y, Bengio Y, Hinton G (2015) Deep learning. Nature 521:436–444

Owhadi H (2015) Bayesian numerical homogenization. Multiscale modeling. Simulation 13:812–828

Owhadi H, Scovel C, Sullivan T et al (2015) Brittleness of Bayesian inference under finite information in a continuous world. Electron J Stat 9:1–79

Raissi M, Karniadakis GE (2017) Hidden physics models: machine learning of nonlinear partial differential equations. arXiv preprint arXiv:1708.00588

Raissi M, Perdikaris P, Karniadakis GE (2017a) Inferring solutions of differential equations using noisy multi-fidelity data. J Comput Phys 335:736–746

Raissi M, Perdikaris P, Karniadakis GE (2017b) Machine learning of linear differential equations using Gaussian processes. J Comput Phys 348:683–693

Raissi M, Perdikaris P, Karniadakis GE (2017c) Numerical Gaussian processes for time-dependent and non-linear partial differential equations. arXiv preprint arXiv:1703.10230

Rasmussen CE, Williams CK (2006) Gaussian processes for machine learning, vol 1. MIT Press, Cambridge

Rudy SH, Brunton SL, Proctor JL, Kutz JN (2017) Data-driven discovery of partial differential equations. Sci Adv 3:1602614

Ethics in Machine Learning and Artificial Intelligence

10

Abstract

Artificial intelligence (AI), machine learning (ML), and deep learning (DL) have progressed tremendously in the last 5 years and have now been applied to the different sectors on human life including agriculture, animal science, plant science, healthcare, and education. However, it is important at this point along with its expansion to evaluate the importance of ethics in AI. We discuss in this chapter highlighting the important role of ethics in AI that is being recognized globally in the current times by tech companies associated with AI.

Keywords

Ethics · Bias · Consequentialism · Deep learning · Machine learning

Abbreviations

AI	Artificial intelligence
ML	Machine learning
DL	Deep learning
NeurIPS	Neural information processing systems
MCM	Moral choice machine

10.1 Introduction

In the last 5 years considerable progress has been made in the field of artificial intelligence (AI), machine learning (ML) and deep learning (DL). With the continuous expansion of application of artificial intelligence in diverse domains of

S. Ghosh, R. Dasgupta, *Machine Learning in Biological Sciences*,
https://doi.org/10.1007/978-981-16-8881-2_10

agriculture, education sector, healthcare and transportation and is being used by the engineers, scientists, designers, policy makers affecting the social, economic, health and daily lives of humans, one needs to understand that algorithms should not be used with the wrong intention of harming the human. Ethics is a matter of debate in AI particularly in health care sector. In 2018 a conference on Neural Information Processing Systems (NeurIPS) in Vancouver, Canada, had a focus on ethics in AI.

Despite tremendous advances technically, in the concepts of accountability and governance of AI, transparency, explainability and bias, fairness, ownership, control, ethics, social impact, policy law, social security are essential issues linked to AI, ML and DL, that needs most discussions at this point of time.

Construction of algorithms needs to be aimed at considering that they do not pose a threat to the human lives and world security and peace. Thus, machines are to be designed with the ability of learning social, ethical, and moral choices.

10.2 Moral Choice

ML has been applied to human texts for extraction of deontological ethical reasoning or understanding an action as right or wrong with template with prompts and responses, including should I or is it ok with answer choices of "Yes/no, as positive/negative response aiming at the design of Moral Choice Machine (MCM).

10.3 Fairness and Concept of Consequentialism

As it is being more realized that AI based systems could potentially create harmful consequences, attempts are being made to introduce fairness in developing algorithms satisfying these criteria. Implications of ethics and fairness find importance in machine learning, with a perspective on consequentialism or outcomes of a matter. Consequentialism or outcomes of actions is a vital concept within the domain of ethics. In mathematical formulation consequentialism includes concepts of act consequentialism, or best action to take in any situation producing best outcomes and rule consequentialism which focusses on action based on set of rules controlling all situations, and adoption of these set of rules leading to the best overall outcomes. Although the concept of consequentialism suffers from challenges arising out of aggregation, subjectivity and uncertainty and does not provide guidance of choosing actions, it enables critique on machine learning fairness.

Fairness criteria based on outcome terms reveals close to consequentialism, as compared to fairness proposals specified without outcomes regard.

10.4 Tackling Biases

Ethical controversies in AI springs in the domains of technologies and their application in predictive policing or facial recognition. Tackling biases in algorithms in patterns of discrimination in data, and avoid affecting already vulnerable populations.

10.5 Business

Tech companies involved in AI research are also now focussed on ethics of their work. Google had presented 12% of papers at NeurIPS. Companies suffer from a lack of diversity. Although staffs and boards seems dedicated to ethics, it is not enough to maintain the ethics in technical solutions. A suggestion involves including of affected groups on ethics boards might enable better practices of ethics by Tech companies associated with AI.

10.6 Discussion

Tech giants including Alphabet, Amazon, Facebook, IBM, and Microsoft and global leaders like Stephen Hawking and Elon Musk have highlighted the important role of ethics and risk assessment in platforms of emerging technology. Issues like unemployment, associated with automation, inequality of distribution of wealth by reduction of human workforce while the owners retaining the wealth, affected human behaviour and interaction by machines, artificial stupidity rising from how to guard mistakes, elimination of AI bias, Security are also major focus of research in this domain.

11 Machine Learning and Life Sciences

Abstract

Machine learning has tremendous applications in diverse domains of Life Sciences, including Health Sciences, basic biology research, clinical research, advances in diagnostics and drug development, advances in radiology and radiotherapy, extraction of information from clinicians notes and Electronic Health Records (HER), animal sciences, welfare, health and industry, plant research, and agriculture industry. This is also expected to bring great advances in the global artificial intelligence based industry and the global market. We discuss in this chapter the major applications of Machine Learning in the major domains of life sciences.

Keywords

SVM · RF · Deep learning · Cancer biology · Epigenetics · Drug

11.1 Introduction

Artificial intelligence (AI) is a branch of cognitive computing combining knowledge and applications of computer, information, and cognitive science by which they are able to learn, reason, audit, and interact with humans. It includes design of machines or robots with computer hardware and software simulating close to human intelligence behavior. The human like behavior of AI includes auditing, creativity, imaging, learning, planning, reasoning, scheduling, writing, and natural language processing (NLP).

Since the last 5–10 years there has been continuous growth and evolution of AI by application of machines containing smart features, endowed with new properties and applications, increasing the functions and productivity of equipment and has found role in both industry and academic research and development. We discuss in

S. Ghosh, R. Dasgupta, *Machine Learning in Biological Sciences*,
https://doi.org/10.1007/978-981-16-8881-2_11

this chapter the various applications of machine learning in (1) health, (2) animal, and (3) agriculture sciences.

11.2 Machine Learning and Life Sciences

Deep learning (DL) method and ML algorithms are capable of combining raw inputs into layers of intermediate features and have been applied to a variety of domains including Biology and medicine.

Methods of DL has been applied to biomedical problems, patient classification, understanding of fundamental biological processes, treatment of patients and it has helped and proved advantageous over conventional methods being fast and less time consuming and far more near accurate.

11.2.1 Applications in Health

Machine learning has diverse applications in the field of Health Sciences, from the analysis of data from research publications patents, clinical trials, and health records by using machine learning techniques including Bayesian inference, Markov chain models, natural language processing (NLP) reinforcement learning, storage and application of hugh data through Cloud, detection of disease, its diagnosis, prognosis, search of biomarkers, extraction of information from Doctors notes and Electronic Health Records (HER) of patients by applications of machine learning including Natural language processing (NLP) and feature extraction tools, advances of understanding of disease biology from analysis of image based studies, digital surveillance of diseases and hence a boon to the public health systems, drug design, better and accurate decision making by the Doctors and clinicians on mode course and protocol of treatment and individual patient care, thereby bettering the healthcare system, enabling personalized care and increasing the quality of life.

11.2.1.1 Applications in Drug Design

Drug discovery for therapies include search for antibodies, compounds, small molecules, peptides, short RNAs or cell based therapies that will react with a molecular target and modulate it to alter the disease state. Chemoinformatics involves extraction, processing, extrapolating, and analysis of data from chemical structures which finds importance in the design of drugs and targets. High throughput screening studies and combinatorial synthesis has generated hugh chemical 'big' data, and thus computational studies and machine learning has found importance in processing of the data, analysis and study of drug design. ML has been applied to chemoinformatics analyses aiding in the drug design and discovery studies.

ML use pattern recognition algorithms to understand mathematical relationships of small molecules and predict their chemical, biological, and physical properties of novel compounds and can be applied to huge datasets.

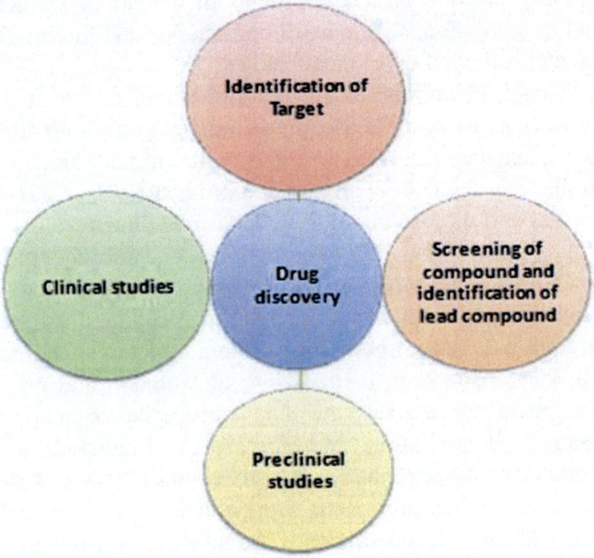

Fig. 11.1 Drug discovery and major areas of applications of machine learning

In the arena of drug discovery ML can enable understanding of relationships between chemical structures and their biological activities, understanding the structure–activity relationships (SARs) of molecules which is primary for any drug discovery. Modeling of QSAR, or quantitative structure–property relationships (QSPR), is the latest application of computational tools in medicinal chemistry for screening of large databases of compounds to understand the biological properties of chemicals from their chemical structure and predict the physicochemical, biological, and environmental properties of compounds, optimization of chemical structure to improve its binding affinity, understand their physiochemical properties and decipher biological functions which is less time consuming, fast, less manpower involving, and predict in silico modifications that influence biological behavior, predicting their toxicity, metabolism, drug–drug interactions, and carcinogenesis. Advanced cheminformatics and ML based approaches can model increasingly complex data (Vamathevan et al. 2019).

Long complex pipelines based on multiple factors are involved in drug discovery and development. Machine learning (ML) tools have enabled better discovery and decision making from analysis of high-throughput and complex data (Fig. 11.1).

Machine learning applications has been effective tools in identification of novel targets, aiding in our understanding of associations between the target and the disease, mechanism of disease, understanding of disease vs. non disease, vs. normal phenotypes, improve design of small-molecule compounds, development of new biomarkers in prognosis, and to study and analyse the effect of drug as treatment protocol, and study its efficacy, improved digital imaging studies and extraction of information from databases and images.

Drug discovery includes stages of development including, target identification and its validation, screening of compounds and finding the lead compound, preclinical and followed by clinical studies.

Target identification studies has enabled our understanding of associations between genes and disease, predictions of targets, identification of alternative targets like splice variants. Data for these studies require high dimensional target association studies based on target disease and drug, analysis of data from omics based studies in normal and disease conditions, data from literature and from failed and successful clinical trials. Screening of compounds and subsequent lead discovery studies using machine learning has enabled design of compounds with desirable properties, synthesis and reactions, ligand based compound screening studies and the data is large in nature with numerous protein structures, and data on properties of ADME (*ADME, absorption, distribution*, metabolism and excretion) of compounds. The preclinical studies using machine learning has been applied on understanding tissue specific identification of biomarkers, prediction of disease-drug responses, biomarkers for prognosis and clinical end points and the data in here largely relies on gene expression data, single cell omics based studies, proteomic and transcriptomic based studies. Clinical development studies using machine learning has enabled, detection of drug response cellular by cell phenotypes, measurement of disease microenvironment like in cancer and the data here is hugh, from laboratory based studies on normal and diseased cell and from clinical based studies and clinical trials (Vamathevan et al. 2019).

Machine learning tools including search, retrieval and extraction tools, together with image mining tools, pattern matching tools are being used in understanding the novel small molecules and predict their chemical, biological and physical properties from the chemical structure, the optimization of its improved binding affinity, biological or physiochemical properties, toxicity, metabolic properties, drug-drug interactions, carcinogenesis can be deciphered.

Machine learning has been applied to diverse stages of drug designing including computer-aided drug synthesis, exploration of chemical space exploration, design of pharmacophore or the part of a molecular structure involved in particular biological or pharmacological interaction and analysis of scaffold or generate, analogue members of core active compound structure and analyze, study and compare with its analog members and design of library. Machine learning converts a compound structure into chemical information by different methods of graph retrieval, generation of descriptor, similarity analysis, and each layer is built upon previous layers (Fig. 11.2).

Although several applications of machine learning are being applied in different stages of drug discovery, there suffer challenges of repeatability, standard of data quality and quantity before they gave desired results in the domain of drug discovery.

Linear Discriminant Analysis (LDA), Principal components analysis (PCA), Support vector Machines (SVM), Random forest (RF), k-Nearest Neighbor (kNN), and Naïve Bayes (NB) have been applied in the design of anticancer drugs (Hu et al. 2019) (Fig. 11.3).

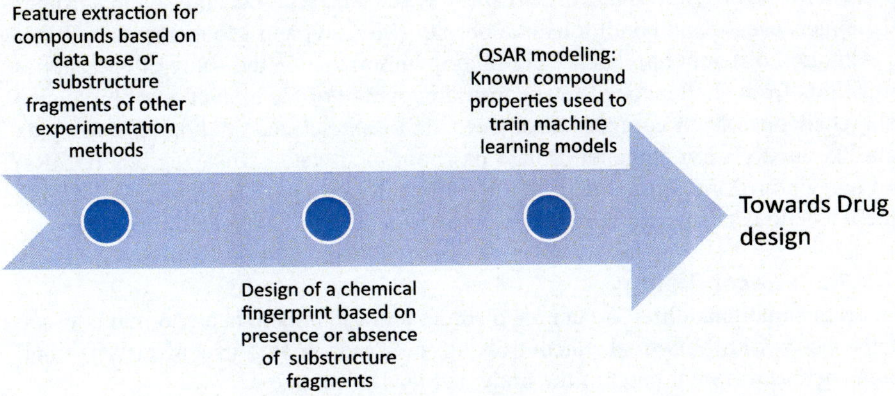

Fig. 11.2 Cheminformatics and QSAR using machine learning

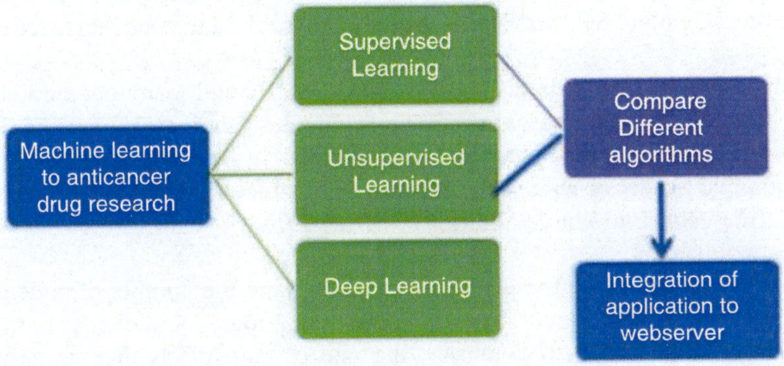

Fig. 11.3 Machine learning in anti-cancer drug research

11.2.1.2 Epigenetics and Disease Biology

Epigenetics involves heritable changes in gene expression without any changes in the DNA sequence leading to changed phenotype without change in genotype.

Since the early inception of the concept of epigenetics in 1942, it has now been known that each cell type has a unique epigenome. Exposure to environmental agent can lead to epigenetic disease state that can be transmitted through generations thus the epidemics of disease is an important domain of study and it is important to identify the portion of genome that is susceptible to epigenetic changes. DNA methylation such as CpG, histone modifications, noncoding RNA (ncRNA), and chromatin structure alteration like nucleosome positioning are different epigenetic modification, influencing gene expression without DNA sequence alteration.

Epigenetic changes have been known to affect the gene expression patterns causing autoimmune disorders, neurological disorders, allergies, cancer, schizophrenia, Fragile X syndrome, Angelman's syndrome, Prader–Willi syndrome,

Rubinstein-Taybi syndrome, Coffin–Lowry syndrome, ATR-X, and Rett syndromes Alzheimer disease and conditions like obesity (Moosavi and Motevalizadeh 2016).

Although different approaches are finding importance in the study of epigenetics but suffers from challenges of time consuming nature of the extraction of biological data, challenges from computational methods to extract and analyze this complex data like methylation epigenetics data has both methylated DNA regions (DMRs) and many non-DMR sites adding to the complexity of data. Thus machine learning on epigenetic data sets are being developed (Moosavi and Motevalizadeh 2016).

11.2.1.3 Cancer Biology

Cancer is a malignant life-threatening disorder and can initiate in any organ or tissue of the body with abnormal, uncontrollable cell growth that can metastasize and spread across different parts of the body and lead to death.

Cancer is globally considered to be the second leading cause of death with an estimated death of 9.6 million people recorded in 2018 of which the colorectal, lung, liver, prostate, and stomach cancer are reported to be common in men, while in women, breast, colorectal, cervical, lung, and thyroid cancer are reported to be common.

The number of cancer patients is on the rise globally and many people living in low and middle income countries are still far from getting early treatment and cannot afford the tremendous associated cost of treatment. In developed countries, with better equipped health systems, and facilities of treatment, however the chances of disease free survival have increased (https://www.who.int/health-topics/cancer#tab=tab_1).

Cancer has also affected the pediatric population and the number of patients are on the rise causing death for children and adolescent population globally. In the low and middle income group of countries, the rate of cure of children is very low of around only 20% due to no access to early diagnosis, of disease low or no access to therapy, death due to toxic effects of therapy, not proper continuation of treatment, unaffordable hugh cost of treatment, while in the high-income countries it is more than 80% (https://www.who.int/news-room/fact-sheets/detail/cancer-in-children).

Unlike adult cancer, the cause for pediatric cancer remains largely unknown, although few studies have hinted at environmental or lifestyle factors. Genetic factors and prenatal exposure to radiation, diethylstilbestrol [DES] and postnatal exposures to radiation and infection by viruses, like infection by HIV, Epstein-Barr virus (EBV) and malaria are reported to increase the risk associated with childhood Cancer. According to the International Classification of Childhood Cancers (ICCC) based on histology, there are 12 major categories of cancer including (1) leukemia, (2) Tumors and cancers of the central nervous system (CNS) and different intracranial and intraspinal neoplasms, (3) lymphomas and other reticuloendothelial neoplasms, (4) carcinomas and other malignant epithelial neoplasms, (5) germ cell, trophoblastic, and other gonadal neoplasms, (6) soft tissue sarcomas, (7) malignant bone tumors, (8) sympathetic and allied nervous system tumors, (9) renal tumors, (10) retinoblastoma, (11) hepatic tumors (12) others and unspecified. According to ICCC childhood cancers are known to occur in children and young adults under the age of 20 years (Hewitt et al. 2003). The most common diagnosed childhood cancers include acute lymphoblastic leukemia (ALL), astrocytoma, neuroblastoma, nephroblastoma and non-Hodgkin lymphoma (Kaatsch 2010) and the trend of Childhood cancer is on the increase (Isaevska et al. 2017).

Childhood acute lymphobastic leukemia, (ALL) is the most common malignancy of the pediatric world accounting for affecting 25% of all childhood cancers (Bhojwani et al. 2015). Different markers including differential expression of 9-*O*-acetylated sialic acids in diagnosis and prognosis of patients are being exploited in understanding the disease biology, diagnosis and monitoring of disease (Pal et al. 2004; Ghosh et al. 2007). Although treatment rates are better in the United States, concerns remain over finding understanding the heterogeneity associated and the minimal residual disease (MRD) exist.

Early diagnosis of Cancer with the help of biomarkers, and effective biochemical histopathological analysis is the only effective way to treat and target Cancer so that specific therapy including surgery, radiotherapy, and chemotherapy protocols can be administered and the body can respond to effective treatment leading to better chances of survival of the Cancer patients and children.

Machine learning (ML) has been applied to identify, classify patterns from large quantities of cancer genomic data leading to identification of disease specific biomarkers, drug targets, and in understanding of important cancer genes. It has been applied in the study of head and neck cancer (HNC), oral cancer and laryngeal cancer employing SVM great accuracy and enabled diagnosis and prognosis of HNC (Patil et al. 2019).

SVM has been applied in different Cancers and domains in Cancer Biology including Classification, typing, biomarker discovery, drug screening, identification of novel targets, driver gene discovery, and gene interactions (Huang et al. 2018) (Fig. 11.4).

Applications of machine learning to gene sequencing data have enabled identification and typing of soft tissue sarcoma subtypes and is used to identify novel diagnostic and prognostic markers and therapeutic targets for soft tissue sarcomas (van Izendoorn et al. 2019). Logistic regression, SVM, and RF have enabled classification of osteosarcoma (Li et al. 2017).

kNN and random forest algorithms applications has enabled identification of gene hubs with implications in tumor proliferation, progression, migration, and metastasis and thereby enable better detection and treatment of pancreatic cancer (Li et al. 2018).

Breast cancers expressing estrogen receptors have been reported as classified as *ER-positive* cancers. Deep learning (DL) and machine learning models, of RF, SVM, recursive partitioning and regression trees (RPART), LDA, prediction analysis for microarrays (PAM), and generalized boosted models (GBM) has enabled detection Estrogen Receptor (ER) status in breast cancer from metabolomics data (Alakwaa et al. 2018).

With huge surge of data from Cancer Research, requirement of new databases is highly emphasised. A Cancer Systems Biology Database (CancerSysDB) has been

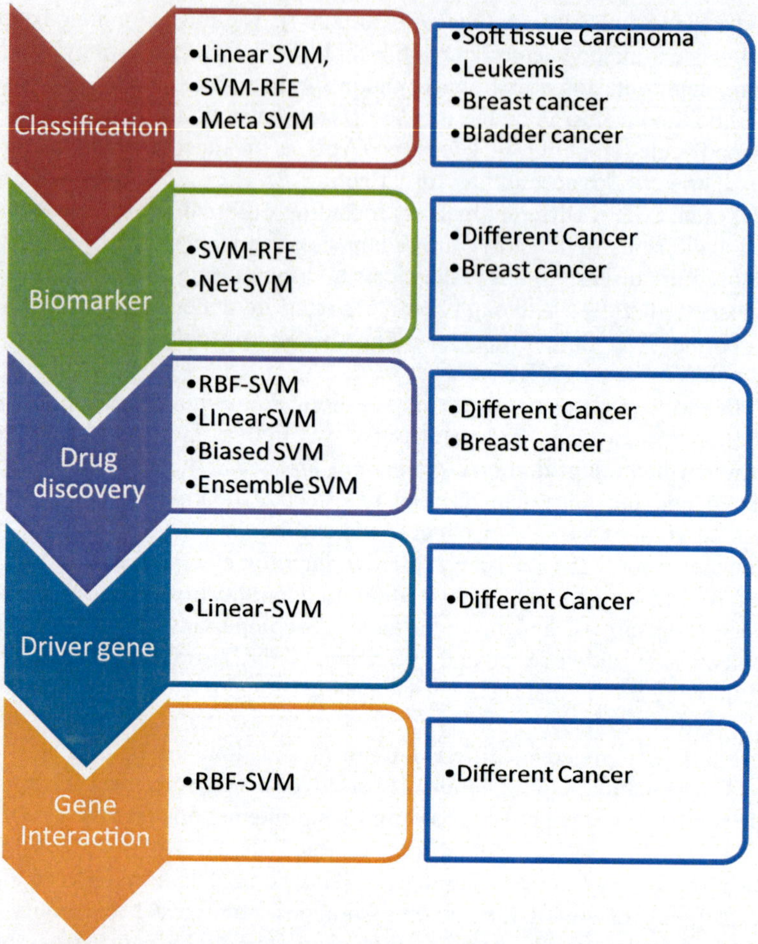

Fig. 11.4 Application in SVM in different aspects of Cancer Biology study: SVM is tool to analyze data and the linear model is used in problem classification and regression in which the algorithm can create a line or a hyperplane used to segregate data into classes. *SVM:RFE* SVM recursive feature elimination is a feature selection tool that can filter required features and eliminate insignificant ones leading to formation of higher classification, *RBF:SVM*: radial basis function SVM, is based on kernel function, *netSVM* network constrained SVM is a classification and network visualization tool, *reg SVM* regression SVM is based on regression method

designed and can enable cancer classification, prediction based on whole-exome data in the Cancer Genome Atlas (Krempel et al. 2018).

11.2.1.4 Infectious Disease Biology

In a study machine learning clinical decision support systems (ML-CDSS) were observed to assist clinicians to address diagnosis, detection, prediction, stratification of sepsis, prediction of prognosis, antibiotic resistance, antibiotic to be chosen and decision on combination antiretroviral therapy in bacterial, viral diseases including tuberculosis, from data collected from intensive care units (ICU), consultation, medical, surgical wards, emergency centres, primary care unit, and application of antimicrobials from diverse socioeconomics settings including high, middle and low-income countries (Peiffer-Smadja et al. 2020). Improvement of treatment in invasive mold disease (IMD) suffers from challenges of lack of cases. NLP tools of machine learning has been used to extract information by screening chest tomography (CT) reports for pulmonary IMD, highlighting the importance of application of antifungal therapy (Baggio et al. 2019).

Rising antimicrobial resistance (AMR) across the globe is a threat to humanity and new approaches to combat AMR are being searched for. Machine learning has been applied to sequencing based data and EHR in the study of antimicrobial susceptibility genotype or prediction of phenotyopes, clinical decision making, search for novel antimicrobials, and use in antimicrobial therapy. However, challenges remain in data quality and interpretability (Macesic et al. 2017). Machine learning has been applied to data from HIV patients and aid in clinical decision making in antiretroviral therapy (ART) and in predicting treatment outcomes of HIV (Bisaso et al. 2017). Infectious diseases, and those transmitted by arthropod vectors has been reported to claim millions of lives and are a global concern. But the dynamics of host-vector-pathogen is extremely complex and machine learning is being applied to decipher the complex relationships between pathogens-vectors and hosts using data mining strategies (Agany et al. 2020).

11.2.1.5 Immunological Disorders

The immune system, plays role in protecting our body from infectious diseases and invading foreign pathogens and can differentiate between foreign and self. However, a dysregulated immune system can lead to diseases.

Autoimmune Disorders

In an autoimmune disorder, the immune system attacks the body parts, as foreign and releases autoantibodies that attack healthy cells and cause the disorder. Some autoimmune diseases like Type 1 diabetes (TID) are organ specific and target only single organ, causing damage to the pancreas while some autoimmune disorder attacks the whole body leading to a systemic disorder like in systemic lupus erythematosus (SLE).

Although factors like genetic, environmental, and infection have been associated to trigger the autoimmune disease, the exact cause of the disorder in the individuals remains largely unknown. Some common autoimmune diseases include celiac disease, TID, Graves' disease, inflammatory bowel disease (IBD), multiple sclerosis (MS), psoriasis, rheumatoid arthritis (RA), and SLE but their diagnosis can be difficult. Treatment may involve administration of nonsteroidal anti-inflammatory drugs (NSAIDs) and immunosuppressants and in severe cases use of intravenous immunoglobulin (Ig) which can lead to improvement of symptoms but cannot provide complete cause to the disorder (Owen et al. 2013; Ghosh 2017, 2019, 2020).

Machine learning tools of Support vector machines (SVMs) and RF have been applied to the study of autoimmune disorder including multiple sclerosis (MS), rheumatoid arthritis (RA), and inflammatory bowel disease (IBD) and have been applied to extract patterns from patient data, to predict treatment outcomes, and better management of disease (Stafford et al. 2020). Machine learning tools of SVM has been applied to analysis of gene expression data and used to understand histologic pattern of rheumatoid arthritis (RA) from analysis of synovial tissue and therefore identify the subtypes, including a high inflammatory subtype with pathophysiology of massive leukocyte infiltration, higher expression of markers for systemic inflammation, auto-antibodies, C-reactive protein (CRP) and expressing severity of pain, low inflammatory subtype revealing transforming growth factor β (TGF- β), glycoproteins, neuronal genes pathways, and another subtypes with mixed pathophysiology (Orange et al. 2018). Machine learning algorithms and natural language processing (NLP) have been used to extract information from electronic health records (EHRs) in diseases including SLE (Jorge et al. 2019). Machine learning has been used in generating effective classification in lupus nephritis applying multilinear regression and RF which finds importance in the assessment of patient prognosis (Tang et al. 2018) and in the classification of Juvenile idiopathic arthritis (JIA) which is the common class of childhood rheumatic diseases (Van Nieuwenhove et al. 2019).

Hypersensitivity Disorders

Hypersensitivity disorders are caused when the immune system generates over reactive responses against innocuous antigens leading to tissue damage and could be fatal also. The Gels and Coombs classification has revealed four types of reactions including (i) Type I hypersensitivity reaction is fast in response, caused by IgE antibody on mass cells and basophils that are crosslinked leading to the release of vasoactive molecules, leading to allergy reactions. Skin test can be done to detect the effect of the allergen. The diseases include allergy, atopy, anaphylaxis, asthma, Churg-Strauss Syndrome.

(ii) Type II hypersensitivity disorders are caused by IgM, IgG antibodies, complement, and membrane attack complex (MAC). IgM or IgG bind to the antigen bound host cell, leading to generation of immune response against host cell causing cell destruction by activation of complement cascade and MAC complex formation. This can be tested by direct or indirect Coombs test. Disorders include Autoimmune hemolytic anemia, Rheumatic heart disease, Thrombocytopenia, Erythroblastosis fetalis, Goodpasture's syndrome, Graves' disease, and Myasthenia gravis.

(iii) Type III hypersensitivity disease is caused by IgG that binds to soluble antigen forming immune complex, that is deposited on walls of vessels, joints, kidney causing local inflammation reactions. Diseases include serum sickness, rheumatoid arthritis, arthus reaction, post glomerulonephritis, membranous nephropathy, reactive arthritis, lupus nephritis, and systemic lupus erythematous (SLE).

(iv) Type IV or delayed type hypersensitivity disorder is caused by T helper (Th) cells. Th1 cells ae activated by antigen presenting cell (APC), leading to the activation of macrophages causing inflammatory reaction that leads to tissue damage. Diseases include contact dermatitis, poison ivy rash caused in contact with poison ivy, chronic rejection of transplant, multiple sclerosis, celiac disease, and Hashimoto's thyroiditis.

Machine learning tools have been used to diagnose complex diseases like Chronic obstructive pulmonary disease (COPD), a chronic inflammatory disease of the lungs caused by obstruction of airflow in the lungs (Spathis and Vlamos 2019), asthma (Spathis and Vlamos 2019; Prosperi et al. 2014) with pathophysiology of inflamed airways, mucus production, difficulty in breathing which could also lead to life threatening condition and eczema (Prosperi et al. 2014) with pathophysiology of inflammation of skin also called as atopic dermatitis and predict clinical outcomes in patents in Emergency with detoriating conditions of asthma or COPD (Goto et al. 2018). Machine learning tools have been instrumental in providing insights into the mechanism in pathogenesis of childhood asthma (Saglani and Custovic 2019), through automated system by the development of a scoring system indicating severity of the disease, time, and effect of treatment (Messinger et al. 2019). Machine learning has been instrumental in early risk prediction and classification of asthma patients (Ullah et al. 2019) using information from Raman spectra and blood sera samples by applying artificial neural network (ANN), SVM, and RF (Ullah et al. 2019). Application of machine learning tools have enables classification in Atopic dermatitis (AD) which is a step towards design of therapy leading to precision medicine (Eyerich et al. 2019).

Airborne pollen like Ambrosia pollen commonly called as ragweed can trigger hypersentitivity reactions causing allergic diseases. Therefore, prediction of pollen in the atmosphere is very helpful in understanding the incidence of pollen related allergic disorders and machine learning tools have been employed in forecasting of daily pollen in environment (Zewdie et al. 2019).

11.2.2 Agriculture

Machine learning is being applied to multi-disciplinary agriculture production systems and finds applications in crop management, improving crop yield prediction, detection of disease and thereby aiding in early decision making, detection of weed and their removal, detection of quality of crop, and identification of species, management and production of livestock, animal welfare, management of resources including water and soil (Liakos et al. 2018).

Machine learning (ML) approaches of data-mining and analysis, analysis of next-generation sequencing (NGS) and phenotyping technologies have enabled breeding and identification of target genes, microRNAs (miRNAs) have been associated in stress conditions (Esposito et al. 2019).

Machine learning has been applied to sensor data (Liakos et al. 2018) collected from sensors including acoustic, bio-sensors, electro-optical and mechanical stored and analyzed daily or hourly (Nayeri et al. 2019) basis monitoring farm animals and is enabling effective and better farm management and decision making for farmers (Liakos et al. 2018).

11.2.3 Animal Sciences

Deep Learning applied to clinical conditions in animals has been able to analyze information from data from databases and veterinary clinical notes (Arguello-Casteleiro et al. 2019).

Veterinary informatics is rather a new branch of developing science focused on animal medicine and machine learning applications of natural language processing (NLP) are enabling steps towards One Health initiatives, techniques and applications in animal health (Lustgarten et al. 2020).

Precision livestock farming has advanced by application of sensors, audio and video recording by cameras, microphones, and accelerometers recording images, sounds, and movements together with the application of machine learning algorithms in non-invasively and even remotely automated monitoring animals to monitor their welfare, detect health, take steps in medication as and when disease is suspected and predict productivity. This has been studied in the welfare of pigs in the swine industry (Lustgarten et al. 2020).

11.3 Discussions

Machine learning has touched almost all spheres of our daily lives. It has particularly found importance in the life sciences. In this chapter we discuss the different applications of the machine learning methods, role of supervised, unsupervised learning, neural network, deep learning methods in different domains of life sciences (1) health, (2) agriculture (3) disease biology and (4) animal sciences.

References

Agany DDM, Pietri JE, Gnimpieba EZ (2020) Assessment of vector-host-pathogen relationships using data mining and machine learning. Comput Struct Biotechnol J 18:1704–1721

Alakwaa FM, Chaudhary K, Garmire LX (2018) Deep learning accurately predicts estrogen receptor status in breast cancer metabolomics data. J Proteome Res 17(1):337–347

Arguello-Casteleiro M, Stevens R, Des-Diz J et al (2019) Exploring semantic deep learning for building reliable and reusable one health knowledge from PubMed systematic reviews and veterinary clinical notes. J Biomed Semant 10(1):22

Baggio D, Peel T, Peleg AY et al (2019) Closing the gap in surveillance and audit of invasive mold diseases for antifungal stewardship using machine learning. J Clin Med 8(9):1390

Bhojwani D, Yang JJ, Pui CH (2015) Biology of childhood acute lymphoblastic leukemia. Pediatr Clin N Am 62(1):47–60

Bisaso KR, Anguzu GT, Karungi SA, Kiragga A, Castelnuovo B (2017) A survey of machine learning applications in HIV clinical research and care. Comput Biol Med 91:366–371

Esposito S, Carputo D, Cardi T, Tripodi P (2019) Applications and trends of machine learning in genomics and phenomics for next-generation breeding. Plan Theory 9(1):34

Eyerich K, Brown SJ, Perez White BE, Tanaka RJ, Bissonette R, Dhar S, Bieber T, Hijnen DJ, Guttman-Yassky E, Irvine A, Thyssen JP, Vestergaard C, Werfel T, Wollenberg A, Paller AS, Reynolds NJ (2019) Human and computational models of atopic dermatitis: a review and perspectives by an expert panel of the International Eczema Council. J Allergy Clin Immunol 143(1):36–45

Ghosh S (2017) Immunology and immunotechnology. Books and Allied Pvt Limited, Kolkata

Ghosh S (2019) Computational immunology, basics. CRC Press, Boca Raton

Ghosh S (2020) Computational immunology, applications. CRC Press, Boca Raton

Ghosh S, Bandyopadhyay S, Mukherjee K, Mallick A, Pal S, Mandal C, Bhattacharya DK, Mandal C, Glycoconj J (2007) O-acetylation of sialic acids is required for the survival of lymphoblasts in childhood acute lymphoblastic leukemia (ALL). Glycoconj J 24(1):17–24

Goto T, Camargo CA, Faridi MK, Yun BJ, Hasegawa K (2018) Machine learning approaches for predicting disposition of asthma and COPD exacerbations in the ED. Am J Emerg Med 36(9): 1650–1654

Hewitt M, Weiner SL, Simone JV (2003) Childhood cancer survivorship: improving care and quality of life. National Academies Press, Washington, DC

Hu Y, Lu Y, Wang S, Zhang M, Qu X, Niu B (2019) Application of machine learning approaches for the design and study of anticancer drugs. Curr Drug Targets 20(5):488–500

Huang S, Cai N, Pacheco PP, Narrandes S, Wang Y, Xu W (2018) Applications of support vector machine (SVM) learning in cancer genomics. Cancer Genomics Proteomics 15(1):41–51

Isaevska E, Manasievska M, Alessi D et al (2017) Cancer incidence rates and trends among children and adolescents in Piedmont, 1967-2011. PLoS One 12(7):e0181805

Jorge A, Castro VM, Barnado A et al (2019) Identifying lupus patients in electronic health records: development and validation of machine learning algorithms and application of rule-based algorithms. Semin Arthritis Rheum 49(1):84–90

Kaatsch P (2010) Epidemiology of childhood cancer. Cancer Treat Rev 36(4):277–285

Krempel R, Kulkarni P, Yim A, Lang U, Habermann B, Frommolt P (2018) Integrative analysis and machine learning on cancer genomics data using the cancer systems biology database (CancerSysDB). BMC Bioinf 19(1):156

Li Z, Reza Soroushmehr SM, Hua Y, Mao M, Qiu Y, Najarian K (2017) Classifying osteosarcoma patients using machine learning approaches. Conf Proc IEEE Eng Med Biol Soc 2017:82–85

Li C, Zeng X, Yu H, Gu Y, Zhang W (2018) Identification of hub genes with diagnostic values in pancreatic cancer by bioinformatics analyses and supervised learning methods. World J Surg Oncol 16(1):223

Liakos KG, Busato P, Moshou D, Pearson S, Bochtis D (2018) Machine learning in agriculture: a review. Sensors 18(8):2674

Lustgarten JL, Zehnder A, Shipman W, Gancher E, Webb TL (2020) Veterinary informatics: forging the future between veterinary medicine, human medicine, and One Health initiatives-a joint paper by the Association for Veterinary Informatics (AVI) and the CTSA One Health Alliance (COHA). JAMIA Open 3(2):306–317

Macesic N, Polubriaginof F, Tatonetti NP (2017) Machine learning: novel bioinformatics approaches for combating antimicrobial resistance. Curr Opin Infect Dis 30(6):511–517

Messinger AI, Bui N, Wagner BD, Szefler SJ, Vu T, Deterding RR (2019) Novel pediatric-automated respiratory score using physiologic data and machine learning in asthma. Pediatr Pulmonol 54(8):1149–1155

Moosavi A, Motevalizadeh AA (2016) Role of epigenetics in biology and human diseases. Iran Biomed J 20(5):246–258

Nayeri S, Sargolzaei M, Tulpan D (2019) A review of traditional and machine learning methods applied to animal breeding. Anim Health Res Rev 20(1):31–46

Orange DE, Agius P, DiCarlo EF et al (2018) Identification of three rheumatoid arthritis disease subtypes by machine learning integration of synovial histologic features and RNA sequencing data. Arthritis Rheumatol 70(5):690–701

Owen JA, Punt J, Stranford SA, Jones PP, Kuby J (2013) Kuby immunology. W.H. Freeman, New York

Pal S, Ghosh S, Bandyopadhyay S, Mandal C, Bandhyopadhyay S, Kumar Bhattacharya D, Mandal C (2004) Differential expression of 9-O-acetylated sialoglycoconjugates on leukemic blasts: a potential tool for long-term monitoring of children with acute lymphoblastic leukemia. Int J Cancer 111(2):270

Patil S, Awan KH, Arakeri G et al (2019) Machine learning and its potential applications to the genomic study of head and neck cancer—a systematic review. J Oral Pathol Med 48:773–779. https://doi.org/10.1111/jop.12854

Peiffer-Smadja N, Rawson TM, Ahmad R et al (2020) Machine learning for clinical decision support in infectious diseases: a narrative review of current applications. Clin Microbiol Infect 26(5):584–595

Prosperi M et al (2014) Predicting phenotypes of asthma and eczema with machine learning. BMC Med Genet 7(Suppl 1):S7

Saglani S, Custovic A (2019) Childhood asthma: advances using machine learning and mechanistic studies. Am J Respir Crit Care Med 199(4):414–422

Spathis D, Vlamos P (2019) Diagnosing asthma and chronic obstructive pulmonary disease with machine learning. Health Informatics J 25(3):811–827

Stafford IS, Kellermann M, Mossotto E, Beattie RM, MacArthur BD, Ennis S (2020) A systematic review of the applications of artificial intelligence and machine learning in autoimmune diseases. NPJ Digit Med 3:30

Tang Y, Zhang W, Zhu M et al (2018) Lupus nephritis pathology prediction with clinical indices. Sci Rep 8(1):10231

Ullah R, Khan S, Ali H, Chaudhary II, Bilal M, Ahmad I (2019) A comparative study of machine learning classifiers for risk prediction of asthma disease. Photodiagn Photodyn Ther 28:292–296

Vamathevan J, Clark D, Czodrowski P et al (2019) Applications of machine learning in drug discovery and development. Nat Rev Drug Discov 18(6):463–477

van Izendoorn DG, Szuhai K, Briaire-de Bruijn IH, Kostine M, Kuijjer ML, Bovée JV (2019) Machine learning analysis of gene expression data reveals novel diagnostic and prognostic biomarkers and identifies therapeutic targets for soft tissue sarcomas. PLoS Comput Biol 15(2): e1006826

Van Nieuwenhove E, Lagou V, Van Eyck L et al (2019) Machine learning identifies an immunological pattern associated with multiple juvenile idiopathic arthritis subtypes. Ann Rheum Dis 78(5):617–628. https://doi.org/10.1136/annrheumdis-2018-214354

Zewdie GK, Liu X, Wu D, Lary DJ, Levetin E (2019) Applying machine learning to forecast daily Ambrosia pollen using environmental and NEXRAD parameters. Environ Monit Assess 191 (Suppl 2):261

Machine Learning and Neglected Tropical Diseases

12

Abstract

A host of different diseases caused by pathogens, virus, bacteria, worms, helminths which may be transmitted by vectors or contaminated food, water, soil or contributed by a lack of healthy, hygienic, practices, improper sanitation, poor living or work condition, predominant in the tropics and subtropical countries are collectively termed as neglected tropical diseases (NTDs). They are a burden to the economy and has a huge bearing on the lives of affected individuals, with a lack of therapy for complete disease eradication, thus claiming large number of lives annually globally. Thus the need for new and effective therapeutic molecules has led to more research in this area of biology and machine learning in contributing towards screening of potential molecules in search of new drug targets.

Keywords

Neglected tropical diseases (NTDs) · World Health Organization (WHO) · Leishmaniasis · Filariasis · Malaria · Dengue · Schistosomiasis

Abbreviations

NTD	Neglected tropical diseases
WHO	World Health Organization
Ae	Aedes
DENV	Dengue virus
CL	Cutaneous leishmaniasis
VL	Visceral leishmaniasis
MCL	Mucocutaneous leishmaniasis
ANN	Artificial neural network

S. Ghosh, R. Dasgupta, *Machine Learning in Biological Sciences*,
https://doi.org/10.1007/978-981-16-8881-2_12

LF	Lymphatic filariasis
NCDS	Noncommunicable diseases
SB	Structure-based
LB	Ligand-based
WNV	West Nile Virus
ZIKV	Zika virus
LC-MS	Liquid chromatography-mass spectrometry
TGR	Thioredoxin glutathione reductase
BU	*Buruli ulcer*
ER	Enoyl reductase
TLR-2	Toll-like receptor 2
HIV	Human immunodeficiency virus
PK	Pyruvate kinase

12.1 Introduction

Neglected tropical diseases (NTDs) include different communicable diseases that appear in tropical and subtropical conditions and are reported from 149 countries as reported by World Health Organisation (WHO) (https://www.who.int/neglected_diseases/diseases/en/). They have been reported to cause disease in more than one billion individuals with a huge burden of cost involved from developing countries. These diseases are predominant in countries which suffer from associated problems of poverty, lack of sanitation, close contact with infectious vectors and domestic animals and livestock.

The list of diseases as enlisted by WHO in this domain includes Buruli ulcer, Chagas disease, Dengue and Chikungunya, Dracunculiasis (guinea-worm disease), Echinococcosis, Foodborne trematodiases, Human African trypanosomiasis (sleeping sickness), Leishmaniasis, Leprosy (Hansen's disease), Lymphatic filariasis, Mycetoma, chromoblastomycosis and other deep mycoses, Onchocerciasis (river blindness), Rabies, Scabies, and other ectoparasites, Schistosomiasis, Soil-transmitted helminthiases, Snakebite envenoming, Taeniasis/Cysticercosis, Trachoma, and Yaws (Endemic treponematoses). In 2017 diseases including chromoblastomycosis, deep mycoses, scabies and other ectoparasites and snakebite envenoming were included (https://www.who.int/neglected_diseases/diseases/en/) in NTD list by WHO.

12.2 Neglected Tropical Diseases

Dengue Dengue is caused by bites of infected *Aedes* species (*Ae. aegypti* or *Ae. albopictus*) mosquito transmitting dengue virus (DENV) and closely related subtypes also called as serotypes 1, 2, 3, and 4 causing fever, body aches, pain, vomiting, nausea, flu-like symptoms, blood in vomit, stools, tiredness, restlessness leading to severe dengue and could be lethal. Dengue is predominant in the tropical climates. With no treatment available for the disease, WHO enlisted dengue as a potential threat for epidemics, being seasonal, peaking at post rainy season bearing a correlation with the increased mosquito population and outbreaks claiming lives.

Rabies Rabies is a vial zoonotoic infection, that is transmitted to human from infected dog bites and which could be fatal causing inflammation of the brain and spinal cord with common symptoms of headache, fever, muscle spasms, excess salivation, paralysis and mental confusion. Forms of rabies include furious rabies revealing symptoms of hyperactivity and hallucinations and paralytic rabies revealing symptoms of paralysis and coma. Rabies is known to claim thousands of lives annually. Other than dogs, the following animals can transmit the disease including foxes, bats, coyotes, skunks, and raccoons.

Trachoma Trachoma is caused by infection of bacteria *Chlamydia trachomatis* and is transmitted by infectious discharges from eye or nose of infected individuals or by exposure to unclean habits and unsafe practices and poor hygiene conditions. If left untreated could lead to blindness. Affects the population globally but is reported to be endemic in countries which are economically poor and in rural settings, like Africa, being most affected continent. But it is also reported from counties of Asia, Australia, Central and South America, and Middle East and is a major public health concern.

Buruli Ulcer Is caused by infections by bacteria Mycobacterium ulcerans releasing the toxin mycolactone causing skin infection, painless nodules, mostly expressed on the arms and legs which may later develop into large ulcers leading to severe damage of the skin, bone and soft tissue. It is predominant in Africa. Early diagnosis and treatment can minimize morbidity, cost of treatment and prevent disability.

Yaws Yaws is a chronic infection caused by the bacteria Treponema pallidum subspecies pertenue affecting the skin and bone, revealing papillomas or noncancerous lumps and ulcers. It is predominant in economically poor countries including countries in Africa, Asia-Pacific, Latin America, and in individuals with poor hygiene conditions.

Leprosy Leprosy, also called as Hansen's disease, is a chronic illness caused *Mycobacterium leprae* bacteria revealing symptoms of affected skin, peripheral nerves, upper respiratory tract mucosa and eyes affecting the range of individuals from infants to the aged. It is a treatable disorder and early detection and treatment can prevent disability. Leprosy is transmitted by nose and mouth droplets, close and frequent contact with infected individuals.

Chagas Disease Chagas disease is called as American trypanosomiasis is caused by *Trypanosoma cruzi*, a protozoan parasite. It is transmitted by bites of infected triatomine bugs, through contaminated food, transfusions of infected blood, transmission at birth, organ transplantation from infected donors causing a life-threatening condition. Chagas disease is reported to be endemic in 21 countries of Latin Americas.

African Trypanosomiasis African Sleeping sickness is commonly called as sleeping sickness is caused by the infection of parasite namely *Trypanosoma brucei. T. b. gambiense* reported from West and Central Africa, and *T. b. rhodesiense* reported from East Africa and are transmitted by bites of infected tsetse flies. The disease has been reported from sub-Saharan Africa and poor countries and in rural areas.

Leishmaniases Leishmaniasis is caused by infection transmitted by *Leishmania* species transmitted by bites of infected female phlebotomine sandflies causing face ulcers, scars and disability causing forms of disease including the common form of cutaneous leishmaniasis (CL), severe form of visceral leishmaniasis (VL), or kala-azar, and disabling form mucocutaneous leishmaniasis (MCL) disease. The treatment of leishmaniasis by first-line and second-line drugs suffers from drawbacks of cost, toxic effects, and routes of administration.

Taeniasis and Neurocysticercosis It is caused by infection by adult tapeworms *Taenia solium* in human intestines causing two forms of the disease including cysticercosis resulting from ingestion of tapeworm eggs leading to the formation of larvae in tissues, and *taeniasis* due to ingestion of raw or improperly cooked, infected pork. Tapeworm eggs after they are ingested by the fecal-oral route, or intake of contaminated food or water infect humans. The developed larval parasites within the host can infect muscles, skin, eyes, and the central nervous system leading to cysticercosis, with detrimental and fatal effects and cause cysts in the brain, leading to neurocysticercosis.

Dracunculiasis Also called as guinea-worm disease is transmitted by ingestion of drinking water from ponds or open wells contaminated with parasitic worm of Dracunculus medinensis containing the cyclops that gets dissolved by stomach acidic pH, leading to the release of the larvae that migrate through the intestinal wall.

Echinococcosis Parasitic disease caused by the infection by larval stages of tapeworms *Echinococcus granulosus* and *Echinococcus multilocularis* forming pathogenic cysts in humans and transmitted by ingesting eggs, in feces of dogs and wild carnivorous animals. Echinococcosis in humans reveals two forms including cystic echinococcosis or hydatidosis and alveolar echinococcosis. Dogs, foxes, and other carnivores species of animals carry the adult worms in their intestine and release parasite eggs in feces. Ingestion of the eggs by humans, lead to development of larvae predominantly in liver and lungs. Both cystic and alveolar echinococcosis can be asymptomatic for many years till the parasite larvae evolve and trigger clinical signs and both can be fatal.

Foodborne Trematodiases Infection mediated by flukes including liver flukes including *Clonorchis sinensis*, *Fasciola gigantica*, *Fasciola hepatica*, *Opisthorchis felineus*, and *Opisthorchis viverrini*, intestinal flukes including *Echinostoma* sp., *Fasciolopsis buski*, and the intestinal trematode heterophyids entering into the human body by ingesting raw or semi or improperly cooked freshwater or brackish water fish and lung flukes including *Paragonimus* sp., that enters the body by eating improperly cooked crab or crayfish and can affect the nervous system causing public health hazard and economic burden. The diseases caused include clonorchiasis, opisthorchiasis and fascioliasis. Foodborne trematodes are a group of diseases that include the parasites *Clonorchis*, *Opisthorchis*, *Fasciola*, and *Paragonimus* claiming 2 million lives annually globally.

Lymphatic Filariasis Also called as elephantiasis is a mosquito borne parasitic infection caused by the nematode roundworms, belonging to family Filariodidea causing enlarged limbs and genitals. The larvae migrates and resides and reproduces in the lymphatic system and produces adult forms and thus continues life cycle and transmission. The disease is painful and causes disfiguring and affects more than 120 million people in the tropical and sub-tropical countries of Asia, Africa, Western Pacific, and Caribbean parts and South America.

Onchocerciasis Also commonly called as river blindness caused by the filaria worm *Onchocerca volvulus* and is transmitted by repeated infected blackflies *Simulium* spp. bites causing itching and eye lesions. The adult worm produces larvae causing visual impairment and permanent blindness. These breeding grounds of blackflies lies along rivers and streams, close to remote villages and agricultural lands. The adult worms produce microfilariae migrating to the skin, eyes, and other organs which are taken up by the female blackfly which are transmitted to a fresh individual passing on the infection.

Scabies It is a parasitic infection caused by mite, *Sarcoptes scabiei var hominis which is* microscopic in a size that burrows into the skin laying eggs, that triggers the host immune response causing itching and rash development. It can be more complex accompanied with bacterial infection, causing skin sores leading to septicemia, heart, and chronic kidney disorders.

Schistosomiasis It is a chronic infection caused by parasitic blood flukes trematodes of genus *Schistosoma*. It is transmitted by the cercariae or larvae released by freshwater snails that penetrate human skin through infected water. The adult worms are found in the urinary tract and intestines veins. Affects 240 million people globally and is predominant in tropical and sub-tropical countries. The diseases caused by *Schistosoma haematobium* include urogenital schistosomiasis and *S. guineensis*, *S. intercalatum*, *S. mansoni*, *S. japonicum*, and *S. mekongi* causes intestinal schistosomiasis.

Soil-transmitted helminthiases or STH is caused by infection by nematode including round worm or *Ascaris lumbricoides*, the whipworm or *Trichuris trichiura* and the hookworms or *Ancylostoma duodenale* and *Necator americanus* that are transmitted by contaminated soil, by human faeces causing anemia, diarrhea, vitamin A deficiency, stunted growth, malnutrition, weakness, pain in the abdomen, intestinal obstruction and impaired development.

Mycetoma is a chronic, inflammatory skin disorder affecting the lower limbs and foot caused by the entry of infectious fungi or bacteria into the subcutaneous tissue. Predominant in poor countries, mostly affecting laborers', farmers in poor countries.

Snakebite Caused by bites of venomous snakes and can be life-threatening caused by toxins. People living and working in rural agricultural set ups such as farmers, fishermen, child laborers are mostly affected.

12.3 Machine Learning in Study of Neglected Tropical Diseases

NTDs cost life and is a hugh burden to the individuals and the countries and majority of them lack proper medical treatments towards their complete eradication and thus search for new and effective therapy finds importance. With less or no therapy available towards complete eradication and treatment of NTDs or side-effects like toxicity, routes of administration in currently used treatment protocols and highly unaffordable cost of treatment are the chief challenges in the treatment of NTDs and calls for more search for new and effective therapeutic molecules.

Candidate therapeutic compounds are available online and machine learning models have been used to identify the candidate molecules for therapy (Hernandez et al. 2018). Therapeutic compounds in Chagas disease suffer from proper understanding of drug targets. Machine learning has been used to screen biological and chemical compounds and their targets from literature, datasets, genomics and metabolic pathways to identify compounds and their targets. A pathway genome data base has been designed for *T. cruzi* (Ekins et al. 2015; Andrade et al. 2019). Detection and measurement of serological antibody levels is indicative of pathogen infection and disease. However, changes in a single antibody IgG level and understanding of humoral immune responses have been applied to detect infection by different pathogens including lymphatic filariasis caused by *Wuchereria bancrofti*, malaria caused by *Plasmodium falciparum*, and protozoans causing enteric diseases including *Cryptosporidium parvum*, *Giardia intestinalis*, *Entamoeba histolytica*, bacteria including enterotoxigenic *Escherichia coli*, *Salmonella* spp., and viruses including norovirus groups I and II analyzed by machine learning approaches enabling surveillance activities (Steel et al. 2017).

Artificial neural network methods (ANNs) have been applied to study ligands against a host of NTDs including Chagas' disease, leishmaniasis (Scotti et al. 2015; Gaba et al. 2014) malaria, sleeping sickness, and tuberculosis (Scotti et al. 2015).

Machine learning has been used to screen compounds with anti-leishmanial activity by virtual screening (Castillo-Garit et al. 2018). Machine learning tools have been used to identify areas under risk and distribution of lymphatic filariasis (LF) across Nigeria (Eneanya et al. 2018).

Unsupervised learning was used to analyze the Hashtag #GlobalHealth, Malaria, #HIV, #TB, (tuberculosis), #NCDS, (noncommunicable diseases), and #NTDS (neglected tropical diseases) tweets towards the understanding of the epidemiology, prevention, control, treatment, and societal impact of NTDs (Fung et al. 2017).

Machine learning tools applied to Structure-Based (SB) and Ligand-Based (LB) modeling have been applied to various NTDs including TB, malaria, dengue, leishmaniasis against different anti-tubercular targets (Aher and Roy 2019).

Machine learning has enabled screening of molecules with Dengue Virus 2 (DENV2) inhibitory and West Nile (WNV) and Zika (ZIKV) virus anti-flaviviral activities (Barigye et al. 2019).

Liquid chromatography-mass spectrometry (LC-MS) analysis and metabolomics of serum samples have enabled screening of Onchocerca volvulus-positive individuals (Denery et al. 2010).

Schistosomiasis is caused by *Schistosoma mansoni*. Thioredoxin glutathione reductase (TGR), enables pathogen survival and is a potential drug target. Machine learning has enabled screening of large chemical compound libraries leading to the discovery of lead molecule (Gaba et al. 2014).

Machine learning has been applied to screen inhibitors of dihydrofolate reductase against Mycobacterium tuberculosis. (Santa Maria et al. 2017).

Attempts are being made to find a vaccines against *Buruli ulcer* (BU) caused by *Mycobacterium ulcerans* by using the enoyl reductase (ER) domain of the *M. ulcerans* mycolactone polyketide synthases electrostatically coupled Toll-like receptor 2 (TLR-2) agonist-based lipopeptide adjuvant, R_4Pam_2Cys and the efficacy of the vaccine was detected by quantitative estimation of the cytokine levels(Mangas et al. 2020).

Anthrax is a NTD affecting humans, livestock, and wildlife globally caused by *Bacillus anthracis* and its distribution has been mapped by using machine learning algorithm (Assefa et al. 2020) thus enabling study of its prediction of its prevention and control.

Treatment of VL in HIV patients suffers from challenges of high relapse and fatal cases by detecting the transcriptome profiles using machine learning approach to select for four genes by pathway analyses (Adriaensen et al. 2020).

Drug resistance in cutaneous leishmaniasis is a major challenge including India. Miltefosine is the only oral drug for the disease and Miltefosine transporter proteins play role in the drug-resistance. Computational modeling has been used to predict its structure. (Kabra et al. 2020). Machine learning models are being used for the study of pyruvate kinase (PK) inhibitors of Leishmania mexicana in targeting of Leishmaniasis (Jamal and Scaria 2013).

Dual RNA-sequencing analysis to *Orientia tsutsugamushi* (Ot), causing scrub typhus has enabled better understanding of the biology of the pathogen and disease (Mika-Gospodorz et al. 2020).

12.4 Discussion

The NTDs are a major global burden but predominant in the tropical and subtropical countries. Diseases like Trachoma can lead to complete blindness and mathematical models are being designed to understand the biology of the disease and research studies and different strategies are being designed towards its cure, prevention, and effective control (Basu and Ghosh 2020).

Many diseases lack proper and complete eradication protocol. Mathematical biology and machine learning have found applications in understanding the biology of disease and attempts are being made in search of suitable and effective molecular drug targets and compounds as effective therapeutic molecules. This area of biology calls in more research in the years to come.

References

Adriaensen W, Cuypers B, Cordero CF, Mengasha B, Blesson S, Cnops L, Kaye PM, Alves F, Diro E, van Griensven J (2020) Host transcriptomic signature as alternative test-of-cure in visceral leishmaniasis patients co-infected with HIV. EBioMedicine 55:102748

Aher RB, Roy K (2019) Computational approaches as rational decision support systems for discovering next-generation antitubercular agents: mini-review. Curr Comput Aided Drug Des 15(5):369–383

Andrade CH, Neves BJ, Melo-Filho CC, Rodrigues J, Silva DC, Braga RC, Cravo PVL (2019) In silico chemogenomics drug repositioning strategies for neglected tropical diseases. Curr Med Chem 26(23):4355–4379

Assefa A, Bihon A, Tibebu A (2020) Anthrax in the Amhara regional state of Ethiopia; spatiotemporal analysis and environmental suitability modeling with an ensemble approach. Prev Vet Med 184:105155

Barigye SJ, García de la Vega JM, Castillo-Garit JA (2019) Undersampling: case studies of flaviviral inhibitory activities. J Comput Aided Mol Des 33(11):997–1008

Basu S, Ghosh S (2020) Mathematical biology and Trachoma. JAZR 31:2

Castillo-Garit JA, Flores-Balmaseda N, Álvarez O, Pham-The H, Pérez-Doñate V, Torrens F, Pérez-Giménez F (2018) Computational identification of chemical compounds with potential activity against leishmania amazonensis using nonlinear machine learning techniques. Curr Top Med Chem 18(27):2347–2354

Denery JR, Nunes AA, Hixon MS, Dickerson TJ, Janda KD (2010) Metabolomics-based discovery of diagnostic biomarkers for onchocerciasis. PLoS Negl Trop Dis 4(10):e834

Ekins S, de Siqueira-Neto JL, McCall LI, Sarker M, Yadav M, Ponder EL, Kallel EA, Kellar D, Chen S, Arkin M, Bunin BA, McKerrow JH, Talcott C (2015) Machine learning models and pathway genome data base for Trypanosoma cruzi drug discovery. PLoS Negl Trop Dis 9(6): e0003878

Eneanya OA, Cano J, Dorigatti I, Anagbogu I, Okoronkwo C, Garske T, Donnelly CA (2018) Environmental suitability for lymphatic filariasis in Nigeria. Parasit Vectors 11(1):513

Fung IC, Jackson AM, Ahweyevu JO, Grizzle JH, Yin J, Tse ZTH, Liang H, Sekandi JN, Fu KW (2017) #Globalhealth Twitter conversations on #Malaria, #HIV, #TB, #NCDS, and #NTDS: a Cross-Sectional Analysis. Ann Glob Health 83(3-4):682–690

Gaba S, Jamal S, Scaria V (2014) Cheminformatics models for inhibitors of Schistosoma mansoni thioredoxin glutathione reductase. Sci World J 2014:957107

Hernandez HW, Soeung M, Zorn KM, Ashoura N, Mottin M, Andrade CH, Caffrey CR, de Siqueira-Neto JL, Ekins S (2018) High throughput and computational repurposing for neglected diseases. Pharm Res 36(2):27

Jamal S, Scaria V (2013) Cheminformatic models based on machine learning for pyruvate kinase inhibitors of Leishmania mexicana. BMC Bioinf 14:329

Kabra R, Ingale P, Singh S (2020) Computationally designed synthetic peptides for transporter proteins imparts allostericity in Miltefosine resistant L. major. Biochem J 477(10):2007–2026

Mangas KM, Buultjens AH, Porter JL, Baines SL, Marion E, Marsollier L, Tobias NJ, Pidot SJ, Quinn KM, Price DJ, Kedzierska K, Zeng W, Jackson DC, Chua BY, Stinear TP (2020) Vaccine-specific immune responses against Mycobacterium ulcerans infection in a low-dose murine challenge model. Infect Immun 88(3):e00753

Mika-Gospodorz B, Giengkam S, Westermann AJ, Wongsantichon J, Kion-Crosby W, Chuenklin S, Wang LC, Sunyakumthorn P, Sobota RM, Subbian S, Vogel J, Barquist L, Salje J (2020) Dual RNA-seq of Orientia tsutsugamushi informs on host-pathogen interactions for this neglected intracellular human pathogen. Nat Commun 11(1):3363

Santa Maria JP, Park Y, Yang L, Murgolo N, Altman MD, Zuck P, Adam G, Chamberlin C, Saradjian P, Dandliker P, Boshoff HIM, Barry CE, Garlisi C, Olsen DB, Young K, Glick M, Nickbarg E, Kutchukian PS (2017) Linking high-throughput screens to identify MoAs and novel inhibitors of mycobacterium tuberculosis dihydrofolate reductase. ACS Chem Biol 12(9): 2448–2456

Scotti L, Ishiki H, Mendonça Júnior FJ, da Silva MS, Scotti MT (2015) Artificial neural network methods applied to drug discovery for neglected diseases. Comb Chem High Throughput Screen 18(8):819–829

Steel C, Kubofcik J, Hamlin KL, Moss DM, Nutman TB, Priest JW, Lammie PJ (2017) Measuring changes in transmission of neglected tropical diseases, malaria, and enteric pathogens from quantitative antibody levels. PLoS Negl Trop Dis 11(5):e0005616

Check for updates

Machine Learning in Cardiovascular Disorders

13

Abstract

Cardiovascular Diseases, affecting human lives and claiming lives globally is a matter of concern globally. Although conventional methods are in use, machine learning applications and advances in radiomics and image based studies have enabled the development of noninvasive methods of detection, better and accurate detection, classification and risk stratification, diagnosis and prognosis in CVD, and the study of disorders with risks. We discuss in this chapter the different applications of machine learning in the domain of CVD.

Keywords

EEG · ECG · CVD · RA

Abbreviations

ECG	Electrocardiography
CV	Cardiovascular
USG	Ultrasonography
TNF-α	Tissue necrosis factor-α
IL-2	Interleukin-2
NT-proBNP	N-Terminal pro-B-type natriuretic peptide
NLP	Natural language processing
HER	Electronic health records
ML	Machine learning
CMR	Cardiovascular magnetic resonance
CT	Computerized tomography
RF	Random Forest
PWV	Pulse velocity wave

S. Ghosh, R. Dasgupta, *Machine Learning in Biological Sciences*,
https://doi.org/10.1007/978-981-16-8881-2_13

RA	Rheumatoid arthritis
CCTA	Coronary CT angiography
AAA	Abdominal aortic aneurysm
iPSc	Induced pluripotent stem cells
LQT	Long QT syndrome (LQTS)
HCM	Hypertrophic cardiomyopathy
ACHD	Adult congenital heart disease

13.1 Introduction

Cardiovascular diseases (CVDs) are the cause for the major number of deaths globally claiming almost 17.9 million lives annually. CVDs involve disorders of the heart and blood vessels causing disorders including coronary heart disease, affecting the blood vessels the supply blood to heart muscle, cerebrovascular disease, affecting the blood vessels and blood supply to the brain, myocardial infarction or heart attack with decreased blood flow or stop of blood flow to heart, damaging heart muscle, peripheral arterial diseases, affecting the blood supply in the arms and legs; rheumatic heart disease causing damage to the heart muscle and valves due to rheumatic fever, by infection caused by streptococcal bacteria, congenital heart disease—due to malformed heart structure since birth, pulmonary embolism and deep vein thrombosis caused by blood clots in the leg veins, can circulate to the heart and lungs (https://www.who.int/health-topics/cardiovascular-diseases/#tab=tab_1; https://www.who.int/en/news-room/fact-sheets/detail/cardiovascular-diseases-(cvds)), aortic atherosclerosis including thoracic and abdominal aneurysms and other conditions (Olvera Lopez et al. 2020) (https://www.who.int/health-topics/cardiovascular-diseases/#tab=tab_1; https://www.who.int/en/news-room/fact-sheets/detail/cardiovascular-diseases-(cvds)). CVD deaths are mostly due to heart attacks and strokes, and almost one-third of the total deaths is premature.

Risk factors for CVD include high blood pressure, glucose, and lipids, overweight and obesity, unhealthy lifestyle, use of alcohol, tobacco, and early detection and proper care can prevent premature deaths (Olvera Lopez et al. 2020) (https://www.who.int/health-topics/cardiovascular-diseases/#tab=tab_1; https://www.who.int/en/news-room/fact-sheets/detail/cardiovascular-diseases-(cvds)).

13.2 Machine Learning and CVD

Artificial intelligence and machine learning tools have been used to find novel genotypes and phenotypes, better patient care, reduce cost, chances of re-admission and mortality rates, risk stratification (Quesada et al. 2019) and is a step towards precision medicine in this domain (Krittanawong et al. 2017).

Machine learning found importance in identifying biomarkers and highlighted the importance of factors including imaging, electrocardiography (ECG), and serum biomarkers in prediction of disease, in contrary to traditional cardiovascular (CV) risk factors. While age factor has been revealed to be most important predictor for mortality, fasting glucose levels and carotid ultrasonography (USG) proved to be stroke predictors. Coronary artery calcium score could be indicative of coronary heart disease and atherosclerosis, left ventricular structure and function and cardiac troponin-T predominantly indicated incident heart failure, creatinine, age, and ankle-brachial index indicated atrial fibrillation, and TNF-α (tissue necrosis factor-α) and IL (interleukin)-2 soluble receptors and NT-proBNP (N-Terminal Pro-B-Type Natriuretic Peptide) levels were important disease indicators. (Ambale-Venkatesh et al. 2017).

Machine-learning is found to be an effective tool on identifying risk factors thereby improve prediction of cardiovascular risk and thus enables decision making for preventive treatment in those individuals only (Weng et al. 2017). Since diabetes and CVD are leading causes of disease, Machine learning tools have been employed for the detection of cardiovascular, pre-diabetic, and diabetic condition. (Dinh et al. 2019). Thus machine learning is enabling cardiovascular healthcare by diverse application of automated imaging interpretation of patient condition from angiography, chest roentgenograms, ECGs, echocardiograms, natural language processing (NLP), and extraction of data from clinical notes and electronic health records (EHR), and helping in diagnosis and prognosis and predicting mortality or complications post-surgery (Kilic 2020). Cardiovascular imaging faces the challenges of storage of hugh volumes of data and their analysis and machine learning methods by its automated approach is enabling storage and analysis of complex and hugh data and holds scope of further developments in cardiovascular imaging (Henglin et al. 2017). Machine learning (ML) has been applied with improvement in automated fast and accurate image acquisition, improved temporal and spatial resolution, refinement and reconstruction, imaging efficiency, quality of output, interpretation and analysis and patient evaluation in cardiovascular magnetic resonance (CMR) and automated quantification of mass and volume of left and right ventricles, myocardial mapping (Leiner et al. 2019; Alonso-Betanzos and Bolón-Canedo 2018), automation with longitudinal strain and 3D ECG (Gandhi et al. 2018) and is helping clinicians in diagnosis and prognosis, predicting stroke (Saber et al. 2019), and fast and accurate decision making (Henglin et al. 2017; Leiner et al. 2019; Johnson et al. 2018). Machine learning has advanced the current applications in radiology with the use of CT scanners and MRI and imaging of lesion and coronary artery assessment, cardiac volumetry and function, and anatomic localization in cardiothoracic and cardiovascular disorders (Retson et al. 2019).

Despite knowledge of cardiovascular Disease (CVD) risk estimation scores in primary prevention of CVD, recently ML tools of k-NN, random forest (RF), and decision tree has enabled CVD prediction more accurately (Dimopoulos et al. 2018).

Machine learning software, mediated analysis of different risk factors of CVD have revealed the importance of albumin/hemoglobin, left ventricular ejection fraction/history of heart failure (HF), acute coronary syndrome (ACS), anti-platelet use history, deceleration time, and intracerebral brain hemorrhage (ICH) or warfarin as risk factors with good prediction possibility (Suzuki et al. 2019). Using machine learning tools for analysis, arterial stiffness parameters detected by arterial tonometry was found to predict development of cardiovascular disease (CVD). And pulse wave velocity (PWV) revealed role in disease prognosis in cardiac disorders (Garcia-Carretero et al. 2019).

EEG data indicates recovery after cardiac arrest; however, to understand the accuracy, machine learning tools were tested on EEG recovery data to predict outcome in hypoxic-ischemic brain injury and outcome post-cardiac arrest (Amorim et al. 2019) and predict survival post-ECG with superior accuracy (Samad et al. 2019).

Autoimmune chronic illness, like rheumatoid arthritis (RA) is associated with increased risk in stroke and irregularities in cardiovascular (CV) functions. However it is difficult to study the risk imposed by conventional methods due to generation of under or overestimates. Advances in medical imaging and application of machine learning tools have enabled proper cardiovascular risk stratification especially in patients with disorders including carotid atherosclerosis, involving building up and accumulation of plaques at injury sites made up of cholesterol clumps, calcium deposits, fibrous tissue, and other cellular debris in arteries delivering blood to the brain and rheumatoid arthritis (RA) study of atherosclerotic plaques in RA patients and appropriate risk stratification (Khanna et al. 2019).

Development in the field of radiomics due to boon from advances from imaging based studies and application in machine learning, coronary CT angiography (CCTA) has been designed as a non-invasive method in the detection of coronary artery disease from medical images based disease phenotype studies (Kolossváry et al. 2019).

Abdominal aortic aneurysm (AAA) involves enlargement of the aorta, at the abdomen and is a life-threatening disorder if it bursts and is predominant in older men and with smoking habit. Database of this disorder has been designed and could be screened using keywords and extract information (Kerut et al. 2019).

Induced pluripotent stem cells (iPScs) have been used to study pathophysiology and develop therapeutic strategies in human cardiac diseases. Cardiac diseases of genetic origin like HCM or Hypertrophic cardiomyopathy caused by defective genes in the heart muscle leading to contraction of left ventricle harder, stiffer and thicker than normal wall and LQT or Long QT syndrome (LQTS) that reveals disorder in heart's electrical activity leading to sudden, dangerous arrhythmias which cannot be controlled due to exercise or stress and can be fatal. This can be identified and separated from healthy individuals by applying machine learning to Ca^{2+} transient signals in induced pluripotent stem cells (iPSC)-derived cardiomyocytes (Joutsijoki et al. 2019).

Machine learning algorithms have been used to estimate prognosis and progress in therapy in adult congenital heart disease (ACHD) or pulmonary hypertension based on data from clinical and demographic sources, ECG, cardiopulmonary exercise testing, and laboratory markers (Diller et al. 2019).

13.3 Discussions

CVDs are a complex groups of diseases affecting the global population across the world and its early prediction and risk factors involved can lead to the control of the disease. Advances in Machine learning algorithms, artificial intelligence applications and advances in radiomics and imaging techniques, extraction of information from EHR, clinical notes by the use of NLP technologies have been boon in early diagnosis and prognosis and treatment protocols to be given to the patients in the disease. Although the applications of machine learning in this domain has just started and is at its infancy, the need of the hour is to find more applications of machine learning in this domain of biology and remains the scope of future day research.

References

Alonso-Betanzos A, Bolón-Canedo V (2018) Big-data analysis, cluster analysis, and machine-learning approaches. Adv Exp Med Biol 1065:607–626

Ambale-Venkatesh B, Yang X, Wu CO et al (2017) Cardiovascular event prediction by machine learning: the multi-ethnic study of atherosclerosis. Circ Res 121(9):1092–1101

Amorim E, van der Stoel M, Nagaraj SB et al (2019) Quantitative EEG reactivity and machine learning for prognostication in hypoxic-ischemic brain injury. Clin Neurophysiol 130(10): 1908–1916

Diller GP, Kempny A, Babu-Narayan SV et al (2019) Machine learning algorithms estimating prognosis and guiding therapy in adult congenital heart disease: data from a single tertiary centre including 10019 patients. Eur Heart J 40(13):1069–1077

Dimopoulos AC, Nikolaidou M, Caballero FF et al (2018) Machine learning methodologies versus cardiovascular risk scores, in predicting disease risk. BMC Med Res Methodol 18(1):179

Dinh A, Miertschin S, Young A, Mohanty SD (2019) A data-driven approach to predicting diabetes and cardiovascular disease with machine learning. BMC Med Inform Decis Mak 19(1):211

Gandhi S, Mosleh W, Shen J, Chow CM (2018) Automation, machine learning, and artificial intelligence in echocardiography: a brave new world. Echocardiography 35(9):1402–1418

Garcia-Carretero R, Vigil-Medina L, Barquero-Perez O, Ramos-Lopez J (2019) Pulse wave velocity and machine learning to predict cardiovascular outcomes in prediabetic and diabetic populations. J Med Syst 44(1):16

Henglin M, Stein G, Hushcha PV, Snoek J, Wiltschko AB, Cheng S (2017) Machine learning approaches in cardiovascular imaging. Circ Cardiovasc Imaging 10(10):e005614

Johnson KW, Torres Soto J, Glicksberg BS et al (2018) Artificial intelligence in cardiology. J Am Coll Cardiol 71(23):2668–2679

Joutsijoki H, Penttinen K, Juhola M, Aalto-Setälä K (2019) Separation of HCM and LQT cardiac diseases with machine learning of Ca2+ transient profiles. Methods Inf Med 58(4):167–178

Kerut EK, To F, Summers KL, Sheahan C, Sheahan M (2019) Statistical and machine learning methodology for abdominal aortic aneurysm prediction from ultrasound screenings. Echocardiography 36(11):1989–1996

Khanna NN, Jamthikar AD, Gupta D et al (2019) Rheumatoid arthritis: atherosclerosis imaging and cardiovascular risk assessment using machine and deep learning-based tissue characterization. Curr Atheroscler Rep 21(2):7

Kilic A (2020) Artificial intelligence and machine learning in cardiovascular health care. Ann Thorac Surg 109(5):1323–1329

Kolossváry M, De Cecco CN, Feuchtner G, Maurovich-Horvat P (2019) Advanced atherosclerosis imaging by CT: radiomics, machine learning and deep learning. J Cardiovasc Comput Tomogr 13(5):274–280

Krittanawong C, Zhang H, Wang Z, Aydar M, Kitai T (2017) Artificial intelligence in precision cardiovascular medicine. J Am Coll Cardiol 69(21):2657–2664

Leiner T, Rueckert D, Suinesiaputra A et al (2019) Machine learning in cardiovascular magnetic resonance: basic concepts and applications. J Cardiovasc Magn Reson 21(1):61

Olvera Lopez E, Ballard BD, Jan A (2020) Cardiovascular disease. In: StatPearls. StatPearls Publishing, Treasure Island

Quesada JA, Lopez-Pineda A, Gil-Guillén VF et al (2019) Machine learning to predict cardiovascular risk. Int J Clin Pract 73(10):e13389

Retson TA, Besser AH, Sall S, Golden D, Hsiao A (2019) Machine learning and deep neural networks in thoracic and cardiovascular imaging. J Thorac Imaging 34(3):192–201

Saber H, Somai M, Rajah GB, Scalzo F, Liebeskind DS (2019) Predictive analytics and machine learning in stroke and neurovascular medicine. Neurol Res 41(8):681–690

Samad MD, Ulloa A, Wehner GJ et al (2019) Predicting survival from large echocardiography and electronic health record datasets: optimization with machine learning. JACC Cardiovasc Imaging 12(4):681–689

Suzuki S, Yamashita T, Sakama T et al (2019) Comparison of risk models for mortality and cardiovascular events between machine learning and conventional logistic regression analysis. PLoS One 14(9):e0221911

Weng SF, Reps J, Kai J, Garibaldi JM, Qureshi N (2017) Can machine-learning improve cardiovascular risk prediction using routine clinical data? PLoS One 12(4):e0174944

Machine Learning and Diabetes 14

Abstract

Diabetes mellitus is a chronic metabolic disease revealing high blood glucose levels that affect the body with associated complications and can cause detrimental fatal effects and is a cause of concern globally. The disease lacks complete cure and cannot be prevented and early diagnosis and proper care and management of the disease is the only hope to prevent mortality. Machine learning tools are being employed in the study of management, risk prediction, diagnosis and prognosis of the disease, and study of its associated complicacies. We discuss in this chapter the different domains of application of machine learning in understanding the biology of diabetes and proper prevention and prediction.

Keywords

Type I · Type II · Diabetes mellitus

Abbreviations

AD	AdaBoost
AMD	Age related macular degeneration
BG	Blood glucose
BMI	Body mass index
CVD	Cardiovascular disease
DLS	Deep learning system
DM	Diabetes mellitus
DR	Diabetic retinopathy
EHR	Electronic health record
GDM	Gestational diabetes mellitus
GTB	Gradient tree boosting

S. Ghosh, R. Dasgupta, *Machine Learning in Biological Sciences*,
https://doi.org/10.1007/978-981-16-8881-2_14

GV	Glycemic variability
GWAS	Genome wide association studies
HbA1c	Glycated hemoglobin
HF	Heart failure
miRNA	MicroRNAs
ML	Machine learning
MLP	Multilayer perceptron
PheWAS	Phenome-wide association study
PWV	Pulse wave velocity
SVM	Support vector machine
T1DM	Type 1 diabetes mellitus
T2DM	Type 2 diabetes mellitus
WHO	World Health Organization

14.1 Introduction

Diabetes mellitus (DM) is a complex chronic metabolic disorder, revealing hyperglycemia or high blood glucose level associated with irregular regulation of glucose metabolism and lead to ill health consequences, including damages in the blood vessels, eyes, heart, kidneys, and nerves associated with various complicacies and could be fatal if not managed well.

Type 2 diabetes mellitus (T2DM), in adults, occurs due to insufficient production of insulin in the body or body is resistant to insulin. The incidence of T2DM has grown dramatically globally in the last two decades. Type 1 diabetes T1DM or juvenile diabetes is also called as insulin-dependent diabetes where pancreas either does not produce any insulin or produces in very less quantity. Symptoms common to T2DM and T1DM include increase in thirst, frequent urine release, hunger, sudden weight loss, ketones in the urine, fatigue, irritation, affected and blurred vision, sores that heal slowly, frequent infections of gums, skin, and vagina.

Associated complications that develop with time include cardiovascular disease (CVD), coronary artery disease, chest pain or angina, heart attack, stroke, atherosclerosis, nerve damage or neuropathy affecting the lower limbs causing a sense of numbness, tingling, burning or pain, loss of sense in limbs. Damage of nerves associated with the digestive system causing nausea, vomiting, diarrhea or constipation, sometimes leads to erectile dysfunction in men, kidney damage or nephropathy and can lead to kidney failure, eye damage or retinopathy, due to damage the retinal blood vessels causing diabetic retinopathy, causing risks of cataracts and glaucoma, damage of foot, due to poor blood flow, cuts, blisters, serious infections, that heal poorly, can lead to leg amputation, is associated with skin infections by bacterial and fungal infections, loss of hearing, Alzheimer's disease, depression.

World Health Organization (WHO) has revealed a 422 million people globally suffering from diabetes, residing in low-and middle-income group of countries, and contributing to 1.6 million deaths annually globally (https://www.who.int/health-topics/diabetes#tab=tab_1).

The diseases suffer from lack of cure and self-care and management are the only steps towards its care. Tracking blood glucose (BG) levels and maintaining physical well-being and good healthy, disciplined, hygienic practices, lifestyles, diet, and insulin intake are the chief steps towards its care and management.

14.2 Machine Learning and Diabetes

Different aspects of the complex metabolic disorder of diabetes mellitus (DM) causing abnormal glucose regulation including its diagnosis, etiopathophysiology and therapy has led to the generation of hugh data and machine learning tools have enabled the analysis of data towards disease prediction (Gruson et al. 2019; Zou et al. 2018), diagnosis, understanding and study of associated complications in diabetes, role of genome and environment and health Care and management (Zou et al. 2018; Kavakiotis et al. 2017).

Application of artificial intelligence in the different aspects of diabetes have enabled generation of a functional taxonomy towards management of diabetes and preventions of complications associated with the disease (Contreras and Vehi 2018).

Modeling the blood glucose levels and the prediction of BG finds importance in diabetes management and machine learning with the help of analysis of hugh data from patients is being instrumental in enabling better and effective diabetes management. Data mining approaches and application of machine learning tools has enabled better prediction of blood glucose particularly in type 1 diabetes and is a step towards generation of development of artificial pancreas (Woldaregay et al. 2019a). Deep learning system (DLS) of machine learning has enabled screening of diabetic retinopathy, glaucoma, and age-related macular degeneration (AMD) in diabetic individuals from retinal examination images (Ting et al. 2017). It has been possible to predict T2DM associated complications like risk of heart failure (Dagliati et al. 2018; Segar et al. 2019) based on electronic health record (EHR) data by applications of machine learning tools from variables of age, body mass index (BMI), diagnosis time, gender, glycated hemoglobin (HbA1c), hypertension, and habit of smoking (Dagliati et al. 2018).

Supervised machine learning has been used to analyze data in EHR from diabetic individuals (Dinh et al. 2019; Nguyen et al. 2019) in prediction of cardiovascular disorders associated with diabetes, condition of pre-diabetes, and detection of diabetes (Dinh et al. 2019).

Machine learning has enabled in finding out the relation between associations between genotype and phenotype in T2DM by genome-wide association study (GWAS) and phenome-wide association study (PheWAS), by analysis of the EHR (Zheng et al. 2017). Machine learning tools to post-liver transplantation data revealed that obese, old male individuals who are recipients of such transplants are a higher risk for new-onset diabetes while donor individuals do not suffer from such risk (Bhat et al. 2018).

Machine learning tools have enabled the classification of pattern and detection of anomalies such as glycemic variability (GV), hyperglycemia, and hypoglycemia in T1D (Woldaregay et al. 2019b). Deep learning algorithm has been used for automated detection of diabetic retinopathy and diabetic macular edema from photographs of retinal fundus examination (Gulshan et al. 2016).

Machine learning (ML) screening, diagnosis, and management of diabetes and its complications (Broome et al. 2020; Rigla et al. 2018) and different machine learning tools are used in precise prediction of disease from datasets (Maniruzzaman et al. 2018). Supervised machine learning tools are used to identify variables for predicting the response of glycated hemoglobin (HbA1c) after insulin treatment in T2DM patients (Nagaraj et al. 2019).

Iridology is an alternative medicine technique applying colors, patterns, and different properties of the iris to understand health and detect disease. Soft computing has been applied to iridology and has enabled automatic diagnosis of T2DM (Samant and Agarwal 2018). Diabetic retinopathy (DR) risk factors could be deduced at using machine learning and data mining approaches and thus is a step towards early prevention for DR in a high-risk population (Tsao et al. 2018).

Although machine learning algorithms have been successfully used in the prediction of hypoglycemia at night, it is still difficult to predict postprandial hypoglycemia reflecting fluctuating glucose levels post-mealtimes. Thus machine learning has been applied to predict postprandial hypoglycemic events (Seo et al. 2019).

Arterial stiffness could predict cardiovascular disease (CVD) detected by Pulse Wave Velocity (PWV) and application of machine learning to the experimental and clinical data enabled application of PWV as prediction factor in prediction of cardiovascular effects in prediabetic and diabetic individuals (Garcia-Carretero et al. 2019).

Early detection and management of T2DM can prevent mortality in high risk areas like China. The use of different machine learning tools including AdaBoost (AD), Gradient Tree Boosting (GTB), Multilayer Perceptron (MLP), Support Vector Machine (SVM), Trees Random Forest (TRF) risk prediction of T2DM was successful (Xiong et al. 2019).

Gestational diabetes mellitus (GDM) detected during late second or early third trimester is associated with complications during pregnancy and is a global concern. Upregulated microRNAs (miRNas) miR-223 and miR-23a were used in diagnosis of GDM and finds importance as biomarkers in the prediction of GDM during the first trimester (Yoffe et al. 2019).

14.3 Discussion

Early diagnosis finds importance in treatment of the disease. With the disease not diagnosed for prolonged period of time or left in untreated condition could be detrimental. Early diagnosis by blood based biochemical test of blood glucose level, and continuous monitoring of blood glucose level are the only available mode of diagnosis.

Maintaining proper control of blood glucose level, by diet, physical activity, medication, if required, blood pressure control, and control of lipid profile to reduce risk of cardiovascular disease (CVD) and other associated complications, regular screening for eyes damage caused due to diabetic retinopathy, regular check of good health of eyes, feet and kidneys enables good management and control of the disease.

Recently machine learning tools have been applied for early, better, and accurate prediction of DM and risk factors associated with the disorder and are facilitating the search of biomarkers for early diagnosis of DM which is a key to prevent the detrimental effects of DM.

More research in these lines is a promising hope for the millions of diabetic individuals across the globe for better and well managed life.

References

Bhat V, Tazari M, Watt KD, Bhat M (2018) New-onset diabetes and preexisting diabetes are associated with comparable reduction in long-term survival after liver transplant: a machine learning approach. Mayo Clin Proc 93(12):1794–1802

Broome DT, Hilton CB, Mehta N (2020) Policy implications of artificial intelligence and machine learning in diabetes management. Curr Diab Rep 20(2):5

Contreras I, Vehi J (2018) Artificial intelligence for diabetes management and decision support: literature review. J Med Internet Res 20(5):e10775

Dagliati A, Marini S, Sacchi L, Cogni G, Teliti M, Tibollo V, De Cata P, Chiovato L, Bellazzi R (2018) Machine learning methods to predict diabetes complications. J Diabetes Sci Technol 12(2):295–302

Dinh A, Miertschin S, Young A, Mohanty SD (2019) A data-driven approach to predicting diabetes and cardiovascular disease with machine learning. BMC Med Inform Decis Mak 19(1):211

Garcia-Carretero R, Vigil-Medina L, Barquero-Perez O, Ramos-Lopez J (2019) Pulse wave velocity and machine learning to predict cardiovascular outcomes in prediabetic and diabetic populations. J Med Syst 44(1):16

Gruson D, Helleputte T, Rousseau P, Gruson D (2019) Data science, artificial intelligence, and machine learning: opportunities for laboratory medicine and the value of positive regulation. Clin Biochem 69:1–7

Gulshan V, Peng L, Coram M, Stumpe MC, Wu D, Narayanaswamy A, Venugopalan S, Widner K, Madams T, Cuadros J, Kim R, Raman R, Nelson PC, Mega JL, Webster DR (2016) Development and validation of a deep learning algorithm for detection of diabetic retinopathy in retinal fundus photographs. JAMA 316(22):2402–2410

Kavakiotis I, Tsave O, Salifoglou A, Maglaveras N, Vlahavas I, Chouvarda I (2017) Machine learning and data mining methods in diabetes research. Comput Struct Biotechnol J 15:104–116

Maniruzzaman M, Rahman MJ, Al-MehediHasan M, Suri HS, Abedin MM, El-Baz A, Suri JS (2018) Accurate diabetes risk stratification using machine learning: role of missing value and outliers. J Med Syst 42(5):92

Nagaraj SB, Sidorenkov G, van Boven JFM, Denig P (2019) Predicting short- and long-term glycated haemoglobin response after insulin initiation in patients with type 2 diabetes mellitus using machine-learning algorithms. Diabetes Obes Metab 21(12):2704–2711

Nguyen BP, Pham HN, Tran H, Nghiem N, Nguyen QH, Do TTT, Tran CT, Simpson CR (2019) Predicting the onset of type 2 diabetes using wide and deep learning with electronic health records. Comput Methods Prog Biomed 182:105055

Rigla M, García-Sáez G, Pons B, Hernando ME (2018) Artificial intelligence methodologies and their application to diabetes. J Diabetes Sci Technol 12(2):303–310
Samant P, Agarwal R (2018) Machine learning techniques for medical diagnosis of diabetes using iris images. Comput Methods Prog Biomed 157:121–128
Segar MW, Vaduganathan M, Patel KV, McGuire DK, Butler J, Fonarow GC, Basit M, Kannan V, Grodin JL, Everett B, Willett D, Berry J, Pandey A (2019) Machine learning to predict the risk of incident heart failure hospitalization among patients with diabetes: the WATCH-DM risk score. Diabetes Care 42(12):2298–2306
Seo W, Lee YB, Lee S, Jin SM, Park SM (2019) A machine-learning approach to predict postprandial hypoglycemia. BMC Med Inform Decis Mak 19(1):210
Ting DSW, Cheung CY, Lim G, Tan GSW, Quang ND, Gan A, Hamzah H, Garcia-Franco R, San Yeo IY, Lee SY, Wong EYM, Sabanayagam C, Baskaran M, Ibrahim F, Tan NC, Finkelstein EA, Lamoureux EL, Wong IY, Bressler NM, Sivaprasad S, Varma R, Jonas JB, He MG, Cheng CY, Cheung GCM, Aung T, Hsu W, Lee ML, Wong TY (2017) Development and validation of a deep learning system for diabetic retinopathy and related eye diseases using retinal images from multiethnic populations with diabetes. JAMA 318(22):2211–2223
Tsao HY, Chan PY, Su EC (2018) Predicting diabetic retinopathy and identifying interpretable biomedical features using machine learning algorithms. BMC Bioinf 19(Suppl 9):283
Woldaregay AZ, Årsand E, Walderhaug S, Albers D, Mamykina L, Botsis T, Hartvigsen G (2019a) Data-driven modeling and prediction of blood glucose dynamics: machine learning applications in type 1 diabetes. Artif Intell Med 98:109–134
Woldaregay AZ, Årsand E, Botsis T, Albers D, Mamykina L, Hartvigsen G (2019b) Data-driven blood glucose pattern classification and anomalies detection: machine-learning applications in type 1 diabetes. J Med Internet Res 21(5):e11030
Xiong XL, Zhang RX, Bi Y, Zhou WH, Yu Y, Zhu DL (2019) Machine learning models in type 2 diabetes risk prediction: results from a cross-sectional retrospective study in Chinese adults. Curr Med Sci 39(4):582–588
Yoffe L, Polsky A, Gilam A, Raff C, Mecacci F, Ognibene A, Crispi F, Gratacós E, Kanety H, Mazaki-Tovi S, Shomron N, Hod M (2019) Early diagnosis of gestational diabetes mellitus using circulating microRNAs. Eur J Endocrinol 181(5):565–577
Zheng T, Xie W, Xu L, He X, Zhang Y, You M, Yang G, Chen Y (2017) A machine learning-based framework to identify type 2 diabetes through electronic health records. Int J Med Inform 97: 120–127
Zou Q, Qu K, Luo Y, Yin D, Ju Y, Tang H (2018) Predicting diabetes mellitus with machine learning techniques. Front Genet 9:515

Machine Learning and Epilepsy

15

Abstract

Epilepsy is a neurological disorder of the brain affecting both children and adult, due to excessive electrical discharges by some brain cells at different sites. Antiepileptic drugs and surgery are the available forms of treatment but seizure free life is the ultimate goal of therapy. Machine learning is enabling early, better and accurate diagnosis and hence enabling decision making in therapy and holds great promise in this field of biomedical research. We discuss in this chapter the different applications of machine learning in understanding the biomedical aspects of epilepsy.

Keywords

Epilepsy · Cognition · Seizures · EEG · AED · GABA · NMDAR

Abbreviations

ADNFLE	Autosomal dominant nocturnal frontal lobe epilepsy
AED	Antiepileptic disease
AMPA	α-Amino-3-hydroxy-5-methyl-4-isoxazolepropionic acid
CAE	Childhood absence epilepsy
CHD	Chromodomain helicase DNA-binding
CNS	Central nervous system
CV	Cardiovascular disease
DEE	Developmental epileptic encephalopathy
DRE	Drug-resistant epilepsy
DS	Dravet syndrome
ECoG	Electrocorticography
EEG	Electroencephalography

S. Ghosh, R. Dasgupta, *Machine Learning in Biological Sciences*,
https://doi.org/10.1007/978-981-16-8881-2_15

EHR	Electronic health record
EIMFS	Epilepsy of infancy with migrating focal seizures
EKG	Electrocardiography
fALFF	Fractional amplitude of low-frequency fluctuation
FIRES	Febrile infection-related epilepsy syndrome
FLE	Frontal lobe epilepsy
fMRI	Functional MRI
FS	Febrile seizures
GABAR	Gamma amino butyric acid
GAD	Glutamic acid decarboxylase
GCPRs	G protein coupled receptors
GEFS+	Generalized epilepsy with febrile seizures plus
GM	Gray matter
GTCS	Generalized tonic-clonic seizures
HFO	High frequency oscillations
ILAE	The International League Against Epilepsy
JME	Juvenile myoclonic epilepsy
kSVM	Kernel support vector machine
MAE	Myoclonic-astatic epilepsy
MCD	Malformations of cortical development
mTOR	Mammalian target of rapamycin pathway (mTOR) pathway
NLP	Natural Language processing
rsfMRI	Resting-state fMRI
SMEI	Severe myoclonic epilepsy in infancy
T solium	*Taenia solium*
TLE	Temporal lobe epilepsy
VGKC	Voltage-gated potassium channel
WHO	World Health Organization

15.1 Introduction

Epilepsy is a chronic non-communicable neurological disorder of the brain, due to excessive electrical discharges by some brain cells at different sites. According to World Health Organization (WHO) it affects around 50 million people globally. Symptoms include recurrent seizures, with involuntary movement of a part or partial or the whole body or generalized and could reveal symptoms of being unconsciousness, disturbed movements, disturbed senses of taste, hearing, vision, mood and cognitive functions, lost control of bowel or bladder function.

Seizures can be for a brief period like jerks in muscles or for prolonged period of convulsions. Frequency of seizures can vary, from less than 1 per year to several seizures in a day but when an individual shows symptoms of two or more unprovoked seizures in a day, he or she is said to be epileptic. Some people reveal fractures, bruises from falling down due to seizures, psychological conditions of anxiety and depression and some may reveal premature death.

Although the exact cause of epilepsy remains largely unknown, yet several factors have been thought to cause the disease including genetic, infectious, immune related, structural, metabolic like brain damage at prenatal or perinatal stage, lack of oxygen supply to brain, trauma during birth, low birth weight, congenital abnormalities in brain malformations, severe head injury, stroke, limiting brain oxygen supply, brain infection like meningitis, encephalitis or neurocysticercosis, genetic disorders, and brain tumor.

Seizures are detected by abnormal patterns in electroencephalography (EEG) and history of seizures of the patient. It can be controlled by the use of antiseizure medicines as per physicians prescription and personal attention and care.

Some steps of preventing post traumatic epilepsy is preventing head injury, adequate perinatal care that can prevent epileptic tendencies associated with birth injury, administration of drugs and use of other methods to reduce fever of child can reduce chance of febrile seizures. Reduction of cardiovascular risk factors can prevent epilepsy associated with stroke, elimination of parasites and avoid infections like neurocysticercosis, which includes cysticercosis infection in the brain and central nervous system (CNS), caused by *Taenia solium* (*T. solium*) leading to epilepsy and forms the most common parasitic infection of the brain in the developing world, including Latin America, Africa, India, and China (Mewara et al. 2013).

There is a need of healthcare as premature death and loss of work productivity is observed and is a financial burden for the developing countries (https://www.who.int/news-room/fact-sheets/detail/epilepsy).

15.2 Epilepsy and Study by Conventional Methods

15.2.1 Cause, Diagnosis and Prognosis

While clinical history, brain imaging highlighting structural and functional causes, are important diagnostic and prognostic features, comorbidities find importance as aetiological and prognostic markers (Thijs et al. 2019). Frontal lobe epilepsy (FLE) reveal psychiatric manifestations of psychosis, altered motor, cognitive and medical, and personality changes (Gold et al. 2016). Very recently genomic studies have revealed genetic predisposition for epilepsy and have enabled a new classification of epilepsies and epileptic seizures (Thijs et al. 2019; Gold et al. 2016). Ion channels mutations are reported to directly or indirectly enhance excitatory neurotransmission or reduce inhibitory neurotransmission thereby leading to hyper excitability responses in the brain that make an individual predisposed to seizures.

Recent development in neuroimaging technology of video-EEG is enabling monitoring and detection of seizure type and diagnosis of epileptogenic zone in the brain. Aetiology of the epilepsies and genetic testing can help in identifying the cause of epilepsy (Guerreiro 2016).

The International League Against Epilepsy (ILAE) has come up with a classification of epileptic seizures and epilepsies, with importance on etiology (Perucca et al. 2018; Pack 2019).

Although the process of epilepsy is dynamic, the prognostic markers include aetiology, abnormal patterns of EEG seizures, types and numbers prior to and post treatment of which aetiology remains as important risk factor associated with premature death. While an early response to treatment is indicative of long-term prognosis, more number of seizures at diagnosis, loss of intellectual ability, and symptomatic aetiology act as negative indicator (Beghi et al. 2015).

Primary brain tumors can cause focal epilepsy, which is a neurological condition revealing predominant symptom of recurring seizures affecting any half of the brain. Leading to new-onset seizures in adults and over 10% of lesional focal epilepsies (Nowell et al. 2015).

15.2.2 Conventional Treatment

Administration of anti-epileptic drugs (AED), in treatment of epilepsy initiated as early as in 1990s and has undergone progressive advances since the last few decades (Manford 2017) but despite the suppression of seizures by AED, it has failed to provide a completely seizure free life.

Surgery has been considered a better option over medical therapy for patients with refractory focal epilepsy (Dewar and Pieters 2015). Epileptic surgery remains the only option for drug-resistant focal epileptic patients (Thijs et al. 2019; Gold et al. 2016) or patients for whom treatment with two AEDs, have failed. Epileptic surgery, however, is affected by factors including epilepsy syndrome, epileptogenic lesion and types, and post-operative follow-up time. For patients where surgery and treatment with AED has failed in control of seizures, alternative AEDs or other non-pharmacological therapies of diet based like ketogenic diet, neurostimulation, may improve seizure control (Perucca et al. 2018) and may be a step towards a better life.

15.2.3 Genetics

The genetic studies have helped in our understanding of the complex condition of epilepsy, its classification, molecular diagnosis, and drug discovery. More than 500 epilepsy-associated genes have been studied to play role in excitation of neurons, synaptic transmission, cortical development. Mutated forms of these genes have been associated in diagnosis, prognosis, and step towards therapy (Weber et al. 2017).

Mutations of different receptors including GABA (Gamma amino butyric acid) – R (receptor), *N*-methyl-D-aspartate NMDA-R, G protein coupled receptors (GCPRs), and ion channels like potassium channels, rapamycin pathway, chromatin remodeling are associated with epilepsy. Gain-of-function mutations in KCNT1,

GRIN2A, GRIN2D and SCN8A genes is finding importance in targeting epilepsy and loss-of-function mutations in non-ion channel epilepsy genes, and gene network and disrupted pathways reveal overlapping with other neurodevelopmental traits including human cognition (Symonds et al. 2017) finds importance in epilepsy. KCNQ2 abbreviated as kv7, is a potassium channel gene in the brain, located in chromosome 20, at 13.3 (20p13.3). Alterations in this gene can lead to developmental and epileptic encephalopathy. KCNT1 gene mutation, can lead to KCNT1-related epilepsy, inherited as autosomal dominant manner including epilepsy of infancy with migrating focal seizures (EIMFS) and autosomal dominant nocturnal frontal lobe epilepsy (ADNFLE, (Gertler et al. 1993–2020)). Early detection of these mutations can be used as effective diagnostic, therapeutic target, and precision medicine treatment (Dewar and Pieters 2015; Steinlein 2008).

15.2.3.1 GABA Receptor Mutation

Recent studies have highlighted the role of inhibitory GABA receptors (R) that bind to neurotransmitter gamma-aminobutyric acid (GABA) and mediate inhibitory synaptic transmission in the central nervous system (CNS). Mutations in GABAA receptor subunit genes including GABRA1, GABRB3, GABRG2 and GABRD have been associated with genetic epilepsy syndromes including childhood absence epilepsy (CAE) occurring at childhood, revealing absence of seizure, in the child who stares blankly, and is non-aware or non-responsive, eyes may blink, show repeated mouth chewing (Macdonald et al. 2010), generalized epilepsy with febrile seizures plus (GEFS+) in which family members reveal seizure ranging from simple febrile seizures (FS) to severe myoclonic-astatic epilepsy (MAE), juvenile myoclonic epilepsy (JME) is the most common epilepsy starting in adolescent period revealing jerks, pure febrile seizures (FS) occur frequently in children, from 6 months to 6 years of age, associated with fever revealing clinical and genetic heterogeneity, and Dravet syndrome (DS) or severe myoclonic epilepsy in infancy (SMEI) is a debilitating childhood encephalopathy with epilepsy characterized by tonic, clonic, or tonic–clonic seizures (Anwar et al. 2019) (Fig. 15.1).

15.2.3.2 NMDA (*N*-Methyl-D-Aspartate) Receptor (NMDAR) Mutation

NMDAR binds to glutamate, the primary excitatory neurotransmitter in the human brain. The receptor is a heteromeric complex of different subunits including GluN1 coded by *GRIN1* gene located at chromosome 9q34.3, GluN2A-2D where *GRIN2A* gene, in chromosome 16p13.2 codes for NMDAR GluN2A subunit., *GRIN2B* gene, located on 12p13.1, codes NMDAR GluN2B subunit,GRIN2C located on 17q25.1 codes for GluN2C, *GRIN2D* gene, on 19q13.3) codes for NMDAR GluN2D subunit and Glu3A-3B are coded by *GRIN3A*, and *GRIN3B* and the mutation of NMDAR subunit has been observed in childhood epilepsy (Xu and Luo 2018).

15.2.3.3 Potassium Channels

Potassuim channels play role in transmembrane transport and mutations in potassium channels have been associated with epilepsy (Köhling and Wolfart 2016).

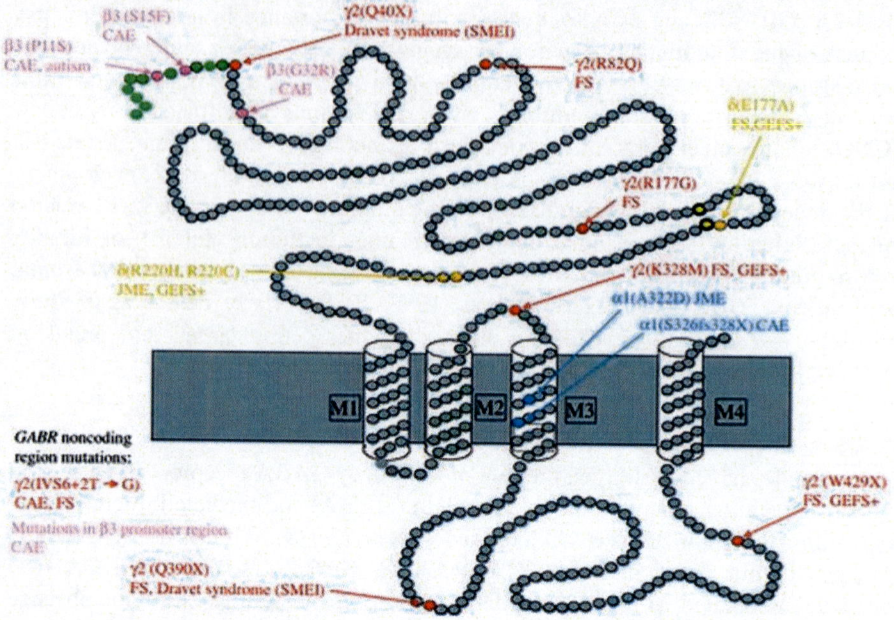

Fig. 15.1 Schematic representation of GABAA R (receptor) subunit mutations associated with genetic epilepsy syndromes. GABAA receptor subunits as precursor protein is translated whose signal sequence (marked in green) is removed forming a mature protein of a large extracellular N terminal domain, four transmembrane domains (represented by M1–M4) and a large cytoplasmic domain. Representation of GABAA receptor α1 (marked in blue), β3 (marked in pink), γ2 (marked in red), and δ (marked in yellow) subunit and their mutations are associated with genetic epilepsy syndromes (Image adapted with permission from Macdonald et al. (2010))

15.2.3.4 G-Protein Coupled Receptors (GCPRs)

G protein-coupled receptors (GPCRs) are seven-transmembrane receptor that binds extracellular substances and transmits signals to the G protein (guanine nucleotide-binding protein). Mutations in GPCR are known to be associated with altered neuronal excitability leading to seizures (Yu et al. 2019).

15.2.3.5 Mammalian Target of Rapamycin Pathway (mTOR) Pathway

The mTOR pathway involves a serine threonine protein kinase that plays a role in the regulation of protein synthesis, mediates cell growth, division, and survival. Mutations in mTOR regulatory genes including TSC1, TSC2, AKT3, DEPDC5 have been associated with epilepsy. Aberrant cascade functions can lead to altered cell size, axon, and dendrite outgrowth leading to epileptogenesis (Crino 2015).

15.2.3.6 Chromatin Remodeling

The chromodomain helicase DNA-binding (CHD) are ATP-dependent chromatin remodeler family of proteins that play role in chromatin reorganization, deposition of histone variants and neurodevelopment. Variant forms of *CHD1*, *CHD2*, *CHD4*,

CHD7, and *CHD8* have been associated with neurological disorders including epilepsy and *CHD2* variant is associated with developmental epileptic encephalopathy (DEE, Lamar and Carvill 2018).

15.2.4 Autoimmune Epilepsies

Autoimmune epilepsies account for around one-third of adults with epilepsies of unknown etiology, seizure disorder and causes, and it is difficult to identify the individuals with autoimmune epilepsy (Dubey et al. 2017). Autoimmune epilepsies associated with anti-*N*-methyl-D-aspartate receptor encephalitis, is associated with antibodies against NR1 and NR2 subunit of NMDA receptor and is known to occur predominantly in women (Yeshokumar and Pardo 2017; Liu et al. 2017), Rasmussen's encephalitis with pathophysiology of inflamed cerebral cortex, epilepsy, and detoriated neurological and cognitive deterioration with role played by autoantibodies (Varadkar et al. 2014), and febrile infection-related epilepsy syndrome (FIRES, (Fox et al. 2017)) of unknown aetiology, affects healthy children of age 3–15 years might involve autoantibodies in the pathogenesis of seizures and can affect healthy children, revealing some or all symptoms of focal seizures, encephalopathy, cognitive deterioration, or other focal neurological deficits.

Studies based on hypothesizing the role of pathogenic auto-antibodies targeting brain against voltage-gated potassium channel (VGKC) complex on neuron surface, NMDAR, γ-aminobutyric acid (GABA)(B)-receptor and the α-amino-3-hydroxy-5-methyl-4-isoxazolepropionic acid (AMPA)-receptor or targeting intracellular antigens like glutamic acid decarboxylase (GAD) and the onconeural antibodies in causing autoimmune epilepsies are being conducted, however the exact pathways they work remains yet to be studied (Bien and Scheffer 2011).

Early detection is important for timely immunotherapy, as appropriate, as conventional AED cannot control seizures and other neurological symptoms (Yeshokumar and Pardo 2017).

15.2.5 Comorbidity

Cardiovascular (CV) comorbidities are commonly observed in epilepsy as arrhythmias occur during and after seizures due to ictal asystole. Studies reveal that genetic factors, drugs, and etiological factors may play role. Seizure activity may lead to transient myocardial ischemia involving obstruction of blood flow to the heart muscle or myocardium due to atherosclerosis which can lead to heart attack and the Takotsubo syndrome also called as broken heart syndrome caused by stressful and emotional situations, also can be triggered by physical illness or surgery (Shmuely et al. 2017).

15.3 Machine Learning and Epilepsy

Machine learning finds application in epilepsy in automated seizure detection from EEG, video, and kinetic data, automated imaging analysis and pre-surgical planning, predicting treatment response and aids in early surgical decision making (Abbasi and Goldenholz 2019). Machine learning methods finds application for predicting patients at low risk by screening from EEG data (Struck et al. 2019).

Machine learning is playing role in prediction of outcome of neurosurgery for survival, recurrence, prognosis, in patients under surgery for various neurological disorders including epilepsy (Senders et al. 2018a), Drug-resistant epilepsy (DRE) patients reveal high risk of morbidity and mortality, and early detection by the machine learning methods enables early decision making for surgery (An et al. 2018).

Machine learning is finding applications in improvement of neurosurgical care in different disorders of the central nervous system (CNS) including epilepsy and has enabled for planning prior to surgical planning, guidance during operation, monitoring of neurophysiological functions, and prediction of neurosurgery (Senders et al. 2018b). Resting-state fMRI or rsfMRI is used to study different neurologic disorders due to neurosurgery and psychiatric disorders. Machine learning tools are used to detect rsfMRI epilepsy networks in temporal lobe epilepsy (TLE), which is the most common form of focal or partial epilepsy (Bharath et al. 2019).

High frequency oscillations (HFOs) of epileptogenic brain are biomarkers used while planning of surgery for patients with medically refractory epilepsy and automated machine learning tools have enabled detection of HFO rate and neuroanatomical localization (Weiss et al. 2019). Machine learning has been used to improve localization of epileptic zones and predicting outcome of epilepsy surgery which is the only option for treatment of drug-resistant patients thus enhancing the chance of living a seizure free life (Nissen et al. 2018).

Machine learning tools have found application in neuroimaging data for diagnosis and prognosis of therapeutic lesion changes within the area of neuroradiology in different neurological disorders including epilepsy (Wissel et al. 2019).

Natural Language processing (NLP) algorithm has been designed to identify potential surgical candidates amongst epilepsy patients from physician notes (Sidhu et al. 2018) and from clinical notes from the electronic health record (EHR) reducing loss of time in decision making (Cohen et al. 2016).

Machine learning methods can detect suicidal tendency, a comorbid condition observed in epilepsy in adults from spoken language (Glauser et al. 2020).

Machine learning is attempting integration of different parameters of imaging and genetic biomarkers to design tailor made epilepsy treatments (Sidhu et al. 2018). Kernel support vector machine (kSVM) has been used to predict localization of epilepsy in accurate ways as compared to standard EEG (Frank et al. 2018).

Machine learning tool of SVM has been used to identify epileptic children with generalized tonic-clonic seizures (GTCS) revealing reduced gray matter (GM) volume, and increased fractional amplitude of low-frequency fluctuation (fALFF) and neuroimaging data revealed the role of right thalamus in the chronic damaging which was not possible by conventional methods (Wang et al. 2018). Machine learning has enabled an automatic detection system identifying normal vs. ictal, and interictal conditions from EEG signals (Wang et al. 2017). Machine learning tools have enabled diagnosis and treatment of temporal lobe epilepsy (TLE) from neuroimaging data (Rudie et al. 2015; Kiran Raj et al. 2018). A supervised machine learning tool was used to predict outcome of AED treatment and improve treatment strategies (Yao et al. 2019). Machine learning tools have enabled accurate detection of seizure preictal state from electrocorticography (ECoG), scalp EEG, and electrocardiography (EKG) and thus solving the major problem of disease identification in epilepsy (Meisel and Bailey 2019).

Machine learning is used to identify atypical language patterns in epileptic individuals separate from normal individuals based on their cerebral activity, detected by functional MRI (fMRI, Torlay et al. 2017).

15.4 Discussions

Machine learning has enabled considerable progress on the understanding, detection of seizures, prediction of surgery (Abbasi and Goldenholz 2019; Struck et al. 2019; Senders et al. 2018a,b; An et al. 2018; Bharath et al. 2019; Weiss et al. 2019; Nissen et al. 2018; Wissel et al. 2019; Sidhu et al. 2018; Cohen et al. 2016; Glauser et al. 2020; Frank et al. 2018; Wang et al. 2017, 2018; Rudie et al. 2015; Kiran Raj et al. 2018; Yao et al. 2019; Meisel and Bailey 2019; Torlay et al. 2017; Sakai and Yamada 2019; Kini et al. 2016; Acharya et al. 2018; Hwang et al. 2019; Kassahun et al. 2014; Armañanzas et al. 2013; Usman et al. 2017). Advances in neuroimaging, automation and machine learning tools have enabled detection of malformations of cortical development (MCDs), in drug resistant epilepsy (DRE, Kini et al. 2016). The identification of epilepsy is difficult in children and machine learning has enabled the detection in children, predict requirement of surgery and is having diverse application in betterment of life and seizure free life. However, with advances in machine learning it is hopeful that it will have better applications in epilepsy in bettering the lives of the patients and remains the scope of future research.

References

Abbasi B, Goldenholz DM (2019) Machine learning applications in epilepsy. Epilepsia 60(10): 2037–2047

Acharya UR, Hagiwara Y, Adeli H (2018) Automated seizure prediction. Epilepsy Behav 88:251–261

An S, Malhotra K, Dilley C et al (2018) Predicting drug-resistant epilepsy - a machine learning approach based on administrative claims data. Epilepsy Behav 89:118–125

Anwar A, Saleem S, Patel UK, Arumaithurai K, Malik P (2019) Dravet syndrome: an overview. Cureus 11(6):e5006

Armañanzas R, Alonso-Nanclares L, Defelipe-Oroquieta J et al (2013) Machine learning approach for the outcome prediction of temporal lobe epilepsy surgery. PLoS One 8(4):e62819
Beghi E, Giussani G, Sander JW (2015) The natural history and prognosis of epilepsy. Epileptic Disord 17(3):243–253
Bharath RD, Panda R, Raj J et al (2019) Machine learning identifies "rsfMRI epilepsy networks" in temporal lobe epilepsy. Eur Radiol 29(7):3496–3505
Bien CG, Scheffer IE (2011) Autoantibodies and epilepsy. Epilepsia 52(suppl 3):18–22
Cohen KB, Glass B, Greiner HM et al (2016) Methodological issues in predicting pediatric epilepsy surgery candidates through natural language processing and machine learning. Biomed Inform Insights 8:11–18
Crino PB (2015) mTOR signaling in epilepsy: insights from malformations of cortical development. Cold Spring Harb Perspect Med 5(4):a022442
Dewar SR, Pieters HC (2015) Perceptions of epilepsy surgery: a systematic review and an explanatory model of decision-making. Epilepsy Behav 44:171–178
Dubey D, Alqallaf A, Hays R et al (2017) Neurological autoantibody prevalence in epilepsy of unknown etiology. JAMA Neurol 74(4):397–402
Fox K, Wells ME, Tennison M, Vaughn B (2017) Febrile infection-related epilepsy syndrome (FIRES): a literature review and case study. Neurodiagn J 57(3):224–233
Frank B, Hurley L, Scott TM, Olsen P, Dugan P, Barr WB (2018) Machine learning as a new paradigm for characterizing localization and lateralization of neuropsychological test data in temporal lobe epilepsy. Epilepsy Behav 86:58–65
Gertler T, Bearden D, Bhattacharjee A et al (1993–2020) KCNT1-related epilepsy, 20 Sept 2018. In: Adam MP, Ardinger HH, Pagon RA, et al. (eds) GeneReviews® [Internet]. University of Washington, Seattle. https://www.ncbi.nlm.nih.gov/books/NBK525917
Glauser T, Santel D, DelBello M et al (2020) Identifying epilepsy psychiatric comorbidities with machine learning. Acta Neurol Scand 141(5):388–396
Gold JA, Sher Y, Maldonado JR (2016) Frontal lobe epilepsy: a primer for psychiatrists and a systematic review of psychiatric manifestations. Psychosomatics 57(5):445–464
Guerreiro CA (2016) Epilepsy: is there hope? Indian J Med Res 144(5):657–660
Hwang G, Nair VA, Mathis J et al (2019) Using low-frequency oscillations to detect temporal lobe epilepsy with machine learning. Brain Connect 9(2):184–193
Kassahun Y, Perrone R, De Momi E et al (2014) Automatic classification of epilepsy types using ontology-based and genetics-based machine learning. Artif Intell Med 61(2):79–88
Kini LG, Gee JC, Litt B (2016) Computational analysis in epilepsy neuroimaging: a survey of features and methods. Neuroimage Clin 11:515–529
Kiran Raj V, Rajagopalan SS, Bhardwaj S et al (2018) Machine learning detects EEG microstate alterations in patients living with temporal lobe epilepsy. Seizure 61:8–13
Köhling R, Wolfart J (2016) Potassium channels in epilepsy. Cold Spring Harb Perspect Med 6(5): a022871
Lamar KJ, Carvill GL (2018) Chromatin remodeling proteins in epilepsy: lessons from CHD2-associated epilepsy. Front Mol Neurosci 11:208
Liu CY, Zhu J, Zheng XY, Ma C, Wang X (2017) Anti-N-methyl-D-aspartate receptor encephalitis: a severe, potentially reversible autoimmune encephalitis. Mediat Inflamm 2017:6361479
Macdonald RL, Kang JQ, Gallagher MJ (2010) Mutations in GABAA receptor subunits associated with genetic epilepsies. J Physiol 588(Pt 11):1861–1869
Manford M (2017) Recent advances in epilepsy. J Neurol 264(8):1811–1824
Meisel C, Bailey KA (2019) Identifying signal-dependent information about the preictal state: a comparison across ECoG, EEG and EKG using deep learning. EBioMedicine 45:422–431
Mewara A, Goyal K, Sehgal R (2013) Neurocysticercosis: a disease of neglect. Trop Parasitol 3(2): 106–113
Nissen IA, Stam CJ, van Straaten ECW et al (2018) Localization of the epileptogenic zone using interictal MEG and machine learning in a large cohort of drug-resistant epilepsy patients. Front Neurol 9:647

Nowell M, Miserocchi A, McEvoy AW (2015) Tumors in epilepsy. Semin Neurol 35(3):209–217

Pack AM (2019) Epilepsy overview and revised classification of seizures and epilepsies. Continuum (Minneap Minn) 25(2):306–321

Park SC, Chung CK (2018) Postoperative seizure outcome-guided machine learning for interictal electrocorticography in neocortical epilepsy. J Neurophysiol 119(6):2265–2275

Perucca P, Scheffer IE, Kiley M (2018) The management of epilepsy in children and adults. Med J Aust 208(5):226–233

Rudie JD, Colby JB, Salamon N (2015) Machine learning classification of mesial temporal sclerosis in epilepsy patients. Epilepsy Res 117:63–69

Sakai K, Yamada K (2019) Machine learning studies on major brain diseases: 5-year trends of 2014-2018. Jpn J Radiol 37(1):34–72

Senders JT, Staples PC, Karhade AV et al (2018a) Machine learning and neurosurgical outcome prediction: a systematic review. World Neurosurg 109:476–486.e1

Senders JT, Zaki MM, Karhade AV et al (2018b) An introduction and overview of machine learning in neurosurgical care. Acta Neurochir 160(1):29–38

Shmuely S, van der Lende M, Lamberts RJ, Sander JW, Thijs RD (2017) The heart of epilepsy: current views and future concepts. Seizure 44:176–183

Sidhu MK, Duncan JS, Sander JW (2018) Neuroimaging in epilepsy. Curr Opin Neurol 31(4): 371–378

Steinlein OK (2008) Genetics and epilepsy. Dialogues Clin Neurosci 10(1):29–38

Struck AF, Rodriguez-Ruiz AA, Osman G et al (2019) Comparison of machine learning models for seizure prediction in hospitalized patients. Ann Clin Transl Neurol 6(7):1239–1247

Symonds JD, Zuberi SM, Johnson MR (2017) Advances in epilepsy gene discovery and implications for epilepsy diagnosis and treatment. Curr Opin Neurol 30(2):193–199

Thijs RD, Surges R, O'Brien TJ, Sander JW (2019) Epilepsy in adults. Lancet 393(10172):689–701

Torlay L, Perrone-Bertolotti M, Thomas E, Baciu M (2017) Machine learning-XGBoost analysis of language networks to classify patients with epilepsy. Brain Inform 4(3):159–169

Usman SM, Usman M, Fong S (2017) Epileptic seizures prediction using machine learning methods. Comput Math Methods Med 2017:9074759

Varadkar S, Bien CG, Kruse CA et al (2014) Rasmussen's encephalitis: clinical features, pathobiology, and treatment advances. Lancet Neurol 13(2):195–205

Wang Y, Li Z, Feng L, Zheng C, Zhang W (2017) Automatic detection of epilepsy and seizure using multiclass sparse extreme learning machine classification. Comput Math Methods Med 2017:6849360

Wang J, Li Y, Wang Y, Huang W (2018) Multimodal data and machine learning for detecting specific biomarkers in pediatric epilepsy patients with generalized tonic-clonic seizures. Front Neurol 9:1038

Weber YG, Biskup S, Helbig KL, Von Spiczak S, Lerche H (2017) The role of genetic testing in epilepsy diagnosis and management. Expert Rev Mol Diagn 17(8):739–750

Weiss SA, Waldman Z, Raimondo F et al (2019) Localizing epileptogenic regions using high-frequency oscillations and machine learning. Biomark Med 13(5):409–418

Wissel BD, Greiner HM, Glauser TA et al (2019) Investigation of bias in an epilepsy machine learning algorithm trained on physician notes. Epilepsia 60(9):e93–e98

Xu XX, Luo JH (2018) Mutations of N-methyl-D-aspartate receptor subunits in epilepsy. Neurosci Bull 34(3):549–565

Yao L, Cai M, Chen Y, Shen C, Shi L, Guo Y (2019) Prediction of antiepileptic drug treatment outcomes of patients with newly diagnosed epilepsy by machine learning. Epilepsy Behav 96: 92–97

Yeshokumar AK, Pardo CA (2017) Autoimmune epilepsies. Semin Pediatr Neurol 24(3):161–167

Yu Y, Nguyen DT, Jiang J (2019) G protein-coupled receptors in acquired epilepsy: druggability and translatability. Prog Neurobiol 183:101682

The IT Industry and Applications in Biology 16

Abstract

The different applications of machine learning tools and Microsoft, Google, Facebook and Pytorch has tremendous application in biology, health, disease, predicting epidemics and disease incidence and in the domain of medical sciences. In this chapter we discuss the different applications of machine learning in applications to health, understanding Trends of disease, analysis of psychological and emotional health, from social networking sites like facebook, analysis of sequencing data using PyTorch algorithm.

Keywords

Google Trends · Lyme disease · Cellulitis · Tick-borne encephalitis (TBE) · Vesicular stomatitis (VS) · COVID-19

Abbreviations

AI	Artificial intelligence
ANN	Artificial Neural Network
DNN	Deep neural networks
DT	Decision Tree
FB	Facebook
ICRISAT	International Crops Research Institute for the Semi-Arid Tropics
URI	Upper respiratory infections

S. Ghosh, R. Dasgupta, *Machine Learning in Biological Sciences*,
https://doi.org/10.1007/978-981-16-8881-2_16

16.1 I. Introduction

Machine Learning and Artificial intelligence (AI) has a promising future with the hugh potential market of growth of softwares by 2025. Development and deployment of powerful AI-based tools is one of the hottest and most happening topics in the technology based industry. A huge investment on Artificial intelligence (AI)-based tools are being done by the industry giants to boost business. This has a tremendous application in human health and life science research. Numerous are the applications including fighting terminal illness, infectious diseases and spreading mass awareness of malignant diseases like Cancer, screening and design of drugs, understanding and predicting trends in epidemics, precision agriculture and farming are areas where artificial intelligence is finding importance and is a boon to the human lives, enabling animal welfare and wellbeing, increased productivity of crops which is on one hand step towards solving the increasing demands on food globally and on the other hand boosting up the agriculture industry and a countries economy. Apple, Amazon, Facebook (FB), Google, and Microsoft are the major industrial leaders who have been playing dominant role in enabling major applications of artificial intelligence in the domain of life sciences (https://www.datamation.com/artificial-intelligence/top-artificial-intelligence-companies.html).

16.2 Google

Google, founded in 1998 is a technology based American based multinational company (MNC) providing services and products in the domains of internet, advertising, search engine, cloud computing, software, and hardware. It is one of the leaders in application of the emerging technology of artificial intelligence. In this chapter we discuss some of the machine learning and artificial intelligence applications with a focus on the area of Life Sciences research.

16.2.1 Artificial Intelligence Applications

AI has been utilized in Google products by Google and finds application in our daily lives from Google Search Engines, Google Maps, Google Photos, YouTube Safe Content, Gmail Smart Reply. Technological products including Tensor Flow and Cloud Auto Machine Learning is being used and others advanced products with diverse applications are being developed by developers.

Google Brain, developed in the 2010, based on deep learning tools can monitor and record the viewing habits of an individual as they surf and visualise the contents of video from different servers. Deep neural networks (DNN) have been used to analyze about viewers and record their choice habits and preferences of viewing making the company profitable.

In 2017, Google Brain added an application of image enhancement system of images with very low resolution by applications of neural networks.

It also led to the development of the Google Translate that can translate words, speech in audio form from one language to text of another language using neural networks by replicating the human brain.

Different other AI based applications of Google including Google's RankBrain, Google's Cloud Video Intelligence segments that can analyse video' stored on Google. Google Jigsaw owned by Google parent Alphabet has applications for a safer internet experience and uses to detoxify the Internet. It is also possible to make applications that can identify hate speech and negative, harmful, hatred and toxic comments by assigning a toxicity score to a content based on different parameters.

The company Alphabet has been reported to file 186 patents within a span of 4 years from 2013 to 2017 in the healthcare industry focused on its AI company DeepMind and Verily Life Science. Reports suggest that Google plans further expand the design of applications in healthcare sector in the USA making use of the AI. AI has been applied in the healthcare sector with major emphasis on disease detection, diagnosis, and treatment (https://medicalfuturist.com/googles-masterplan-for-healthcare/).

A Health IT Analytics report mentions that Google's machine learning (ML) program can detect breast cancer with 89% accuracy compared to a human pathologist who can detect it with just 73% accuracy (https://healthitanalytics.com/news/google-deep-learning-tool-99-accurate-at-breast-cancer-detection).

16.2.2 In Search of Biomedical Literature

The application of computation has enabled availability of different medical databases on the World Wide Web (www), like Google Scholar, PubMed, Scopus, and Web of Science that provides search facilities that extracts biological and biomedical scientific data and thereby accelerates research. Although PubMed is most widely used search engine for extracting information in literature for biomedical research, Google Scholar, finds importance in finding the most relevant and obscure (Falagas et al. 2008; Nourbakhsh et al. 2012) information and articles with higher citation numbers (Nourbakhsh et al. 2012). Studies have revealed no significant differences in searching answers for questions related to drug related literature within Google Scholar and PubMed (Freeman et al. 2009) and that PubMed proves to be a better search tool than Google Scholar with the Clinical Queries filter for topics related to respiratory care (Anders and Evans 2010).

16.2.3 Google Trends and Disease Forecast

Google Trends is another application created by Google that enables graphical analysis of diverse queries over time ranging from election to football matches to the trends by the latest havoc created by the pandemic COVID-19.

16.2.3.1 Trends in Disease Biology

One of the recent free and novel applications of Google Trends enables access to users of search data in the internet, indicating trends and variations aiding in searches related to population behavior and health (Nuti et al. 2014).

Disease Surveillance

When Integrated Disease Surveillance Programme (IDSP) data reporting from India and Google Trends where compared in predicting epidemics outbreaks for different infectious diseases in India including (i) Chikungunya, an *Aedes aegypti* (*Ae. aegypti*) and *Aedes albopictus (Ae. albopticus)* mosquito borne viral disease caused by RNA virus of alphavirus genus, family Togaviridae, (ii) Dengue, caused by Dengue virus (DENV), transmitted by mosquitoes *Ae. aegypti* and *Ae. albopictus*, (iii) malaria transmitted by female *Anopheles* mosquitoes, caused by *Plasmodium vivax* and *Plasmodium falciparum* and (iv) enteric fever or Typhoid fever caused by *Salmonella typhi* or *Salmonella paratyphi* there revealed a temporal correlation, indicating its application in predicting disease outbreaks (Verma et al. 2018). Google Trends have found important applications in the newly developing branch of science called as infodemiology, the term coined in the year 2000, by Gunther Eysenbach (2002) involving description and analysis of data on health information and web based communication patterns with role in public health with aims to improving public health. Infoveillance is another new branch of science in the field of public health research, a term coined in 2004 by Gunther Eysenbach including syndromic surveillance, focused on public health related issues, from information and content available online. This led to the development of Google Flu, use of other social media like Twitter to predicting disease outbreak patterns (Eysenbach 2006, 2009, 2011). Machine learning have led to further development of different tools in this domain including Google Flu Trends where one can querry and get information on trends in flu disease and can compare with a countries data by official surveillance (Ginsberg et al. 2008). Later Google Dengue Trends was developed that could provide trends in spread or occurrence of dengue from queried terms and could enable comparison with countries' official surveillance data (Chan et al. 2011). Using Machine learning tools Flu Detector tool was developed by University of Bristol with advanced feature selection tool to extract flu related terms from twitter, then giving a flu-score for UK regions based on geolocated tweets (Vasileios and Nello 2010).

16.2.3.2 Prediction of Disease Incidence and Trend

Lyme Disease

Lyme disease is tick borne disease in humans caused by the bacteria belonging to class spirochete under genus Borrelia, including *Borrelia burgdorferi (B burgdorferi)* and *Borrelia mayonii (B mayonii)* transmitted by infected black-legged tick or *Ixodes scapularis (I scapularis)*, commonly called as deer tick, in the East and Midwest, *Ixodes pacificus* (*I pacificus*) on the Pacific Coast, *Ixodes ricinus* (*I ricinus or* sheep tick) and *Ixodes persulcatus* (*I persulcatus* or taiga tick) in Europe and Asia, respectively. It reveals symptoms of fever, headache, fatigue, skin rash also termed as erythema migraines, swollen lymph nodes, joint and muscle aches (Shapiro 2014). Google Trends data has been reported to correlate well with the reported incidence of Lyme disease in Germany, however, further research on accuracy on forecasting models developed on traditional data was also reported (Kapitány-Fövény et al. 2019).

Cellulitis

Cellulitis is a bacterial infection primarily thought to be contributed by β-hemolytic *Streptococcus* and *Staphylococcus aureus* affecting the deep dermis and subcutaneous tissue, revealing symptoms of erythema, warmth, tenderness, and swelling and affects the world population largely predominantly reported from the United States. The chief challenges in this disease lies in effective and early diagnosis, administration of appropriate first-line antibiotics for drug resistant varieties, secondary states of imunosuppression, chronic liver or kidney disease (Laico 2018) that adds to the complicacy in the disease.

Google Trends, has been able to find out trends in cellulitis in human population globally reported from 31 countries on six continents explaining the observed seasonality and peaking effect in summer months (Zhang et al. 2018).

Disease Related to Pollution

Pollution is a major cause of disease and mortality across the globe. World Health Organization (WHO) in 2016, reported 4.2 million deaths due to air pollution through different diseases including lung cancer, chronic obstructive pulmonary disease (COPD), ischemic heart disease and stroke and respiratory infection (https://www.who.int/gho/phe/outdoor_air_pollution/burden_text/en/) affecting even the stem cells (Ghosh and Ansar 2015). Google Trends has been instrumental in providing a mechanism to track trends in pollution parameters (Laico 2018) and thus has been instrumental in spreading awareness and is a step towards controlling pollution mediated health hazards.

Tick-Borne Encephalitis (TBE)

TBE is a zoonotic infectious disease caused by tick-borne encephalitis virus (TBEV), belonging to Flaviviridae family including Far eastern Tick-borne encephalitis virus, European or Western tick-borne encephalitis virus and Siberian tick-borne encephalitis virus. It is transmitted by infected tick bites affecting the central nervous system (CNS) that can also spread by taking contaminated food, water, milk, or contact with skin farm animal including cow, sheep. The disease reveals symptoms of fever, achiness, appetite loss, headache, nausea, and vomiting (Bogovic and Strle 2015). It has been possible to apply Google Trends to predicting incidence and monitoring of TBE in Germany (Sulyok et al. 2020).

Vesicular Stomatitis (VS)

VS is caused by negative-sense RNA *virus*, Vesicular stomatitis virus (VSV) belonging to Rhabdoviridae family and *Vesiculovirus genus with* immunological classes including New Jersey (NJ) and Indiana (IND) affecting the livestock including, cow, cattle, horses, and swine causing blisters and lesions on gums, lips, tongue, nasal mucosa, oral mucosa, hoof crown, nipple and teats, coronary bands and may affect the agriculture industry adversely. Humans develop meningitis on being infected with vesicular stomatitis (Rozo-Lopez et al. 2018). Google Trends has been instrumental in predicting the trend and forecast of VS (Wang et al. 2018) and thus can be step towards controlling the infection.

Measles

Measles is caused by virus belonging to paramyxovirus family and genus *Morbillivirus*, known to infect humans, being transmitted by contact, respiratory droplets by coughing, sneezing and is a air borne disorder, infecting the respiratory tract spreading throughout the body affecting the skin, respiratory and immune system. It is a highly contagious disease, revealing symptoms of fever and rash on face, upper neck, hands and feet affecting the world population globally every year (Naim 2015). Google Trends has been able to predict the incidence and spread of the infection of measles from Italy (Santangelo et al. 2019).

COVID-19

Coronavirus infectious disease or COVID-19 (Ge et al. 2020; Ye et al. 2020; https://covid19.who.int/) is a zoonotic infection caused by the novel β coronavirus, termed as severe acute respiratory syndrome coronavirus-2 (SARS-CoV-2) and was reported in December 2019 in the Wuhan district of China initially that later spread across the globe, causing global pandemic claiming millions of lives. WHO reported 24,822,800 confirmed cases of COVID-19, including 838,360 deaths, on 30 August 2020. Most people revealed mild to moderate symptoms of fever, dry cough, feeling tired, aches and pains, sore throat, diarrhea, conjunctivitis, headache, loss of taste or smell, skin rash, or discoloration of fingers or toes, while some were asymptomatic. The infection spreads through droplets form coughs, sneezes, or exhaled breath of infected person, or by touching an infected fomite or contaminated surface and then touching the mouth, eyes, or nose. The acute symptoms of the disease include acute respiratory distress syndrome (ARDS) and multiple-organ failure leading to death within short time from cytokine storm (Ge et al. 2020; Ye et al. 2020; https://covid19.who.int/). Google Trends (GT) has been able to predict importance of loss of smell during COVID-19 infection correlating with observed mortality from data from eight countries including France, Germany, Iran, Italy, Netherlands, Spain, United Kingdom (UK), United States of America (USA), highlighting the importance of this symptom of loss of smell in COVID-19 (Walker et al. 2020). A study from Taiwan revealed that GT could predict timing and location and thereby play role in risk communication in affected population where infection by COVID-19 was rampant (Husnayain et al. 2020).

Zika Virus (ZIKV)

ZIKV disease is caused by single stranded (s-s) RNA virus belonging to genus flavivirus, under family Flaviviridae with two lineages namely Asian and African, transmitted by infected *Aedes* mosquito. The African ZIKV revealed life cycle and sylvatic transmission within monkeys and apes with infection in humans hosts as occasional, but the Asian lineage of ZIKV revealed main host as humans. The infection was first reported from Uganda in 1947 in monkeys and in 1952 in humans in 1952, followed by disease outbreaks in countries in Africa, America, and the Asia-Pacific region from 1960s to 1980s. In 2007 it was reported from Yap Federated States of Micronesia, French Polynesia in 2013 and other countries in the Pacific, in 2015 in Brazil accounting for a total of incidence reports from 86 countries globally. Symptoms of the disease include fever, body rash, headache, conjunctivitis, muscle and joint pain, and malaise and the infection is reported to be associated with neurological disorders of Guillain–Barré syndrome (GBS) and malformations in fetus and microcephaly (Rawal et al. 2016; https://www.who.int/news-room/fact-sheets/detail/zika-virus). This serious nature of the disease has made it important for their frequent updates of emergence and outbreaks.

Google Trends have been used to generate a model predict confirmed Zika cases with studies from Brazil and Colombia. Such early prediction could be helpful information for awareness of masses and so that necessary alert and precautionary measures can be taken by the healthcare personnel to prevent the outbreak or combat it and prevent the further spread (Morsy et al. 2018).

Autoimmune Disease

Systemic lupus erythematosus (SLE, (Gurevitz et al. 2013; Fortuna and Brennan 2013; Adinolfi et al. 2016)) is an autoimmune, chronic inflammatory disease, caused by generation of autoantibodies in the body of the individual that are against nuclear and cytoplasmic antigens. Thus the body's immune system attacks healthy tissue affecting the skin, joints, kidneys, brain, heart, blood cells and vessels, lungs and other organs, affecting the women more as compared to the men and can make its incidence at any age. Although the exact cause of the disease is not known, the factors associated with the incidence includes, genetic, environmental, hormonal, some medicines. Infodemiology and infoveillance and application of Google Trends by Big Data monitoring approach has enabled follow of studies related to SLE and its symptoms (Radin and Sciascia 2017).

Cancer

Due to increased number of incident cases of cancer reported from Brazil awareness in the mass regarding breast and prostate cancer found importance. Online campaigns were found to be effective in spreading awareness of cancer information as indicated by The Google Trends database (Quintanilha et al. 2019). Research studies have revealed that Google Trends have been instrumental in serving information to the masses and mass education on the malignant disorder, of throat cancer (Faoury et al. 2019). However, more applications of the disease detection and prognosis need to be worked out as a future scope of research.

16.3 Amazon

Another major user and applicant to machine learning and AI is Amazon. The company has employed applications and patented an AI based technological application of anticipatory shipping that enables the company to get the order to the nearest possible location of the address even prior to the actual purchase made by the customer. AI is the core technology finding applications in Amazon's forecasting and Prime Amazon in dynamic pricing algorithm on its marketplace (https://www.financialexpress.com/industry/sme/amazon-wants-to-send-your-order-even-before-you-place-it-jeff-bezos-firm-working-on-this-tech/1664299/).

16.4 Microsoft

Microsoft is deploying AI solutions across different industries including healthcare, agriculture, retail, education, banking, and financial services. Microsoft uses (AI) to fight against hackers by learning from the data of each company using its systems. Using AI initiative Project Hanover, Microsoft is working with the aim of finding applications for the best cancer treatment with algorithms to understand the development of cancer and prediction of the best drug combinations to fight against it. Together with International Crops Research Institute for the Semi-Arid Tropics (ICRISAT), Hyderabad, India it uses Artificial Intelligence (AI) to help farmers increase their crop yields by 10–30% by predicting the best sowing date for crops (https://www.accenture.com/t00010101t000000z__w__/gb-en/_acnmedia/pdf-75/accenture-techvision-2018-tech-trends-report.pdf; https://www.icrisat.org/icrisat-awarded-microsofts-ai-for-earth-grant-to-promote-environmental-sustainability/).

Use of advanced sensors and analytics together with applications of AI has enabled find and predict flavors. This has been used in The Danish brewer Carlsberg the world's oldest industries by the researchers in the Beer Fingerprinting Project (https://news.microsoft.com/transform/can-ai-help-brewers-predict-how-new-beer-varieties-will-taste-carlsberg-says-probably/).

16.5 Facebook

Facebook is a data-rich company extensively using Artificial intelligence (AI) for projects of today and tomorrow to connect every person on the planet. Using deep neural networks, the platform decides on type of advertisements to be shown to the in a more insightful way. Facebook is working on building AI-powered systems that will understand images at the pixel level and not classify the entire image (https://in.springboard.com/blog/google-vs-amazon-vs-microsoft-vs-facebook-who-is-leading-the-ai-race%EF%BB%BF/).

Many software industry giants are working towards storing of data and development of software for medical industry and hospitals, with products and services in the area of cloud-computing focussed towards health-care.

Social media has brought major changes how people communicate, react to incidents, share experiences with their own defined community. Recent studies show that the use of Facebook, Twitter, and Instagram has impacted relationship between people, promoted ideas positively and also otherwise. In the new world of progress in interactions by social media reveals a new dimension of reports of cases of depression. In many events, every moment people are evaluating their life and achievement with others in the same community. Recent studies by a group of researchers at University of Italy, Padova on the associations between problematic Facebook use, psychological distress and well-being in the population of adolescents and young adults, shows significant correlation between face book use and psychological well-being. Results of random effects meta-analysis confirmed a positive correlation between problematic Facebook use and psychological distress. Moderation analysis revealed that effect sizes were larger in older samples. Moreover, a negative correlation between problematic Facebook use and well-being was observed (Marino et al. 2018).

Facebook can also be used to promote ideas and missions. Research groups in Alberta, Canada used facebook to recruit human sample for medical testing purposes. They did research on evaluation of the effectiveness of paid advertisements on Facebook as a platform for recruiting pregnant women to a randomized controlled trial in comparison with traditional recruitment approaches. Some of the women studied through Facebook, and some through traditional approaches. This study revealed that Facebook advertisements were effective platform for reaching pregnant women (Adam et al. 2016). Facebook chooses their target samples through complex Artificial intelligence-based models, specially the Artificial Neural Network (ANN). Internet-based social networks like Facebook also has been recently implicated in causing chronic psychological stress affecting health negatively and can impact the incidence of upper respiratory infections (URIs) and was studied using online questionnaires (Campisi et al. 2012).

Social network data finds applications in the study of a user's feelings, emotions, and sentiments based on their online communication and finds importance in the detection of depression in users. Decision Tree (DT) and machine learning methods have been instrumental in providing accuracy in detection of depression and enable the finding of solutions to mental health problems in Facebook (FB) users (Islam et al. 2018).

16.6 Pytorch

With more and more data on high-throughput sequencing of proteins, genomics, transcriptomics, and metabolomics being available and being very complicated, it has become more important to analyze the data properly to gain insights into the information in biological system and understand the biological processes.

PyTorch, is a deep learning based library for Python programmes that enables study and building of deep learning projects and finds application in biology. It is flexible, easy to use, with accelerated computation using graphical processing units

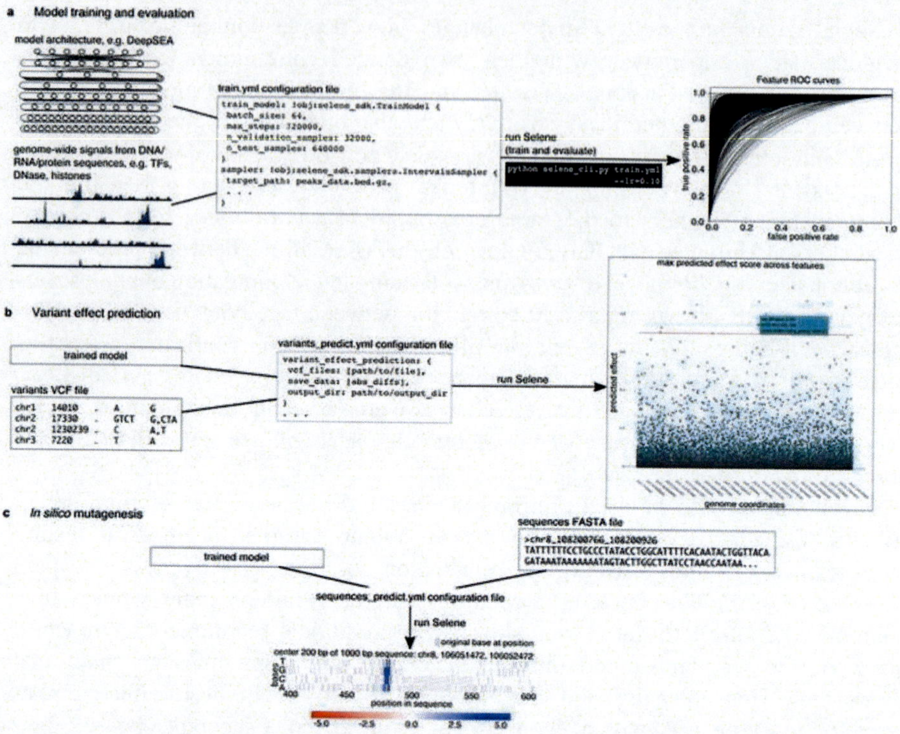

Fig. 16.1 Use of Selene in understanding and analysis of sequencing data. (Reproduced with permission from Chen et al. (2019)). **(a)** Selene enables training and evaluation of new deep learning models, **(b)** Selene supports trained models to interpret variants. **(c)** Selene supports to get mutation effect scores and understand it by heatmap

(GPUs) and enables numerical optimization on mathematical expressions (Stevens et al., 2020). Selene (https://selene.flatironinstitute.org/), based on PyTorch has been used for fast and easy development, training, and application for any biological DNA RNA, or protein sequence data (Chen et al. 2019). DeepSEA (http://deepsea.princeton.edu/) is a deep learning-based algorithmic framework that can predict the chromatin effects of sequence alterations with single nucleotide sensitivity, epigenetic state of a sequence, including transcription factors binding, DNase I sensitivities, histone marks, chromatin effects of sequence variants (Zhou and Troyanskaya 2015). Selene is designed with data sampling and training for a trained model using the trained model, analysis, prediction and visualization (Fig. 16.1).

16.7 Discussion

Machine Learning in biology has now expanded applications in identifying coding regions of genes, analyzing genomic data obtained by next-generation sequencing (NGS) approaches, prediction of structure of protein from data obtained from proteomics based studies and proves a promise to the animal, health, and agriculture industry. Machine learning applications in text-mining tools have enabled identification of novel drug targets from published articles and secondary databases.

Deep learning and neural network can enable to understand the thinking of human brain. AI in healthcare has enabled personalized medications, predict accurately and improve the quality of life. It holds promise in prediction of mental health and detection and probable cure of depression by analysis of social network data.

References

Adam LM, Manca DP, Bell RC (2016) Can Facebook be used for research? Experiences using Facebook to recruit pregnant women for a randomized controlled trial. J Med Internet Res 18(9): e250

Adinolfi A, Valentini E, Calabresi E et al (2016) One year in review 2016: systemic lupus erythematosus. Clin Exp Rheumatol 34(4):569–574

Anders ME, Evans DP (2010) Comparison of PubMed and Google Scholar literature searches. Respir Care 55(5):578–583

Bogovic P, Strle F (2015) Tick-borne encephalitis: a review of epidemiology, clinical characteristics, and management. World J Clin Cases 3(5):430–441

Campisi J, Bynog P, McGehee H et al (2012) Facebook, stress, and incidence of upper respiratory infection in undergraduate college students. Cyberpsychol Behav Soc Netw 15(12):675–681

Chan EH, Sahai V, Conrad C, Brownstein JS, Aksoy S (2011) Using Web search query data to monitor dengue epidemics: a new model for neglected tropical disease surveillance. PLOS Negl Trop Dis 5(5):e1206

Chen KM, Cofer EM, Zhou J, Troyanskaya OG (2019) Selene: a PyTorch-based deep learning library for sequence data. Nat Methods 16(4):315–318

Eysenbach G (2002) Infodemiology: the epidemiology of (mis)information. Am J Med 113(9): 763–765

Eysenbach G (2006) Infodemiology: tracking flu-related searches on the web for syndromic surveillance. AMIA Ann Symp Proc 2006:244–248

Eysenbach G (2009) Infodemiology and infoveillance: framework for an emerging set of public health informatics methods to analyze search, communication and publication behavior on the Internet. J Med Internet Res 11(1):e11

Eysenbach G (2011) Infodemiology and infoveillance tracking online health information and cyber behavior for public health. Am J Prev Med 40(5 suppl 2):S154–S158

Falagas ME, Pitsouni EI, Malietzis GA, Pappas G (2008) Comparison of PubMed, Scopus, Web of Science, and Google Scholar: strengths and weaknesses. FASEB J 22(2):338–342

Faoury M, Upile T, Patel N (2019) Using Google Trends to understand information-seeking behaviour about throat cancer. J Laryngol Otol 133(7):610–614

Fortuna G, Brennan MT (2013) Systemic lupus erythematosus: epidemiology, pathophysiology, manifestations, and management. Dent Clin N Am 57(4):631–655

Freeman MK, Lauderdale SA, Kendrach MG, Woolley TW (2009) Google Scholar versus PubMed in locating primary literature to answer drug-related questions. Ann Pharmacother 43(3): 478–484

Ge H, Wang X, Yuan X et al (2020) The epidemiology and clinical information about COVID-19. Eur J Clin Microbiol Infect Dis 39(6):1011–1019

Ghosh S, Ansar W (2015) Indoor air pollution: impact of health and stem cells. J Stem Cells 9(4): 269–281

Ginsberg J, Mohebbi MH, Patel RS, Brammer L, Smolinski MS, Brilliant L (2008) Detecting influenza epidemics using search engine query data. Nature 457(7232):1012–1014

Gurevitz SL, Snyder JA, Wessel EK, Frey J, Williamson BA (2013) Systemic lupus erythematosus: a review of the disease and treatment options. Consult Pharm 28(2):110–121

Husnayain A, Fuad A, Su EC (2020) Applications of Google Search Trends for risk communication in infectious disease management: a case study of the COVID-19 outbreak in Taiwan. Int J Infect Dis 95:221–223

Islam MR, Kabir MA, Ahmed A, Kamal ARM, Wang H, Ulhaq A (2018) Depression detection from social network data using machine learning techniques. Health Inf Sci Syst 6(1):8

Kapitány-Fövény M, Ferenci T, Sulyok Z et al (2019) Can Google Trends data improve forecasting of Lyme disease incidence? Zoonoses Public Health 66(1):101–107

Laico SA (2018) Insights and tracking of pollution awareness using Google Trends. J Health Pollut 8(18):180609

Marino C, Gini G, Vieno A, Spada MM (2018) The associations between problematic Facebook use, psychological distress and well-being among adolescents and young adults: a systematic review and meta-analysis. J Affect Disord 226:274–281

Morsy S, Dang TN, Kamel MG et al (2018) Prediction of Zika-confirmed cases in Brazil and Colombia using Google Trends. Epidemiol Infect 146(13):1625–1627

Naim HY (2015) Measles virus. Hum Vaccin Immunother 11(1):21–26

Nourbakhsh E, Nugent R, Wang H, Cevik C, Nugent K (2012) Medical literature searches: a comparison of PubMed and Google Scholar. Health Inf Libr J 29(3):214–222

Nuti SV, Wayda B, Ranasinghe I et al (2014) The use of google trends in health care research: a systematic review. PLoS One 9(10):e109583

Quintanilha LF, Souza LN, Sanches D, Demarco RS, Fukutani KF (2019) The impact of cancer campaigns in Brazil: a Google Trends analysis. Ecancer Med Sci 13:963

Radin M, Sciascia S (2017) Infodemiology of systemic lupus erythematous using Google Trends. Lupus 26(8):886–889

Rawal G, Yadav S, Kumar R (2016) Zika virus: an overview. J Family Med Prim Care 5(3): 523–527

Rozo-Lopez P, Drolet BS, Londoño-Renteria B (2018) Vesicular stomatitis virus transmission: a comparison of incriminated vectors. Insects 9(4):190

Santangelo OE, Provenzano S, Piazza D, Giordano D, Calamusa G, Firenze A (2019) Digital epidemiology: assessment of measles infection through Google trends mechanism in Italy. Ann Ig 31(4):385–391

Shapiro ED (2014) Borrelia burgdorferi (Lyme disease). Pediatr Rev 35(12):500–509

Stevens E, Antiga L, Viehmann T (2020) Deep learning with PyTorch. Manning Publications, Shelter Island. ISBN 9781617295263

Sulyok M, Richter H, Sulyok Z, Kapitány-Fövény M, Walker MD (2020) Predicting tick-borne encephalitis using Google Trends. Ticks Tick Borne Dis 11(1):101306

Vasileios L, Nello C (2010) Tracking the flu pandemic by monitoring the social web. In: Second international workshop on cognitive information processing, pp 411–416. ISBN 978-1-4244-6459-3

Verma M, Kishore K, Kumar M, Sondh AR, Aggarwal G, Kathirvel S (2018) Google search trends predicting disease outbreaks: an analysis from India. Healthc Inform Res 24(4):300–308

Walker A, Hopkins C, Surda P (2020) Use of Google trends to investigate loss-of-smell-related searches during the COVID-19 outbreak. Int Forum Allergy Rhinol 10(7):839–847

Wang J, Zhang T, Lu Y, Zhou G, Chen Q, Niu B (2018) Vesicular stomatitis forecasting based on Google trends. PLoS One 13(1):e0192141

Ye Q, Wang B, Mao J (2020) The pathogenesis and treatment of the 'Cytokine Storm' in COVID-19. J Infect 80(6):607–613

Zhang X, Dang S, Ji F et al (2018) Seasonality of cellulitis: evidence from Google Trends. Infect Drug Resist 11:689–693

Zhou J, Troyanskaya O (2015) Predicting effects of noncoding variants with deep learning–based sequence model. Nat Methods 12:931–934

Applications and Software of Machine Learning and Artificial Intelligence (AI) in Medical Knowledge and Health

17

Abstract

Deadly diseases claim millions of life every year and therefore the key focus in Health Research lies in early and accurate detection, better medication, patient centric health policy, post therapy care to combat the disease and increase the quality of life. Modern life has been impacted greatly by big data and machine learning. Machine learning algorithms and statistical methods are enabling revolution in healthcare sector also with algorithms that perform like human physicians. It has been possible to predict diseases from imaging data far more accurately and early phases of disease progression. Big data has been generating in an increasing scale with applications in the Health Science. It has been possible to apply machine leaning tools towards data extraction from Electronic Health Records (EHR). It is also a step towards predicting high risk individuals, prepare well ahead of emergency viral/pathogen attacks that would prevent chances of epidemics, predict wounds, trauma and injury in car accidents, detect Cancer at early stages and also extract accurate information from ECG which forms one of the few major applications of Machine Learning in the Health Sciences. In this chapter we highlight (1) study in diabetes, (2) predicting Cancer using machine learning and deep learning algorithm, (3) interpreting ECG, (4) detection of glaucoma through machine learning algorithm.

Keywords

Bigdata · Machine learning · Electronic Health Records · ECG · CVD · Diabetes

Abbreviations

3D	3 Dimensional
AD	Alzheimer's disease

S. Ghosh, R. Dasgupta, *Machine Learning in Biological Sciences*,
https://doi.org/10.1007/978-981-16-8881-2_17

ADR	Adverse drug reactions
AI	Artificial intelligence
CHD	Coronary heart disease
CNN	Convolutional neural networks
CT	Computed tomography
CVD	Cardiovascular disease
ECG	Echocardiography
EHR	Electronic Health Records
LDCT	Low-dose computed tomography
LUMAS	Lung malignancy scores
ML	Machine learning
NLP	Natural language processing
PAD	Peripheral arterial disease
PCA	Principal component analysis
PD	Parkinson's disease
ROI	Region of interest
SL	Supervised learning
SSL	Semi-supervised learning
T2D	Type 2 diabetes
TID	Type I diabetes
USL	Unsupervised learning
WHO	World Health Organization

17.1 Introduction

It is the biggest challenge and heart touching reasons to humanity to search for effective ways in the detection of deadly diseases that claim lives every year and is a matter of great concern globally. Diseases include a full host of manifestations with causes ranging from malfunction and or failure of the systems, organs of the body, to infection by pathogens and environmental agents, mediated by trauma and accidents, malignancy. For most cases an early diagnosis, targeted therapy and prevention of infection, search for agents to target these disease causing agents, targeting malignant/cancerous cells, effective post-therapy care, and patient management remains the key areas of research in Health Science (Cheng and Hu 2018; Somvanshi and Venkatesh 2014; Hamidi and Kretzler 2018).

Conventional methods in disease diagnosis and treatment have developed considerably since the last more than five decades focused on increasing the quality of life and life expectancy. Yet deadly diseases such as Cancer, cardiovascular diseases, neurodegenerative disorders such as Alzheimer's Disease (AD), Parkinson's Disease (PD), autoimmune disorders (AID) like rheumatoid arthritis (RA), allergy and related disorders suffers from appropriate early diagnosis and early stages of disease

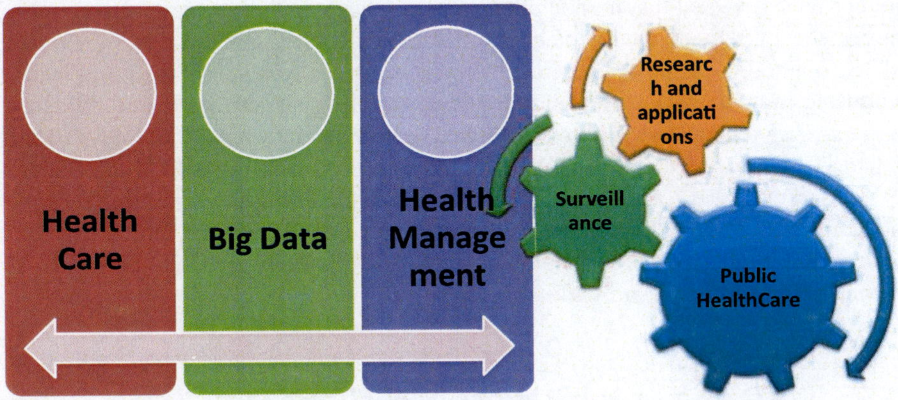

Fig. 17.1 Applications of Big data in Public Health

management. Thus it is very important to find suitable markers in early diagnosis towards targeting such disorders.

Approaches from biochemical studies, studies from proteomics, genomics, molecular biology, histology, *in situ* hybridizations, recombinant DNA technology (rdt) are few areas are generating high-throughput data which have advanced our knowledge in the domains of health and disease biology. Machine learning algorithms are helping us in analyzing the complex data from the experimental sources.

The healthcare industry faces two major challenges as regard to reduction in healthcare costs and initiatives towards personalized tailor-made therapy and healthcare and therefore challenges in this industry lie with the management of electronic record data, their appropriate integration, improved diagnoses aided by computers and predictions and detections of disease. Machine Learning tools have been instrumental in resolving these challenges.

Big data, however, encompasses large quantities of data or information, like data from biobanks from large number ranging from more than 100 s to millions of patients which may have been collected for billing, etc. purposes but which can also serve as an information source on diverse parameters of health, like genomes, disease indexes, socioeconomic parameters such as economic status, address, pin codes, etc. as well. This data of multiple parameters finds importance in revealing health status, monitoring, and surveillance of these individuals.

Bigdata, recently therefore finds importance in health care and management. It has been instrumental in the domains of epidemiology, surveillance, and causal inference ((Cheng and Hu 2018), Fig. 17.1) from the point of view of Public Health. Machine learning has been instrumental in enabling the process. Application of machine learning to analyze Big data has enabled validation of patterns of diverse parameters related to Health and analysis of complex relationships between variables like relation between clinical or environmental factors and its association with risk for disease. Application of Big data has enabled trafficking affected individuals.

Studies have revealed that mobile phone data has been applied successfully to track movements in individual with infectious disorders including cholera, dengue, Ebola, AIDS, malaria, rubella, Zika virus infection and schistosomiasis and has been promising in tracking the affected individuals travelling by air by use of GPS data-loggers, magnetometers, Twitter, and web based search engines (Dolley 2018; Tatem 2014; Finger et al. 2016; Wesolowski et al. 2014, 2015, 2016; Isdory et al. 2015; Mari et al. 2017; Huff et al. 2016). This finds importance in predicting disease incidence and outbreaks like epidemics and hence enables better tackling of such epidemics and preventive measures can be taken by the individual and the state and prevent precious human life losses.

17.2 Big Data and Applications in Public Health

Big data used by public health researchers and practitioners includes genomic, proteomic, transcriptomic or metabolomic datasets, geospatial analyses, collected electronic data on medical record, measurements taken automatically by FitBit or a global positioning system (GPS) device, records from postings in social media, or records in cell phone of the participants wherein, each data has separate implications and challenges for public health practice and research (Table 17.1). Big data has enabled precision in different domains on public health including disease surveillance, signal detection, risk prediction, targeting interventions and understanding the pathophysiology, biology, etiology spread, transmission of disease.

Using Natural Language processing (NLP) and tools of Machine learning tools like Entity Extraction attempts are being made to extract information in text from handwritten prescriptions which could be stored in database for the use of many (Fig. 17.2).

17.3 Public Health Surveillance Using Tools of Machine Learning

Surveillance systems applied to public health find importance in maintaining healthy populations by monitoring rates of disease incidence, health conditions, and environmental effects. Google Flu Trends uses big data for surveillance related to effluent data. However, managing a wide data set poses challenges and Machine learning models encompassing tools are being applied for hypothesis generation and prediction and testing (Mooney and Pejaver 2018).

17.4 Machine Learning and Big Data and Human Health Science

Machine learning includes techniques that can enable fitting of models algorithmically based on data patterns. These techniques include the following.

Table 17.1 Types of big data related to public health

Source	Example	Bigness of data	Major technical concerns	Application
-omics based or biological	• Whole exome profiling • Metabolomics	Wide	• Lab effects • Informatics	• Etiolological research • Screening
Geospatial Medical Geographic information system or medical (GIS)	Neighborhood features		Spatial autocorrelation	• Etiolological research • Surveillance
Electronic Health Records (EHR)	Records of patients	• Tall • Wide	• Data refining • Natural language	• Clinical research • Surveillance • Decision making
Effluent data	• Google Search Results • Reddit	• Tall	• Selection bias • Natural language	• Screening • Surveillance • Locating hidden social network

Table adapted with modifications from (Mooney and Pejaver 2018) Under Creative Commons Attribution 4.0 International License
Wide data Sets: With Many columns, tall Datasets with many rows

1. *Supervised learning (SL):* Involves identification of patterns formed by relating variables to outcomes which can be measured with maximum accuracy while prediction of outcomes. Like, an automatically fitted regression model finds importance as a supervised learning (SL) technique.
   In this SL, model is trained by using a range of inputs or features that are associated with an outcome that is known. Like in studies related to health, training of a model is done to relate a person's characteristics including, height, weight, habit like smoking to a certain outcome like onset of diabetes within duration of period of 5 years. After successful training of the algorithm it will be capable of predicting on applications to new data set.

   Predictions based on models trained by the application of Supervised learning either can be discrete being positive or negative representing either benign or malignant form of tumor or continuous with an allotted score ranging from 0 to 100. A model producing classes or discrete categories is termed as a classification algorithm. This has been used in predicting benign or malignant form of a tumor. Here the probability of a class falls between 0 and 1 in which 0 represents impossible and 1 represents the definite. Any probability of greater than 50 is classified into class of 1, however this threshold may vary with algorithm.

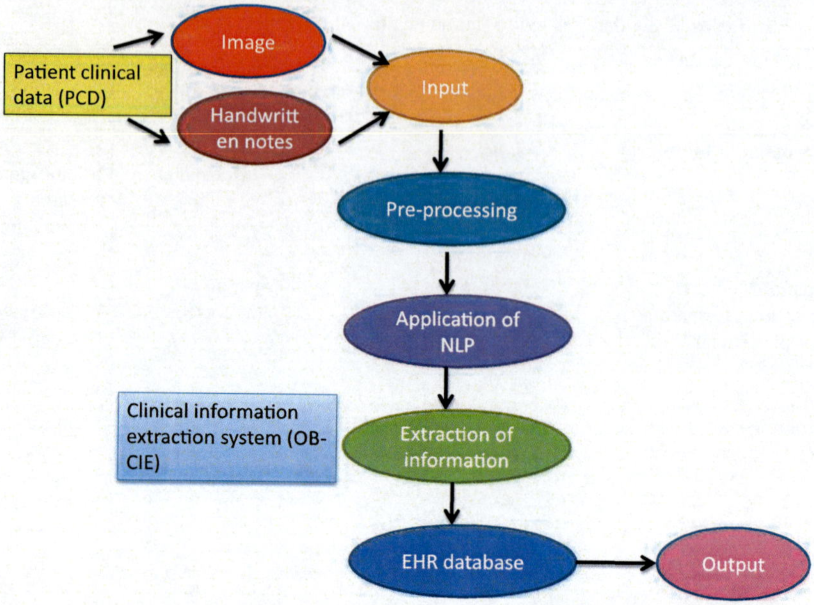

Fig. 17.2 Using a ML Vision model, handwriting to text. An Ontology-based Clinical information extraction system (OB-CIE) is used to extract the clinical information from the physician's handwritten clinical notes and convert them into patient clinical data (PCD) in the (OB-CIE) system to be recognized in the EHR database. (Image and legend adapted from Yehia et al. (2019) with permission)

A model that returns a continuous value prediction is termed as a regression algorithm. This finds application like predicting an life expectancy of an individual or the detection of dose of chemotherapy (Mooney and Pejaver 2018; Sidey-Gibbons and Sidey-Gibbons 2019) that he can tolerate.

Supervised Learning finds important application in prediction of re-admission in a hospital, transmission of disease like tuberculosis (tb), prediction of serious injuries due to accidents.

2. *Unsupervised Learning (USL)* involves detection of properties of trends and patterns characteristics of the input data set to and involves unlabeled data.
Like principal component analysis (PCA), used to identify the covariance structures. With relation to medical applications, USL has been applied to spatial and spatio-temporal profiling, detection of disease outbreak and surveillance studies, relating patient clinical features with clinical outcomes and environmental monitoring.

In opposed to SL, USL does not include a outcome that is predefined. Patterns in USL are searched by the application of algorithms in absence of any input from the user and therefore finds importance in finding undefined patterns or clusters in datasets also is called as techniques for dimension reduction. Examples include Latent Dirichlet Analysis (LDA), PCA, and *t*-Distributed Stochastic Neighbor Embedding (t-SNE).

Table 17.2 Supervised and unsupervised machine learning approaches in big data application in the domain of public health

Learning	Method	Some public health domain application
Unsupervised	Content analysis	• Surveillance
	K-means clustering (KMC)	• Detection of hot spot
	Retrospective event detection	• Case ascertainment
Supervised	K-nearest neighbors (k-NN) clustering	• Hot spot detection by spatiotemporal means • Clinical outcomes based on inputs of genetic data
	Naive Bayes	• Surveillance
	Neural networks	• Predicting disease • Predicting vaccination outcome • Predicting mortality
	Support vector machines (SVMs)	• Diagnosis • Detection of disease from social media posts
	Decision trees	• Identifying risk factor • Comparing cost effectiveness of treatments

Table adapted with modification from Mooney and Pejaver (2018) Under Creative Commons Attribution 4.0 International License

3. *Semi-supervised learning (SSL)* is a hybrid model, with prediction as a goal. However here the most of the data points miss the outcome information. Semi-supervised methods (SSM) and unsupervised methods (USM) include the involvement of data mining as precursors to supervised.

   Some applications of SSL algorithms (Table 17.2) lies in the detection of adverse drug reactions (ADR) and associated warning from the data from social media, identification of air pollutants from smart-phone data.

17.5 Machine Learning and Bigdata: Applications in Human Health

Machine learning is enabling integration of data types towards better understanding of complex relations between genetical make up of an individual, its interaction with the environment, and development of disease. Data from the biobanks are being used to decipher new genetic variants, from genome, environment, family history association studies, thereby enabling inference of health parameters associated with disease risk.

17.5.1 Study in Diabetes

Diabetes mellitus (DM) commonly referred to as diabetes, is metabolic deadly disorder. It is, chronic in nature characterized by elevated levels of blood glucose as the pancreas fails to synthesize the requisite insulin, which is the hormone that is involved in regulation of blood sugar level. It can also occur when the body fails to utilize the insulin that is synthesized in the body. Diabetes results from increased blood sugar levels also called as hyperglycemia. This can cause serious damage to the blood vessels, eyes, heart, kidneys, nerves claiming lives.

The types of diabetes include (1) Type 1 diabetes (TID) or insulin-dependent, juvenile diabetes wherein the pathophysiology involves non-production of insulin and the individual has to take daily insulin from outside. The common symptoms that can occur suddenly include excessive urination or polyuria, kidney failure, increased thirst or polydipsia, constant hunger, weight loss, affected vision, leading to blindness, fatigue heart attacks, stroke and lower limb amputation. (2) Type 2 diabetes (T2D) or non-insulin-dependent occurs when the body fails to use the produced insulin. It is the most common type globally. They reveal similar but may be less pronounced symptoms as those in type 1 diabetes. (3) Gestational diabetes is hyperglycaemia observed in some mothers during pregnancy. According to global statistics there is an number of diabetic individuals has increased from 108 million in 1980 to 422 million in 2014 and the number has been rapidly increasing in middle- and low-income countries and the cause for 1.6 million deaths in 2016 (World Health Organization 2016; https://www.who.int/health-topics/diabetes#tab=tab_1). Thus detection and prediction of diabetes is an area of active research globally.

Machine learning has been applied to improvise the accuracy of diagnosis and prognosis of disease though analysis of medical images, like automated screening of individuals for diabetic retinopathy. This would enable screening, diagnosis, preventive care and treatment and post treatment monitoring of patients. Scientist in Google and in collaboration with scientist across the globe reveals the design of new techniques applying machine learning and deep learning to automate image analysis like that of physicians.

17.5.2 Predicting Cancer

The World Cancer statistics by the World Health Organization (WHO) reveals Cancer as the second largest cause of death, claiming 9.6 million lives in 2018 globally with an estimated high rate of 70% morbidity reported from in countries with low and middle income (https://www.who.int/news-room/fact-sheets/detail/cancer; WHO 2020). An estimated 160,000 deaths in 2018 has been reported from lung cancer in the United States of America (USA). Therefore its early diagnosis and techniques for population screening is a most important requisite. Recently, Deep learning algorithm and models have been proposed that can use a current and previous patient's computed tomography (CT) data towards prediction of the risk

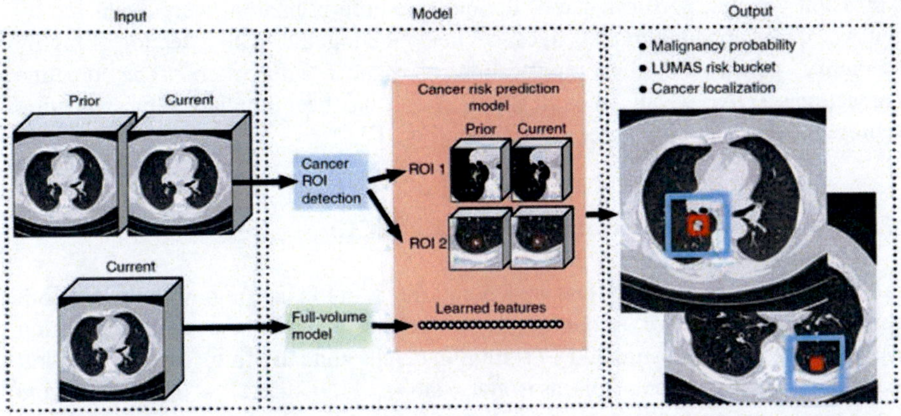

Fig. 17.3 A modeling framework was designed to predict cancerous nodules for each patient, using LDCT volume and a prior LDCT volume as input. The model can indicate and predict overall malignancy for LUMAS. (Image and legend reproduced with permission from Ardila et al. (2019))

of lung cancer and reveals to be a potential method with increased accuracy, consistency, and adoption for screening of lung cancer globally.

A radiologist greatly depends on full assessment of LDCT volume, in screening high lung cancer risks. Models like Lung-RADS and PanCan finds importance in determining risk ratings associated with malignancy.

However bottlenecks in the current modes of screening rises from poor sensitivity and specificity, missed or diagnosis that was late and unnecessary biopsy procedures that might have been done from false positive and false negatives. Therefore firstly a three-dimensional (3D) deep convolutional neural networks (CNN), models were used to perform analysis of whole-CT volumes, using low-dose volumetric computed tomography (LDCT) volumes from pathologically cases of confirmed cancer in patients used as data for training. Secondly a convolutional neural network model or CNN region of interest also called as ROI detection model was designed to detect region for 3D cancer candidate from the CT volume data. Thirdly a CNN was used to predict cancer risk based on outputs from the cancer ROI detection model and full-volume model. The final models could include regions from a patient's earlier scans, indicating cancer candidate regions in the current scan, followed by designating a malignancy score called as 'case' which may contain multiple CT volumes. To compare with the data and prediction of the radiologist, a threshold of the model's predictions were designed including three different cutoffs values/scores that eventually produced four different lung malignancy scores (LUMAS, Ardila et al. 2019) (Fig. 17.3).

Gene extraction has been used for years to detect cancer effectively. It poses a challenge because of the high dimensional and complex nature of the data. A group of scientists from University of Oregon, Corvallis, used deep learning techniques to identify the genes related to breast cancer. Using a Stacked Denoising Auto encoder (SDAE) for deep extraction of functional features from high dimensional gene expression profiles, performance of the extracted representation were evaluated by supervised classification models in cancer detection together with detection of highly interactive genes and their application as cancer biomarkers. The pipeline representing SDAE, (Danaee et al. 2017) model has been applied to classification of breast cancer and biomarkers extraction (Fig. 17.4).

17.5.3 Interpreting Echocardiography (ECG)

Cardiovascular disorders (CVDs) are a major threat to life causing mortality for both men and women globally. According to the statistics of WHO, they are the claiming lives globally, with an estimated 17.9 million deaths annually (https://www.who.int/health-topics/cardiovascular-diseases/#tab=tab_1; Institute of Medicine (US) Committee on a National Surveillance System for Cardiovascular and Select Chronic Diseases 2011). CVDs affect heart, blood vessels and include an array of disorders including (1) coronary heart disease (CHD) that affects the blood vessels carrying blood supply to the heart muscle, (2) cerebrovascular disease arising from damaged blood vessels that supply blood to the brain, (3) rheumatic heart disease caused by damaged heart muscle and valves from rheumatic fever, as a consequence of infection by streptococcal bacteria and other conditions, (4) peripheral arterial disease (PAD) or disorder of blood vessels carrying blood to the arms and legs, (5) congenital heart disease occurring since birth due to disease malformed heart structure, (6) deep vein thrombosis initiated by a sudden blood clot or thrombus that happen suddenly in one or more deep veins especially in those of legs and pulmonary embolism involving a blockage by blood clot in one of the pulmonary arteries. Strokes and heart attacks occur due to blocked blood supply preventing blood flow to the heart or brain. This may be due to excessive deposition of fats in blood vessels inner lining walls carrying blood to heart or brain. Strokes occur by bleeding from a blood vessel in the brain or from blood clots. Risk factors for stoke include diabetes, tobacco, obesity, lack of physical activity, unhealthy diet, alcohol usage, hypertension, and hyperlipidaemia.

Therefore, early and accurate detection of cardiovascular disease is very much required. ECG uses ultrasounds for capturing high temporal and spatial resolution heart images and finds application in the treatment of cardiovascular disorder. However, deep learning and CNN model named EchoNet has been designed and applied to ECG to identify cardiac structures, function, predict age, sex, weight and height, anatomic structures and their local functions including pacemaker leads, enlarged left atrium, left ventricular hypertrophy, end systolic and diastolic volumes, and ejection in much better and with higher resolution and greater clarity as opposed to conventional methods (Ghorbani et al. 2020) (Fig. 17.5).

A CNN model EchoNet was trained on a data images of 2.6 million ECG from more than 2500 patients used to identify cardiac structures and function so that it could predict and identify cardiac anatomy and systemic risk factors (Fig. 17.1) for medical classification or carry out prediction task from ECG of new patients (Ghorbani et al. 2020).

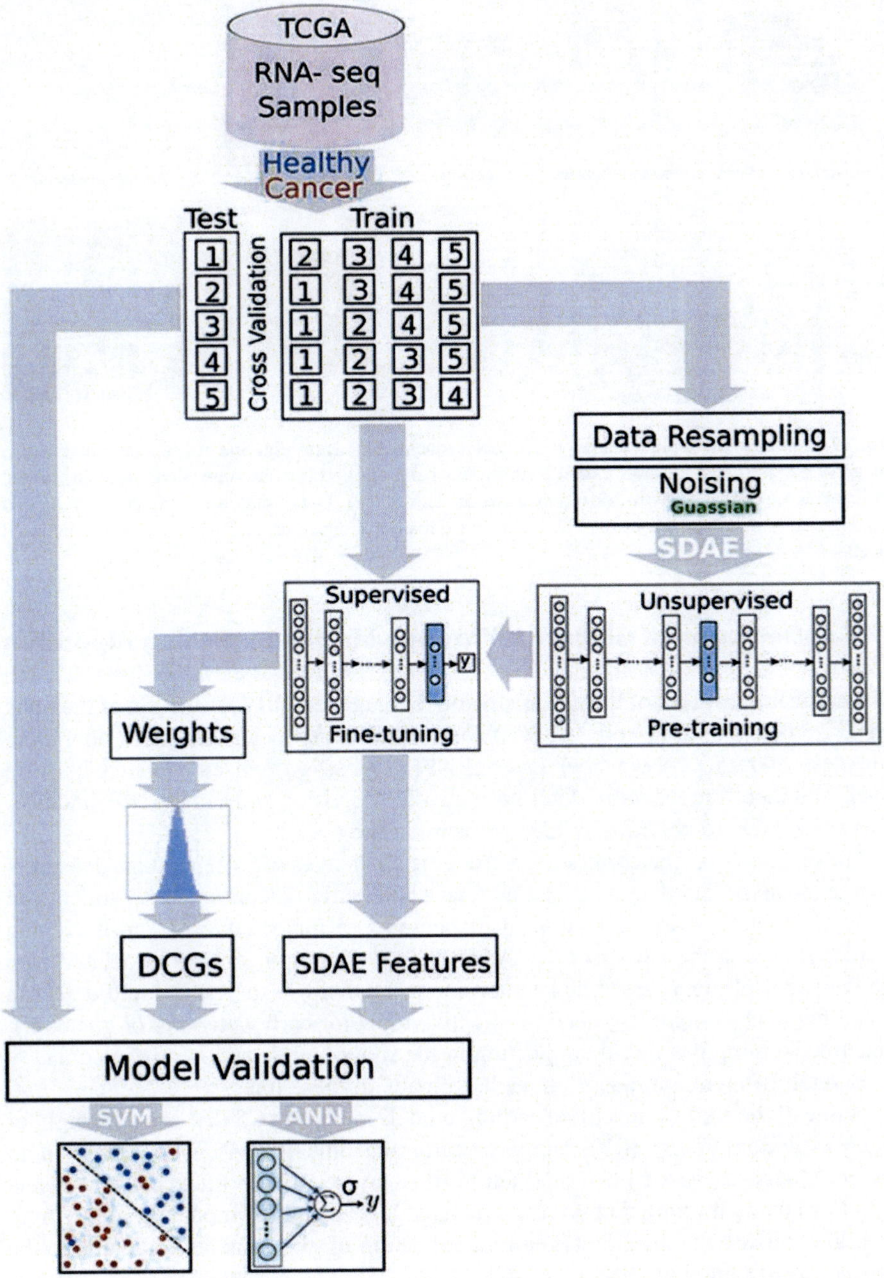

Fig. 17.4 Machine Learning model to detect Breast Cancer by marker extraction. (Image adapted with permission from open access article (Danaee et al. 2017))

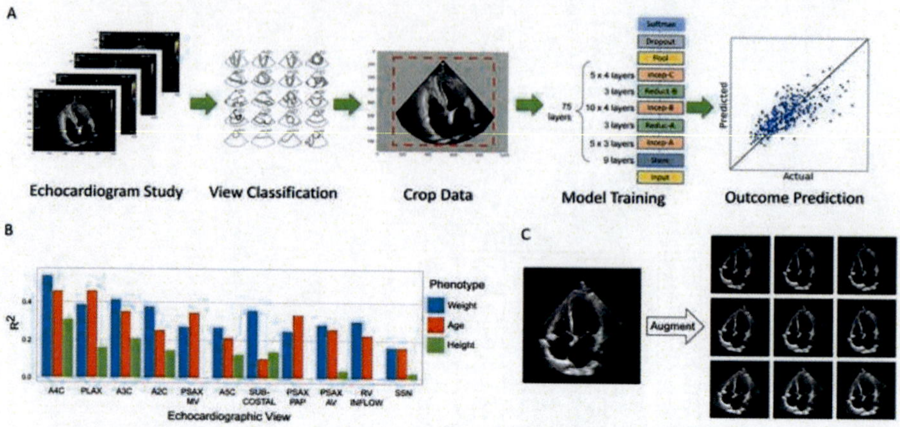

Fig. 17.5 Schematic representation of EchoNet machine learning pipeline has been applied for (**a**) image selection, cleaning, and model training has been applied for outcome prediction. (**b**) model performance compared with different cardiac views. (**c**) Data analysis. (Image and legend reproduced from Ghorbani et al 2020, under a Creative Commons Attribution 4.0 International License)

17.5.4 Detection of Glaucoma Through Machine Learning Algorithm

Glaucoma is a disorder of the eye arising out of damages and dysfunction of the optic nerve, retinal ganglion cells (RGCs) loss, optic nerve head structural alteration, thickness of retinal nerve fiber layer (RNFL), ganglion cell and inner plexiform layers and therefore the individuals have an affected vision and can lead to blindness. This occurs due to abnormally high pressure in the eye.

The diagnosis of glaucoma suffers from the challenges of its early stage detection. Misdiagnosis or late diagnosis can be fatal while early detection and treatment can delay vision loss. Early detection of glaucoma is a major concern and intriguing problem to researchers in this filed. Added levels of complicacy are associated with the detection of glaucoma, brain tumor due to altered eye disc shape and defect in visual field. Thus there is a need of effective tools for early detection of glaucoma. Machine learning has found application in the study and detection of glaucoma.

Several different approaches have been made in these lines.

In one of the studies, machine learning models including C5.0, *k*-nearest neighbor (KNN), random forest (RF), support vector machine (SVM), were designed to predict glaucoma based on retinal nerve fiber layer (RNFL) thickness and visual field (VF) using training dataset and evaluated it by using RF models enabling high accuracy, sensitivity, specificity, and identification of glaucoma as compared to the normal eyes (Kim et al. 2017).

Recently a group of researchers worked on machine learning algorithm techniques on automated detection of glaucoma through clinical data with more accuracy (An et al. 2019).

Codes are available at GitHub, written in Python (An et al. 2019):One of the results are as shown (Fig. 17.6):

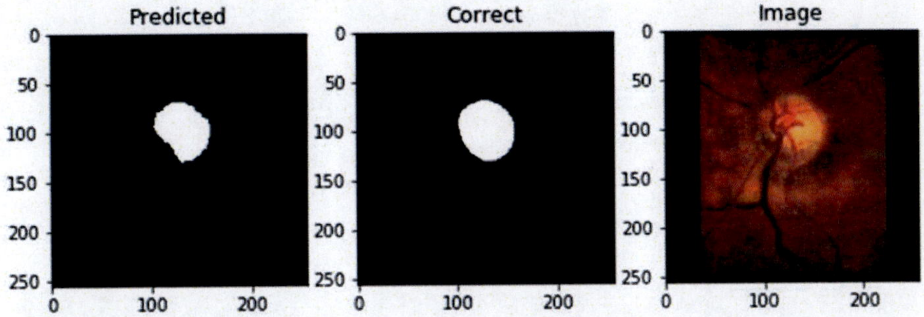

Fig. 17.6 Predicting glaucoma. (Image with permission from Sevastopolsky, Artem. "Optic disc and cup segmentation methods for glaucoma detection with modification of U-Net convolutional neural network." *Pattern Recognition and Image Analysis* 27.3 (2017): 618–624.[github repository] Artem Sevastopolsky. (2017, April 15). seva100/optic-nerve-cnn: First version of the software. (Version v1.0.0). Zenodo. https://doi.org/10.5281/zenodo.546338)

Another group is aiming at the development of an algorithm CNN for diagnosis of open-angle glaucoma, based on 3D optical coherence tomography (OCT) data and color fundus images, trained to differentiate the disc fundus images of eyes of normal and glaucomatous individuals and enabled accurate diagnosis in glaucoma over and above the conventional methods (An et al. 2019).

17.6 Machine Learning and Health Policy

Machine learning (ML) and artificial intelligence (AI) technology has found application in designing of public health policy with an overall patient focused approach involving patient related information like medical imaging, health records, whole genome studies, the workforce and resource management involved and is thus promising step towards maintenance of good health and social care mediated by predicted clinical outcomes, population screening, preventive strategies, disease protection, maintaining healthy lifestyles, and prediction of high risk individuals. With the application of machine learning to health records it has been possible to monitor the spread of influenza. It has been also possible to design vaccine supplies timely (Mooney and Pejaver 2018) (Fig. 17.7).

Environmental data obtained from climate sensors can predict climate changes, pollution trends and shifts in ecosystem and thereby enabling preparation steps to be taken for emergency responses against sudden climate changes and manage ecological crisis situations (Mooney and Pejaver 2018).

Application of Machine Learning Algorithm to data from city transportation has been used in prediction of locations of injuries or trauma experienced by people by car accidents and could be a step towards prevention of car accidents on one hand and attending to emergency situations of people on the other.

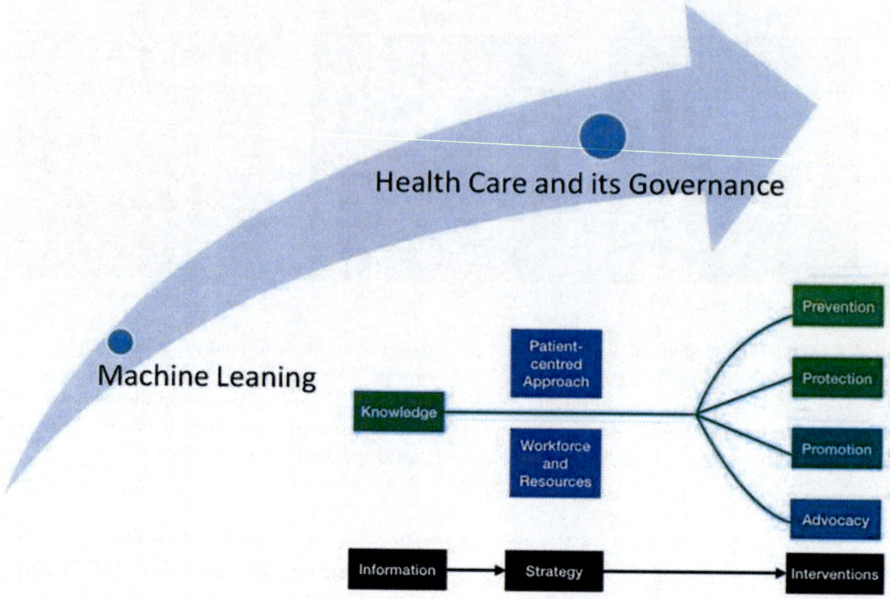

Fig. 17.7 Machine learning in health care and its governance. (Image adapted with modification from Ashrafian and Darzi (2018) under open access article distributed under the terms of the Creative Commons Attribution License)

Machine Learning finds application in the promotion of health. This can be done from data stored in electronic health records (EHRs) to identify unhealthy behaviors like procurement of tobacco, alcohol- or sugar-based drinks.

Machine learning has enabled steps towards disease protection by design of screening strategies. Patients with high risk rate of colorectal cancer has been detected by applying Machine learning tools to blood count test. Deep learning image screening has enabled screening data from mammography.

Application of Machine learning has enabled patient advocacy together with maintaining ethics to well-being in population.

17.7 Novel Corona (COVID-19) Virus Infection and Machine Learning: A Recent Most Application

While the book is authored, the whole world is witnessing an epidemic due to corona virus induced flu and pneumonia. Zoonotic Coronaviruses (CoV) causes disorders ranging from the common cold to infectious diseases including Middle East Respiratory Syndrome (MERS-CoV), Severe Acute Respiratory Syndrome (SARS-CoV) with symptoms of, cough, respiratory distress, fever, shortness of breath, breathing difficulties, pneumonia infection, kidney failure and even death. Corona virus (2019-nCoV) has been reported to cause a recent outbreak in China and is an

Fig. 17.8 NCorona virus infection: WHO https://experience.arcgis.com/experience/685d0ace521648f8a5beeeee1b9125cd, reproduced with permission

alarming concern globally. Till the year 2020 early January as many as 4631 confirmed cases claiming 106 deaths, from 11 countries or regions were reported. Transmission of SARS-CoV from civet cats to humans and transmission of MERS-CoV from dromedary camels to humans have been reported followed by a human to human rapid transmission (Fig. 17.8).

A group of scientists created a tool using genomic match algorithm and the tool can identify the corona virus infection (https://www.genomedetective.com/app/typingtool/cov/).

17.8 Big Data and Future Direction of Public Health Through Machine

In future major machine learning and big data will be of application to the public health research in a major way.

- Machine Learning (ML) has started confronting outcome measurement error specially for low sensitivity data sets.
- There are several machine learning algorithm working on real-time prediction using a logic system of reward and penalty, rather than model fitting on past data. This system is called Reinforcement Learning System. Reinforcement Learning System will change the way tools and data surveillance work now.
- Deep-learning approaches, uses huge data volumes and computation to identify image classification and finds increasing application in analysis of health related data.

Besides, Public Health organizations and researchers now increasingly using neural network to identify the probability and pattern of spread of a disease and its chances of turning to an epidemic.

- Targeting infectious disease and neglected tropical diseases remains a major concern. Early diagnosis of cancer through CNN model on MRI remains as one of the most successful implementations of machine learning and neural network.

17.9 Discussions

Alan Turing in 1950 said that "*AI, or 'machine intelligence', will have a widespread and pervasive role within our society.*"

With the advanced applications of AI in health sector, It has also become possible by application of Machine Learning towards framing of a patient centric Health policy keeping in mind the well-being of the individual in every aspect of life. We have discussed in this chapter the various applications of machine learning in Human Health.

Problems related to optimization of data collection in healthcare is being alleviated by the application of improved tools of machine learning. Application of natural language processing (NLP) with appropriate design will enable data extraction from clinician notes for effective clinical and research use culminating to the application by Epidemiologists to suggest for better effective efficient protocols in public health management (Mooney and Pejaver 2018).

It is the advancement of machine learning that AI has been applied in generation of the health policy.

However, there are thoughts and predictions that application of machine learning may lead to job losses in Healthcare sector like those related to diagnostic radiology or pathology. However, there are also contradictory thoughts that it will also generate new jobs and roles.

References

An G, Omodaka K, Hashimoto K, Tsuda S, Yukihiro Shiga Y, Takada N, Kikawa T, Yokota H, Akiba M, Nakazawa T (2019) Glaucoma diagnosis with machine learning based on optical coherence tomography and color fundus images. J Healthc Eng 2019:4061313, 9p. https://doi.org/10.1155/2019/4061313

Ardila D, Kiraly AP, Bharadwaj S et al (2019) End-to-end lung cancer screening with three-dimensional deep learning on low-dose chest computed tomography. Nat Med 25:954–961

Ashrafian H, Darzi A (2018) Transforming health policy through machine learning. PLoS Med 15(11):e1002692. https://doi.org/10.1371/journal.pmed.1002692

Cheng L, Hu Y (2018) Human disease system biology. Curr Gene Ther 18(5):255–256

Danaee P, Ghaeini R, Hendrix DA (2017) A deep learning approach for cancer detection and relevant gene identification. Pac Symp Biocomput 22:219–229

Dolley S (2018) Big Data's role in precision public health. Front Public Health 6:68

Finger F, Genolet T, Mari L, de Magny GC, Manga NM, Rinaldo A et al (2016) Mobile phone data highlights the role of mass gatherings in the spreading of cholera outbreaks. Proc Natl Acad Sci U S A 113(23):6421–6426

Ghorbani A, Ouyang D, Abid A et al (2020) Deep learning interpretation of echocardiograms. NPJ Digit Med 3:10

Hamidi H, Kretzler M (2018) Systems biology approaches to identify disease mechanisms and facilitate targeted therapy in the management of glomerular disease. Curr Opin Nephrol Hypertens 27(6):433–439

Huff A, Allen T, Whiting K, Breit N, Arnold B (2016) FLIRT-ing with Zika: a web application to predict the movement of infected travelers validated against the current Zika virus epidemic. PLoS Curr 8. https://doi.org/10.1371/currents.outbreaks.711379ace737b7c04c89765342a9a8c9

Institute of Medicine (US) Committee on a National Surveillance System for Cardiovascular and Select Chronic Diseases (2011) A nationwide framework for surveillance of cardiovascular and chronic lung diseases, vol 2: Cardiovascular disease. National Academies Press (US), Washington, DC

Isdory A, Mureithi EW, Sumpter DJ (2015) The impact of human mobility on HIV transmission in Kenya. PLoS One 10:e0142805. https://doi.org/10.1371/journal.pone.0142805

Kim SJ, Cho KJ, Oh S (2017) Development of machine learning models for diagnosis of glaucoma. PLoS One 12(5):e0177726

Mari L, Gatto M, Ciddio M, Dia ED, Sokolow SH, De Leo GA et al (2017) Big-data-driven modeling unveils country-wide drivers of endemic schistosomiasis. Sci Rep 7:489. https://doi.org/10.1038/s41598-017-00493-1

Mooney SJ, Pejaver V (2018) Big Data in public health: terminology, machine learning, and privacy. Annu Rev Public Health 1(39):95–112

Sidey-Gibbons J, Sidey-Gibbons C (2019) Machine learning in medicine: a practical introduction. BMC Med Res Methodol 19:64

Somvanshi PR, Venkatesh KV (2014) A conceptual review on systems biology in health and diseases: from biological networks to modern therapeutics. Syst Synth Biol 8(1):99–116

Tatem AJ (2014) Mapping population and pathogen movements. Int Health 6(1):5–11

Wesolowski A, Buckee CO, Bengtsson L, Wetter E, Lu X, Tatem AJ (2014) Commentary: containing the Ebola outbreak - the potential and challenge of mobile network data. PLoS Curr 6. https://doi.org/10.1371/currents.outbreaks.0177e7fcf52217b8b634376e2f3efc5e

Wesolowski A, Qureshi T, Boni MF, Sundsøy PR, Johansson MA, Rasheed SB et al (2015) Impact of human mobility on the emergence of dengue epidemics in Pakistan. Proc Natl Acad Sci U S A 112:11887–11892

Wesolowski A, Buckee CO, Engø-Monsen K, Metcalf CJE (2016) Connecting mobility to infectious diseases: the promise and limits of mobile phone data. J Infect Dis 214(suppl 4):S414–S420. https://doi.org/10.1093/infdis/jiw273

WHO report on cancer: setting priorities, investing wisely and providing care for all. World Health Organization, Geneva. 2020. Licence: CC BY-NC-SA 3.0 IGO

Global report on diabetes, World Health Organization (2016). https://apps.who.int/iris/bitstream/handle/10665/204871/9789241565257_eng.pdf;jsessionid=05FF75E89AD714779A54F039EA3B38A4?sequence=1

Yehia E, Boshnak H, Gaber SA, Abdo A, Doaa S, Elzanfaly DS (2019) Ontology-based clinical information extraction from physician's free-text notes. J Biomed Inform 98:103276

Cloud Computing Infrastructure in Healthcare Industry

18

Abstract

Cloud Computing has become a default infrastructure in the healthcare industry. In UK, there is a nationwide backbone of Electronic Health Record (EHR). Whether it is a General Practice (GP) or a Primary Care Unit or Hospital or Diagnostic Center, every establishment needs to access the EHR over the Cloud Infrastructure. Patient records are stored in the NHS Cloud over internet. In North America, especially in the US, while there is no nationwide EHR, most of the Independent Practice Association (IPA) uses software solutions available over cloud, utilizing a Software As A Service (SAAS) model. Large hospitals use private clouds to manage different units and systems such as Emergency, inpatient and outpatient services, Hospital ERP, and Revenue Cycle Management (RCM). Most of the Business Intelligence (BI) systems are available over the cloud infrastructure. In this chapter we will address different cloud system infrastructure and models.

Keywords

Healthcare · Cloud computing · SAAS · PAAS · IAAS

18.1 Introduction

Healthcare and medical sector of industry has the problem of generation and maintenance of huge data. Cloud Technology with the advantages of robust computing power, with high flexibility and scalability of resources, robust storage capacities, ubiquitous access to pooled resources, across wide geographic range through a broad network has been instrumental in storing hugh data, its transfer and easy access in cost effective way with minimal infrastructure.

S. Ghosh, R. Dasgupta, *Machine Learning in Biological Sciences*,
https://doi.org/10.1007/978-981-16-8881-2_18

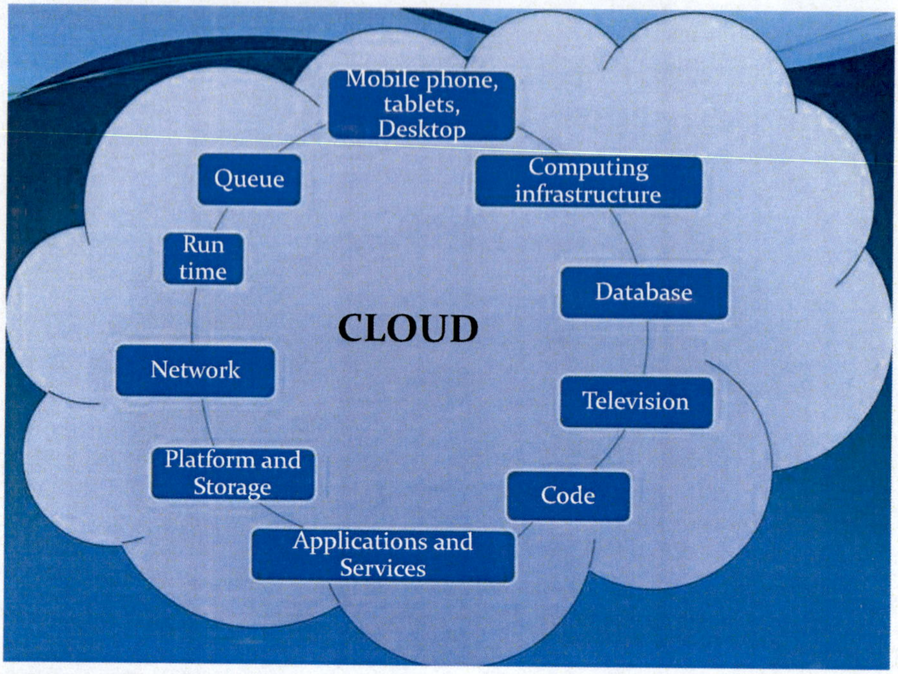

Fig. 18.1 Cloud computing

The term Cloud computing owes its origin by the National Institutes of Standards and Technology (NIST) and is known model for ubiquitous, easy, flexible, on-demand access to shared computing resources. Mell and Grance (Griebel et al. 2015; Rostrom and Teng 2011; Mell and Grance 2011) enlisted the following criteria for Cloud computing technology.

Cloud Technology has the characteristics of providing service network, through different communication media including Ethernet, Integrated Service Digital Network (ISDN), Fiber Network, multi-protocol Label Switching (MPLS), Shared or Dedicated Access with Wireless-Fidelity (Wi-Fi) based system, Very Small Aperture Terminal (VSAT), and Virtual Private Network (VPN). Services are flexible, scalable, and elastic in features, with scale up options, capable of pool of shared resources with different organizations with broader utilization of services. Service Models for clouds include Software as a Service (SAAS), Infrastructure as a Service (IAAS), and Platform as a Service (PAAS).

Cloud Services can be controlled and automatically optimized (Mu-HsingKuo 2011; Oren 2013; Shao et al. 2015; Kharat et al. 2012; Wang and Alexander 2014; Geoff 2012; Regola et al. 2013; Fusaro et al. 2011; Griebel et al. 2015) (Fig. 18.1).

18.2 Cloud Transformation Models

The different cloud computing transformation include private, public, hybrid, and Community Cloud which can be selected based on its use.

1. Public Cloud
   Finds effective application in public domain providing services on applications, software and storage with unlimited storage space huge size of infrastructure and can provide services to small, medium enterprise networks. Few of the type includes, App Engine by Google, Azure Services platform by Windows, Elastic Cloud Computer by Amazon, BlueCloud by IBM.
2. Community Cloud
   It can be used for a large group in collaboration which can be controlled either internally or by other third party, used by internal or external organization.
3. Hybrid Cloud
   Hybrid cloud is so named as it includes features of both public and private cloud with flexibility of using multiple models with advantages of optimal utilization, data-centre consolidation, risk transfer and all time availability and security over and above public cloud.
4. Private Cloud
   Designed by an individual organization with restricted usage outside the organization, less elastic in nature and is managed internally by the host organization and is considered advantageous over other cloud models in terms of data security risk and reliability. Examples are Amazon Virtual Private Cloud (Amazon VPC), Microsoft Private, Eucalyptus Cloud Platform, and IBM Smart-Cloud Foundation (Fig. 18.2)

18.3 Cloud Service Models

Cloud service models include Software as a Service (SAAS), Platform as a Service (PAAS), and Infrastructure as a Service (IAAS).

SAAS Cloud service enables access through Public Internet and can offer healthcare organizations, quick access to business applications, maintaining customer relationship management (CRM), and enabling on-demand hosted services.

PAAS offers advantage for infrastructure platform to run the application and can be used for digital MRI analysis, X-rays dashboard built on Microsoft Azure and Digital Pathology System for pathological slides in a enhanced security environment for web-based services.

IAAS offering more permissions and liberty of usage finds application in diagnostics test and medical devices and can store large amount of medical facilities (Fig. 18.3).

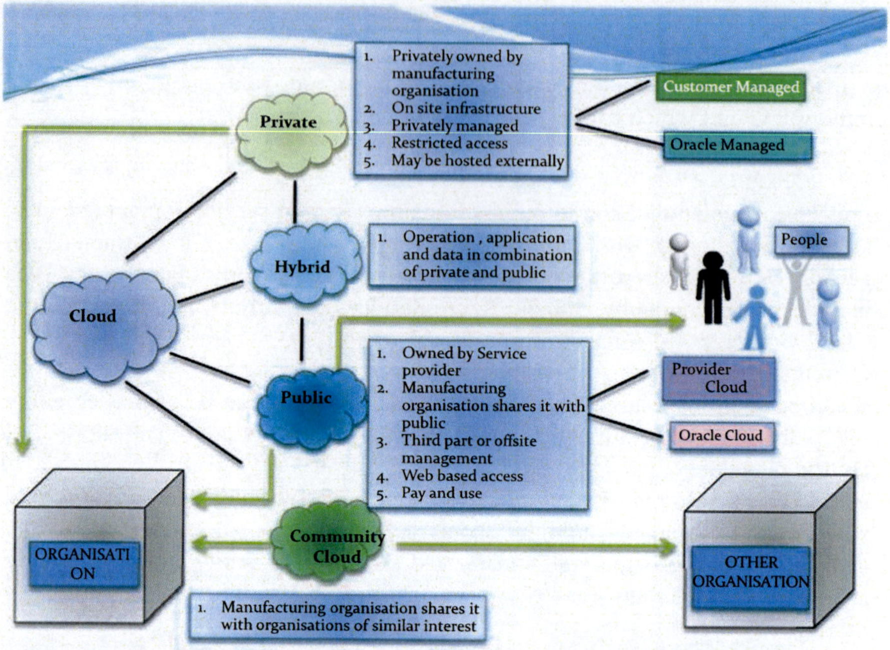

Fig. 18.2 Cloud deployment models

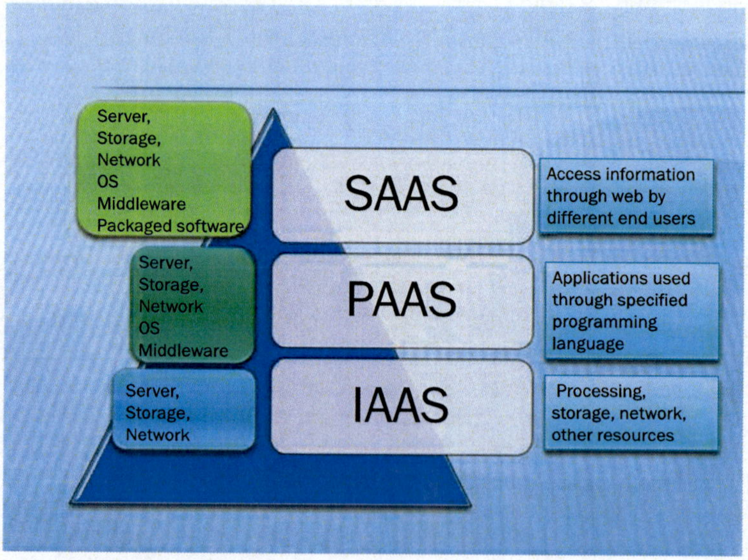

Fig. 18.3 Cloud service models

18.4 Cloud Computing and Healthcare

Healthcare applications for cloud includes medical applications of telemedicine, Electronic medical records (EMR), medical imaging like Xrays, radiology, USG, CT Scans, Digital Pathology, Tele-dermatology on one hand and administrative applications including booking of appointments, registration of patient details, billing, scheduling and reimbursement like functions. In telemedicine cloud storage includes storing or sharing of patient data while tele-consultation, follow-up and education consisting of video conferencing, medical consultation by telemedicine software and medical attachments including recording of heartbeat, Electrocardiography (ECG), pulse rate, saturation of peripheral oxygen (SPO_2) Blood Glucose, human non-invasive blood pressure (NIBP), Pathological Slides, X-Ray Scanner and Dermatology Camera. Video Cloud services for Telemedicine & Tele-health Services enables video conferencing from the different device, location and network. EMR is being facilitated by cloud-based medical records and medical image and is also helping in collaborative solutions. Medical Imaging includes storage, sharing and computation of images.

Adoption of Cloud technology is beneficial both for the healthcare providers and the patient care. On one had it lowers the cost of operation, relying more on data storage and use of computing resources rather than depending on the frequent need of purchasing the hardware and servers thereby improving business. Provides efficient working environment and promotes easy exchange of information by computer systems or software by means of data integrations across healthcare system enabling giving timely medications and advises towards betterment of health. Recently medical research has been greatly helped by the application of Big Data analytics and artificial intelligence algorithms (AIA) in cloud-based storing of patient data. It has improved telemedicine and applications in the domain of telehealth (Table 18.1).

The pay per use model of Cloud based technology is enabling developing and using of services. Cloud computing has been instrumental in handling, processing, and storing of data from OMICS based studies including studies from genomics, proteomics, and molecular medicine in the domain of biology and health (Wang and Alexander 2014) which offers advantage of flexible and on-demand services over conventional methods of text-or data mining algorithms, with temporary usage and costly maintenance of permanent infrastructure. Commercial providers including Amazon and Microsoft have enabled easy and effective Cloud Technology.

Table 18.1 Criteria of cloud computing (Griebel et al. 2015; Rostrom and Teng 2011; Mell and Grance 2011)

	Criteria
1.	On demand services
2.	Access through broad network
3.	Pooling of resources
4.	Rapid elasticity
5.	Flexibility
6.	Measured services

18.5 Challenges in Cloud Technology in Healthcare

Healthcare organizations depend on different collaborations of patient history files, prescriptions, diagnostic test reports, X-rays, radiological data, histological data & pathological images, inter-department collaboration and communication, wards, An outpatient department (OPD) clinics and is really challenging. Maintaining the confidentiality of patient data is another challenge. Arranging such an infrastructure is a challenge more so in the rural sectors also entailing cost, education of healthcare professionals in such technology and need of trained manpower to run such application. Although Cloud technology provides solution to hospital's data center, there is a risk of maintaining patient data confidentiality, security, loss, and misuse of sensitive data (Tables 18.2 and 18.3).

18.6 Discussion

Cloud applications are transforming the Healthcare Industry by better information management and reduced operating risks, flexible, involves less cost and less infrastructure requirement and is a benefit to patients with improved healthcare and management of healthy life. Together with application of Big Data Analytics, Artificial intelligence algorithms and internet has improved its efficient application in the domain of Healthcare Sector. The healthcare industry would be greatly benefitted by adoption of such technology at large.

Table 18.2 Advantages of Cloud computing

Domain	Advantage
Business	1. Cut down of operational cost 2. Providing instant patient services 3. Bringing efficiency in work output 4. Interoperability
Patients	1. Easy access to patient data 2. Leading to ease of consultation and getting advises on health well in advance even beyond geographical locations 3. Patient can have control over his own health data aiding in patient education 4. Enable providing of high quality personal care 5. Improved patient treatment and better quality of life 6. Redundancy of data can be reduced
Insurance	1. Easy access to patient data across pharmaceuticals, insurance and makes payment process easier and less cumbersome
Medical research	1. Application of big data analytics and AIA has improved closed based storing of patient data 2. Ease handling of hugh data sets
Telemedicine	1. Remote access of data 2. Improved telemedicine, post-hospitalization care, and virtual medication 3. Ease in data sharing, improved data accessibility

Table 18.3 Challenges in cloud Technology in Healthcare Sector

Domains	Criteria	Potential solutions
Security	1. Security issues associated with patient data 2. Sensitive matter and needs to be guarded against potential external threats 3. Data-breach	1. Data encryption 2. Use of security keys to access patient specific sensitive data 3. Block-chain to secure data
Regulations compliance	Patient specific healthcare data needs to follow regulations like Health Insurance Portability and accountability Act, 1996 (HIPAA) in USA, under US Department of Health and Human Services (HHS), Health Information Technology for Economic and Clinical Health Act (HITECH Act) law, under American Recovery and Reinvestment Act (ARRA) bill, 2009, General Data Protection Regulation, under the European Data Protection Regulation, dated May 25th, 2018	Data hosted in the cloud domain needs to comply to all regulations.
Downtimes of system	Reliable yet suffers disadvantage of system downtime	1. Plan ahead 2. Maintenance of strategy and keep a system at place to overcome downtime or shorten it at the lowest minimum time

However, there are still challenges and risks associated with confidentiality of patient, reducing the system downtime and maintenance of existing regulations and calls in for more research in this domain.

References

Fusaro VA, Patil P, Gafni E, Wall DP, Tonellato PJ (2011) Biomedical cloud computing with Amazon web Services. PLoS Comput Biol 7(8):e1002147. https://doi.org/10.1371/journal.pcbi.1002147

Geoff W (2012) Making the cloud work for healthcare. Health Manage Technol 33:2; ProQuest

Griebel L, Prokosch H, Köpcke F et al (2015) A scoping review of cloud computing in healthcare. BMC Med Inform Decis Mak 15:17. https://doi.org/10.1186/s12911-015-0145-7

Kharat AT, Safvi A, Thind S, Singh A (2012) Cloud computing for radiologists. Indian J Radiol Imaging 22(3):150–154

Mell P, Grance T (2011) The NIST definition of cloud computing (draft). NIST Spec Publ 800(145):7

Mu-HsingKuo A (2011) Opportunities and challenges of cloud computing to improve health care services. J Med Internet Res 13(3):e67

Oren H (2013) How to maintain security in the cloud. Health Manage Technol 34:6; ProQuest

Regola N, Nitesh V, Chawla NV (2013) Storing and using health data in virtual private cloud. J Med Internet Res 15:3

Rostrom T, Teng CC (2011) Secure communications for PACS in a cloud environment. Conf Proc IEEE Eng Med Biol Soc 2011:8219–8222

Shao Z, Yang B, Zhang W, Zhao Y, Wu Z, Miao M (2015) Secure medical information sharing in cloud computing. Technol Health Care 23(suppl 1):S133–S137

Wang L, Alexander CA (2014) Medical applications and healthcare based on cloud computing. Int J Cloud Comput Serv Sci 2(4):217–225

Artificial Intelligence Industry and the Domain of Life Sciences

19

Abstract

Cloud computing services have found applications as low cost, effective to analyze, store data and have been applied to health, biomedical, and life sciences domain. Both Microsoft Azure and Amazon Web Services (AWS) have been instrumental in facilitating research in the domain of life sciences. We discuss in this chapter the application of AZURE and AWS in research in the domain of Life sciences.

Keywords

AWS (Amazon Web Services) · Azure · COVID-19 · Chatbots · NLP · HPC

Abbreviations

AI	Artificial intelligence
AML	Azure Machine Learning
ASSS	Amazon Simple Storage Service
DA	Data analytics
DDI	AWS Diagnostic Development Initiative
DL	Deep learning
DRMAA	Distributed Resource Management Application API
EC2	Elastic Compute Cloud
EHR	Electronic health records
FINRA	Financial Industry Regulatory Authority
HPC	High performance computing
JIST	Java Image Science Toolkit
MIPAV	Medical-image processing, analysis, and visualization
ML	Machine learning

S. Ghosh, R. Dasgupta, *Machine Learning in Biological Sciences*,
https://doi.org/10.1007/978-981-16-8881-2_19

NGS	Next-generation sequencing
NL	Natural language
NLP	Natural language processing
OS	Operating system
PDB	Protein Data Bank
VMs	Virtual machines

19.1 Introduction

Cloud computing has various applications which are found to be easy to implement, low cost, scalable, elastic, and user-friendly. These are being applied to solving problems in the complex domain of biology including neurological imaging studies, genomics, bioinformatics, structural biology. Amazon Web Services (AWS) and Microsoft AZURE are two such major systems that are allowing research in pharmaceutical, managing health data with necessary compliance to current regulations, maintaining data security and privacy as well as enabling extraction of information and analysis of data from omics based studies and protein structural biology data, study of vast complexity of the microbiome that is enabling understanding biology in health and disease. We discuss in this chapter the different applications of AZURE and AWS in the different domains of biological sciences.

19.2 Artificial Intelligence Industry

Amazon Web Services (AWS)

AWS has been into the health care industry for some time now. Recently while the world is witnessing the havoc death and millions of affected people due to the global pandemic COVID-19 with no cure or vaccination strategy in place and needed a collaborative effort from industry, academia, and healthcare sector. During the pandemic, the world has also witnessed the complete collapse of health care industry due to the impact of COVID-19 pandemic which touched adversely almost every aspect of our lives, from our daily lives, to our livelihood, to the restructuring of economies, loss of jobs, and also affected the mental and emotional well-being of the people globally. Any information related to the infection, transmission, healthcare support, medication are needed to be transmitted to the masses very quickly. In the healthcare sector they have provided diverse applications to healthcare providers, insurers, and payers by handling patient data with security and compliance to regulations, pharmaceutical and biotechnology industry by providing effective and quick solutions, enabling faster clinical trials, manufacture and commercialization of therapeutics in strict compliance to security and regulations and in the area of Genomics in analysis of hugh volumes of sequencing data, data from clinical trials and is a step towards design of personal and precision medicine.

At this juncture AWS has been instrumental in initiating collaborative relations with healthcare sectors, public health organizations, government agencies, and business in life science domain globally in support of collaborative research in tackling the global pandemic of COVID-19 that has touched badly in almost every sector of our lives from health to economy to our daily livelihood.

19.2.1 Chatbots

This includes providing technical support to people and organizations to help maintain operations in clinical setup, providing increased number of contact center and their increased efficiency of receiving and sending information like with web chatbots which are like virtual assistants, created by applications of artificial intelligence (AI) software enabling a chat by conversation with a user by a virtual assistant in natural language (NL) through messaging applications including websites, mobile apps, or by voice calls through telephone.Chatbots or rather question answer system through application of natural language processing (NLP) is of profound importance in interaction between humans and machines and finds application in different organizations and end-use applications, including healthcare sector.

19.2.2 Voice Conversational Agents

Application of voice conversational agents enabled supplementation of administrative staff to help patients, enabled clinical and medical staff in providing services in telemedicine, enabled global research collaboration with cloud desktop, streaming solutions, more so finds importance in challenging times of COVID-19 pandemic when communication and working itself are difficult.

Amazon Connect is one such contact center with web chatbots and voice conversational agents that enabled management of high volume of incoming calls of patients and anxious relatives to help take decision on treatment of patients. AWS end user computing cloud desktop and application streaming solutions are enabling workers to function in an effective and compliant manner.

19.2.3 High Performance Computing (HPC)

AWS enables application of elastic, scalable cloud infrastructure with unlimited capacity and room for experimentation for HPC applications in cloud for efficient and faster work of engineers and researchers, across the globe and applications could vary from, sequencing studies in Biology, computational biology Cancer Research, patient specific clinical trials, climate forecast, geophysical and seismic imaging. This is further advanced with applications from other cloud-based services including machine learning (ML), deep learning (DL), data analytics (DA), artificial intelligence (AI) and offers advantages of shorter or no queues, and flexible infrastructure.

This technology is being applied by different companies across the world including Bristol-Myers Squibb, a global pharmaceutical company, FINRA or Financial Industry Regulatory Authority, BP which is a multinational oil and gas company London, United Kingdom and Autodesk trust AWS (https://www.autodesk.com/trust/overview) that can scale designs to run many simulations in an hour instead of several hours or days with security of data, assets, and data protection.

During COVID-19, AWS launched HPC for Research and AWS Diagnostic Development Initiative (DDI) to facilitate diagnosis, treatment, and vaccine studies. AWS Data Exchange now hosts COVID-19 research dataset from John Hopkins University, USA.

We discuss in subsequent section some of the recent applications of AWS in biological sciences.

1. Neuroimaging studies
   The field of neuroimaging is vital in understanding the structure and function of human brain in the cloud but suffers from limitation of trained researchers in parallel computation and increased need in computational methods of data-sharing, scanning of images with high resolution and use of statistical methods. Recently AWS has been used to test workflows in neuroimaging through cloud computing and cluster-on-demand type of service (Madhyastha et al. 2017).
2. Study of microbiome
   AWS cloud has been studied for application in storing data of the microbiome followed by its computation and analysis and 19 vaginal and 50 gut microbiome samples could be analyzed revealing their diversity, taxonomy, and abundance using Amazon Simple Storage Service (ASSS) for microbiome sequence storage, Linux Elastic Compute Cloud (EC2) servers for data computation and analysis, and security keys to create and manage encryption, together with application of bioinformatics and statistical tools including Quantitative Insights Into Microbial Ecology 2 and R Studio within the Linux EC2 for statistical analysis of microbiome in Linux or Mac based operating system (OS, (Bai et al. 2019)) (Fig. 19.1).
3. Processing of medical images
   HPC of AWS has been used in medical image processing due to its affordable, scalable, reliable, on-demand, and inexpensive, easy-to-use cloud computing services with integration of Java Image Science Toolkit (JIST), a plugin for Medical-Image Processing, Analysis, and Visualization (MIPAV), Distributed Resource Management Application API (DRMAA) compliant system (Bao et al. 2016).
4. Biomedical domain and genomics
   Using the ultrafast aligner for short human DNA sequences such as Bowtie could index the around 2.2 GB genome with a Burrows-Wheeler index and enable mapping of human genome and AWS services have been instrumental in storage of data (http://aws.amazon.com/importexport/; Fusaro et al. 2011).

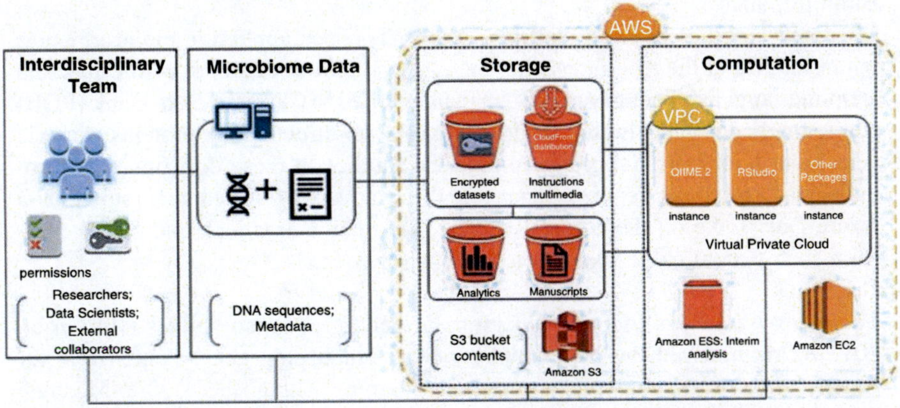

Fig. 19.1 The microbiome data analysis pipeline using AWS (Amazon Web Services), S3: Simple Storage Service; VPC: virtual private cloud; QIIME: Quantitative Insights Into Microbial Ecology; EBS: Elastic Block Store; EC2: Elastic Compute Cloud. Image reproduced with permission from (Bai et al. 2019) ©Jinbing Bai, Ileen Jhaney, Jessica Wells. Originally published in open access article in JMIR Medical Informatics (http://medinform.jmir.org), distributed under the terms of the Creative Commons Attribution License

Globus Genomics was developed to rapidly analyze hugh volumes of data obtained from next-generation sequencing (NGS) with automation, encompassing data analysis specification, configuration, and multi-step processing pipelines by Galaxy workflow, creating Amazon Machine Images and on-demand resource provider on Amazon EC2, and scheduling the pipelines by HT Condor scheduler, open source software used in the draft assembly of human genome (Madduri et al. 2014).

Spark clusters on Google Cloud Platform and AWS and Hail (https://doi.org/10.5281/zenodo.2646680) have been used to whole-genome analysis in exploration of genomic variants dataset and is a step towards improved research in the domain of genetics; however, studies on its speed and cost remain a domain of future research (Krissaane et al. 2020).

19.3 Industry and Biological Sciences

Microsoft (https://www.microsoft.com/en-in/industry/health/pharmaceuticals) has designed AZURE that has application in the health and life science industry in maintaining electronic health records (EHR), application of medical devices, analysis of research data. Azure platforms abstracts hardware into Virtual Machines (VMs) and runs on them with scalability and portability, and works on a Windows based Operating system (OS) with PAAS which gives liberty to users of not continuous updating of OS.

We discuss the applications in life sciences in the subsequent sections.

1. Bioinformatics
   Microsoft cloud computing platform, Azure, has been applied in bioinformatics, with usability of the resource, using the PAAS platform of Azure with Linux and scripting language background (Shanahan et al. 2014). Protein Data Bank (PDB) stores hugh data and high-resolution images on structure of proteins, nucleic acids, and nucleic acid–protein complexes. Studies in these domains of macromolecular structures require parsing and extraction of stored data stored and Azure Data Lake on the Clouds enabled such data extractions and calculations from such structures of macromolecules (Mrozek et al. 2019).
2. Genomics
   Microsoft Azure has found applications in genomics based studies (Kim et al. 2012). This finds all the more importance in identifying genes responsible for genetic disease. A supervised machine learning using the Azure Machine Learning (AML) Platform was used to analyze the literature from databases including Entrez of NCBI, Gene2Pubmed (http://www.ncbi.nlm.nih.gov/gene), Clinical Genomics Database of NHGRI (http://research.nhgri.nih.gov/CGD/) for study of literature related to basic, clinical, and management of disease (Bornstein et al. 2016).

19.4 Discussion

Technology of Cloud computing is perhaps developed to a considerable extent and is having applications in handling analysis, calculation of hugh high-throughput biological data. The ease of performance, low cost, scalability, user-friendliness are important parameters that are being provided by both AZURE and AWS in the supplications of different biological research areas together with applications of machine learning. Researchers across the globe are gradually working in this developing domain of integrated sciences towards better understanding of problems in life sciences and managing health.

References

Bai J, Jhaney I, Wells J (2019) Developing a reproducible microbiome data analysis pipeline using the Amazon Web Services Cloud for a Cancer Research Group: proof-of-concept study. JMIR Med Inform 7(4):e14667

Bao S, Damon SM, Landman BA, Gokhale A (2016) Performance management of high performance computing for medical image processing in Amazon Web Services. Proc SPIE Int Soc Opt Eng 9789:97890Q. https://doi.org/10.1117/12.2217396

Bornstein AT, McLoughlin MH, Aguilar J, Wong WS, Solomon BD (2016) Tracking medical genetic literature through machine learning. Mol Genet Metab 118(4):255–258

Fusaro VA, Patil P, Gafni E, Wall DP, Tonellato PJ (2011) Biomedical cloud computing with Amazon Web Services. PLoS Comput Biol 7(8):e1002147

Kim I, Jung JY, Deluca TF, Nelson TH, Wall DP (2012) Cloud computing for comparative genomics with windows azure platform. Evol Bioinformatics Online 8:527–534

Krissaane I, De Niz C, Gutiérrez-Sacristán A et al (2020) Scalability and cost-effectiveness analysis of whole genome-wide association studies on Google Cloud Platform and Amazon Web Services. J Am Med Inform Assoc 27:1425

Madduri RK, Sulakhe D, Lacinski L et al (2014) Experiences building globus genomics: a next-generation sequencing analysis service using Galaxy, Globus, and Amazon Web Services. Concurr Comput 26(13):2266–2279

Madhyastha TM, Koh N, Day TKM et al (2017) Running neuroimaging applications on Amazon Web Services: how, when, and at what cost? Front Neuroinform 11:63

Mrozek D, Dąbek T, Małysiak-Mrozek B (2019) Scalable extraction of big macromolecular data in azure data Lake environment. Molecules 24(1):179

Shanahan HP, Owen AM, Harrison AP (2014) Bioinformatics on the cloud computing platform Azure. PLoS One 9(7):e102642

Toxicity: An Introduction

20

Abstract

Toxicity induced by drug or chemicals may be life-threatening to human lives. Therefore a drug prior to human use should be tested for its toxic effects in preclinical studies in animals. In this chapter we discuss the types of toxicity, pathophysiology associated with drug induced toxicity, and the molecular basis involved. This forms the introductory chapter towards the next chapter which highlights the application of machine learning in detecting toxicity.

Keywords

TEN · SJS · Toxicity

Abbreviations

ARBs	Adverse drug reactions
Cd	Cadmium
CO	Carbon monoxide
CMI	Cell mediated immunity
CNS	Central nervous system
CNTs	Carbon nanotubes
HT	Hyperthermia
LD_{50}	Lethal dose
MWCNTs	Multiwalled carbon nanotubes
R-S	Receptor–substrate
SJS	Stevens–Johnson syndrome
SWCNTs	Single walled carbon nanotubes
TEN	Toxic epidermal necrolysis
WHO	World Health Organization

S. Ghosh, R. Dasgupta, *Machine Learning in Biological Sciences*,
https://doi.org/10.1007/978-981-16-8881-2_20

20.1 Introduction

Toxicity indicates the degree of damage a substance can inflict on an organism. It could be inflicted on a whole or part of an organism, including an animal, bacterium, or plant, or cell or an organ like liver. The nature of toxic agents could be

1. Chemical, including inorganic elements of lead, mercury, hydrofluoric acid, nitrogen dioxide, cyanide and chlorine gas, and organic matter including methyl alcohol, drugs, and toxins.
2. Radioactive substances, like uranium, harmful ionizing radiations generated from the substance, gamma radiation, microwaves.
3. Biological toxins include pathogens including microbes, plants, fungus, venom on infection may release toxins, tetanus, botulinum, cholera, ricin, mushroom toxin, snake venom.
4. Physical toxicants including coal dust, asbestos fibers, or silica can be toxic, interfering with biological activity when inhaled and corrode tissues, water when taken excessively can act as a physical toxicant due to decrease in concentration of vital ions, asphyxiant gases act as physical toxicants by displacing oxygen.

20.2 Types of Toxicity

20.2.1 Acute Toxicity

Acute exposures to a toxic substance may not last long but may be toxic enough to cause severe harm or death. Like, for example, acute toxicity with suicidal intent in taking of cyanamide through oral or dermal route is known that affects the parasympathetic nervous system in both animals and in humans. Cyanamide exposure through dust or liquid can cause eye irritation and form ulcers in moist skin. In 1989, leakage of methyl isocyanate from industry in India killed 5000 people and permanently disabled 30,000 people. Inhalation of poisonous gas carbon monoxide (CO) from industry can also claim lives.

Radiation therapy can lead to acute toxicity causing death of actively dividing cell populations, in skin epithelium, oral tumors on mucous membranes, reddening of skin or erythema, formation of blisters or vesicles or vesiculation, peeling of upper skin layer or desquamation, severe exfoliative dermatitis, or erythroderma manifested by **severe** inflammation of skin surface, followed by necrosis at a later stage, localized hair loss due to damaged hair follicle epithelium.

Acute toxicity could develop from adverse drug reactions (ADRs) and has been reported from studies on the drugs. Therefore, testing of acute toxicity finds importance in detection and measure of drug or chemical use safety. LD_{50} or lethal dose 50 is a test which is a measure of the lethal dose of an agent including toxin, radiation, or pathogen and the LD_{50} value indicates the dose of the agent that kills half of the tested population members in a specified test. It is used to measure acute toxicity.

Acute toxicity can also arise from intestinal or renal disease. Paralysis, cardiac disorders, nausea, depression of nervous or respiratory system bowel hypomotility, hyporeflexia have been associated with magnesium toxicity. Exposure to silicones has been reported to cause mild skin, eye irritations and corrodes the mucous membranes (Shionogi, and Co., Ltd. 1990; Olson and Smollin 2008; Peach et al. 2015; Laurence 1982; Institute of Medicine (US) Committee on the Safety of Silicone Breast Implants et al. 1994).

20.2.2 Subchronic Toxicity

Repeated exposure of a toxic substance of human exposure to toxic drugs and pharmaceuticals and environmental agents for several weeks to months can lead to subchronic toxicity. Like warfarin (Coumadin®) tablets can reduce blood clots, thereby reducing risk of stroke, heart attack but can act as subchronic toxic substance, exposure to lead in workplace over several weeks can lead to anemia (Mückter 2003; Deaton and Nappe 2020).

20.2.3 Chronic Toxicity

However sometimes continuous exposure to a toxicant over extended months or years can lead to irreversible side effects, these are called as chronic toxic inducing agents causing continuous damage to specific organ systems finally leading to disease. This includes cirrhosis in alcoholic individuals taking ethanol for years, chronic kidney disease caused in workmen due to lead exposure, chronic bronchitis caused in cigarette smokers over long periods of time, pulmonary fibrosis of black lung disease caused in coal miners.

Turpentine is an occupational hazard and its chronic exposure can lead to anemia, cerebral atrophy, changes in behavior, damage of bone marrow, glomerulonephritis, and skin disorders including dermatitis, eczema, and even lung cancer. Paints and varnishes when continuously exposed to workers can reveal disturbances in urination, albuminuria, and urinary casts. Chronic toxicity on exposure of carbon nanotubes (CNTs) of single walled carbon nanotubes (SWCNTs) and multiwalled carbon nanotubes (MWCNTs) can lead to inflammation and inflammatory responses, angiogenesis or fibrosis (Saeidnia 2014; Ghiazza and Fenoglio 2014; Naas 2005). Chronic toxicity has symptoms of gastrointestinal irritations, nausea, vomiting, and diarrhea. Overdose of exposure or intake of cinchona can lead to cinchonism in chronic toxicity with symptoms of headache, fever, visual problems, mydriasis, affected hearing or tinnitus, nausea, vomiting, hot flushed skin, rash, central nervous system (CNS) impairment, lethargy, memory loss, hallucinations.

20.2.4 Carcinogenicity

Carcinogenicity is caused by series of complex processes by exposure to a carcinogen that leads to abnormal cell proliferation and differentiation leading to cancer. The process of transformation includes steps of initiation when carcinogen induces irreversible changes and mutations of normal genes of tumor suppressors in normal cells leading to abnormal cell growth, followed by promotion stage when these cells progress towards cancer. A neoplasm or tumor reveals uncontrolled growth induced mass of cells, benign tumors remain confined to their site of origin but do not spread and are not metastatic and are curable but malignant tumors also called as cancer reveal a metastatic tumor capable of migrating into adjacent tissues and are more difficult to treat and often lead to death (Cohen et al. 2019; Wolf et al. 2019; Benigni and Bossa 2011; Swenberg and Beauchamp 1997).

20.2.5 Developmental Toxicity

Developmental toxicity affects the developing embryo or fetus and can result from exposure of toxic substance to either of the parents before the fetus is conceived or direct exposure of toxic substance to the mother and her developing embryo or fetus. These agents act on embryo cells, causing cell death or cell damage, causing abnormal development, they may also cause a mutation in a parent's germ cell, and being transmitted to the fertilized ovum, lead to development of abnormal embryos. This may lead to lethality of the embryo manifested by failure to conceive, abortion, and stillbirth, embryotoxicity causing retardation or delayed growth of specific organ systems, teratogenicity which is an irreversible effect causing permanent congenital defects in live offspring, like cleft palette or missing limbs (Baker et al. 2018; Hazelden 2013; Hougaard and Hansen 2007).

20.2.6 Genetic Toxicity

Caused by damaged DNA and affected gene expression including mutation, chromosomal aberration, aneuploidy, or polyploidy caused by mutagen through a process called mutagenesis. The mutation in the germ cell, is heritable and passed on to the next generation, while if it is the somatic cell or somatic mutation, it can lead to altered cell growth as in cancer or cell death as in teratogenesis (EFSA Scientific Committee et al. 2018; Rencuzogullari and Aydin 2018).

20.3 Routes of Exposure of Toxicants

Exposure to toxic substances could be through skin, ingestion, inhalation, and by injection with brief exposure, of a single dose or multiple doses over time. It may be in different physical states of solid, liquid, gas and can affect adversely the genetic makeup and overall health of an individual (Behal et al. 2016).

20.4 Drug Induced Toxicity

20.4.1 Pathophysiology

Adverse drug reactions (ADR) is a wide domain and can lead to patient death and tackling it properly remains a major challenge to the global healthcare industry. The main challenge to an anesthetist is anaphylaxis and malignant hyperthermia (HT) related to the administration of drugs in the peri-operative period of anesthetized patient delaying the diagnosis and treatment and adversely affects treatment, it could activate IgE mediated immune responses (Patton and Borshoff 2018; Paton 1975). Tools for detection of drug toxicity leading to IgE mediated hypersensitivity reactions lie in skin testing and *in vitro* testing (Patel and Khan 2017).

The adverse severe cutaneous reactions, involving serious systemic immunologic disorders with high rates of mortality observed include Stevens–Johnson syndrome (SJS), is a life-threatening mucocutaneous syndrome caused by drugs including sulfonamides, penicillin, aspirin, barbiturates, isoniazid, phenytoin and toxic epidermal necrolysis (TEN) is drug induced skin disorder revealing skin peeling in sheet layers and mucosal surfaces and needs immediate management (Araujo and Flowers 1984; Lerch et al. 2018; Harr and French 2012).

The syndromes of ADR hold back to 1922, when American pediatricians, Albert Mason Stevens and Frank Chambliss Johnson, first identified the disease in two young boys revealing symptoms of continued fever, inflamed oral mucosa, and acute conjunctivitis; however, controversy over identifying features in SJS and TEN remained (Table 20.1). Only in 1993, SJS has been identified as a disease on the TEN spectrum revealing pathophysiology of blisters on erythematous or purpuric maculae, on face and trunk, with less than 10% affected body surface area, TEN with more than 30% affected cutaneous surface area and is separate from erythema mulitforme, but both disorders revealed manifestations of mucosal reactions with distinctive patterns of cutaneous lesions and different precipitating factors (Seagrave and Bamba 2017).

Table 20.1 SJS and TEN features

	Pathophysiology		Inducing agents	
	Common	Differences	Drugs	Infectious agents
SJS	1. Severe hypersensitivity (HS) reactions 2. Affecting skin and mucous membranes 3. Keratinocyte death 4.Subepidermal layer splitting 5. Epidermal necrosis 6. Painful lesions: small blister on purpuric macules 7. Fever, malaise 8. Appearance within 0–3 weeks	1. Less than 10% affected cutaneous area 2. Rare area of confluent erythema	1. Sulfonamides 2.Anticonvulsants 3. Non-steroidal anti-inflammatory drugs (NSAIDs) 4. Allopurinol 5. Antibiotics	1. Viruses 2. Bacteria 3. Fungi and 4. Protozoa
TEN		1. More than 30% affected cutaneous area 2. Common confluent erythema 3. Necrotic epidermis in large sheets 4.Outer epidermal layer separates easily from basal layer with lateral pressure		

20.4.2 Molecular Basis

Drugs can induce hepatic injury, renal injury and disorders, nephrotoxicity as liver, and kidney are involved in metabolism of drugs and chemicals (Mitchell et al. 1977).

Exposure to cadmium (Cd) is toxic to human and animals as it accumulates in organs and kidney cortex and Cd toxicity inhibits protein synthesis, carbohydrate metabolism, drug metabolizing enzymes in liver of animals, leading to manifestations of hypertension in animal models, proteinuria, affected nervous system, affected lungs, humoral immune (HI), and cell mediated immune (CMI) responses in animals (Nath et al. 1984). Some drugs are known to affect the lungs causing respiratory disorders (Lei et al. 2017).

To understand drug induced toxicity it is important to understand receptor–substrate (R-S) interactions. The Ah receptor of a protein-small molecule interaction has been associated with extreme acute toxicity in experimental studies in animal models (McKinney 1985). Drug toxicities lead to functional, biochemical, and/or structural alterations and receptor systems are involved in toxic responses. Structural toxicities produce changes in the structure of cell, organ, tissue, or cell like fatty infiltration of the liver cells in chloroform anesthesia or alcohol consumption, phenothiazine toxicity is associated with cataracts (Loomis and Hayes 1996).

Therefore it is important to understand the effect of the drug induced toxicity prior to its administration in patients and evaluation of toxicity finds importance prior to any preclinical trial to access its safety as new drug related to human health and the fate of drugs. The study involves extremely costly and laborious time-consuming experimental work in animals, involved animal ethics. Recently computational models and machine learning has been instrumental in mediating the toxicity studies and detecting endpoints as carcinogenicity, mutagenicity, and hepatotoxicity which we discuss in detail in the subsequent chapter.

20.5 Discussion

World Health Organization (WHO) has enlisted Adverse Drug Reaction (ADR) under International Classification of Diseases and classified into six types including dose-related or augmented, non-dose-related or bizarre, dose-related and time-related or chronic, time-related or delayed, withdrawal or end of use and failure of therapy as failure (Edwards and Aronson 2000). Thus testing of a drug finds importance in preclinical studies. This involves costly experimentation and validation of studies in animal models in preclinical studies. Recently machine learning and computational biology have enabled development of algorithms that has enabled easy, effective, less time-consuming, and less costly methods in the detection of the toxicity of the proposed drug prior to its use in human population. We discuss the impact and importance of machine learning on the drug toxicity studies in subsequent chapters.

References

Araujo OE, Flowers FP (1984) Stevens-Johnson syndrome. J Emerg Med. 2(2):129–135

Baker N, Boobis A, Burgoon L et al (2018) Building a developmental toxicity ontology. Birth Defects Res 110(6):502–518

Behal N, Wong A, Mantara R, Cantrell FL (2016) Human poisoning through atypical routes of exposure. J Community Health 41(1):105–108

Benigni R, Bossa C (2011) Mechanisms of chemical carcinogenicity and mutagenicity: a review with implications for predictive toxicology. Chem Rev 111(4):2507–2536

Cohen SM, Boobis AR, Dellarco VL et al (2019) Chemical carcinogenicity revisited 3: risk assessment of carcinogenic potential based on the current state of knowledge of carcinogenesis in humans. Regul Toxicol Pharmacol 103:100–105

Deaton JG, Nappe TM (2020) Warfarin toxicity. [Updated 2020 Jan 10]. In: StatPearls [Internet]. StatPearls Publishing, Treasure Island. https://www.ncbi.nlm.nih.gov/books/NBK431112/

Edwards IR, Aronson JK (2000) Adverse drug reactions: definitions, diagnosis, and management. Lancet. 356(9237):1255–1259

EFSA Scientific Committee, More S, Bampidis V et al (2018) Genotoxicity assessment of chemical mixtures. EFSA J 17(1):e05519

Ghiazza M, Fenoglio GV (2014) Chapter 8 - Carbon nanotubes: properties, applications, and toxicity. In: Health and environmental safety of nanomaterials polymer nanocomposites and other materials containing nanoparticles. Woodhead Publishing, Sawston, pp 147–174

Harr T, French LE (2012) Stevens-Johnson syndrome and toxic epidermal necrolysis. Chem Immunol Allergy 97:149–166

Hazelden KP (2013) The developmental toxicity testing of biologics. Methods Mol Biol 947:31–36

Hougaard KS, Hansen AM (2007) Enhancement of developmental toxicity effects of chemicals by gestational stress. A review. Neurotoxicol Teratol 29(4):425–445

Institute of Medicine (US) Committee on the Safety of Silicone Breast Implants (1994) In: Bondurant S, Ernster V, Herdman R (eds) Safety of silicone breast implants, vol 4: Silicone toxicology. National Academies Press (US), Washington, DC. https://www.ncbi.nlm.nih.gov/books/NBK44789/
Laurence DR (1982) Adverse drug reactions and acute poisoning reviews, vol 1, No. 1. J R Soc Med 75(8):683
Lei T, Chen F, Liu H et al (2017) ADMET evaluation in drug discovery. Part 17: development of quantitative and qualitative prediction models for chemical-induced respiratory toxicity. Mol Pharm 14(7):2407–2421
Lerch M, Mainetti C, Terziroli Beretta-Piccoli B, Harr T (2018) Current perspectives on Stevens-Johnson syndrome and toxic epidermal necrolysis. Clin Rev Allergy Immunol 54(1):147–176
Loomis TA, Hayes AW (1996) Chapter 7 - Classification of harmful effects of chemicals. In: Loomis's essentials of toxicology, 4th edn. Elsevier, pp 101–106
McKinney JD (1985) The molecular basis of chemical toxicity. Environ Health Perspect 61:5–10
Mitchell JR, McMurtry RJ, Statham CN, Nelson SD (1977) Molecular basis for several drug-induced nephropathies. Am J Med 62(4):518–526
Mückter H (2003) What is toxicology and how does toxicity occur? Best Pract Res Clin Anaesthesiol 17(1):5–27
Naas DJ (2005) Quinidine. In: Encyclopedia of toxicology, 2nd edn. Elsevier, Oxford, pp 593–594
Nath R, Prasad R, Palinal VK, Chopra RK (1984) Molecular basis of cadmium toxicity. Prog Food Nutr Sci 8(1-2):109–163
Olson K, Smollin C (2008) Carbon monoxide poisoning (acute). BMJ Clin Evid 2008:2103
Patel SV, Khan DA (2017) Adverse reactions to biologic therapy. Immunol Allergy Clin North Am. 37(2):397–412
Paton WD (1975) The molecular basis of drug toxicity. J Clin Pathol Suppl (R Coll Pathol). 9:1–6
Patton K, Borshoff DC (2018) Adverse drug reactions. Anaesthesia 73:76–84
Peach MS, Showalter TN, Ohri N (2015) Systematic review of the relationship between acute and late gastrointestinal toxicity after radiotherapy for prostate cancer. Prostate Cancer 2015:624736
Rencuzogullari E, Aydin M (2018) Methodology of genotoxic and teratogenic studies in rats. Methods Mol Biol 1797:555–575
Saeidnia S (2014) Turpentine encyclopedia of toxicology, 3rd edn. Academic, Boston, pp 860–865
Seagrave Z, Bamba S (2017) Adverse drug reactions. Dis Mon 63(2):49–53
Shionogi & Co., Ltd. (1990) Summary of toxicity data on methyl isothiocyanate (MITC). J Pestic Sci 15(2):297–304. Released 5 Aug 2010. Online ISSN 1349-092
Swenberg JA, Beauchamp RO Jr (1997) A review of the chronic toxicity, carcinogenicity, and possible mechanisms of action of inorganic acid mists in animals. Crit Rev Toxicol 27(3): 253–259
Wolf DC, Cohen SM, Boobis AR et al (2019) Chemical carcinogenicity revisited 1: a unified theory of carcinogenicity based on contemporary knowledge. Regul Toxicol Pharmacol 103:86

21 Machine Learning (ML) and Toxicity Studies

Abstract

Drug or chemicals induced toxic changes in the body can lead to mortality and therefore it is mandatory to test the proposed drug for its toxic effects in animal models by preclinical studies. However this method entails cost and is time-consuming. Recently applications of experiments based on high-throughput studies and computational biology, it has been possible to predict the toxicity of drugs and screen them and release them to the market at lower cost and time. We discuss in this chapter the different machine learning applications that have been applied to toxicity studies.

Keywords

Machine learning · Toxicity

Abbreviations

ADRs	Adverse drug reactions
ANN	Artificial neural networks
DL	Deep learning
DNNs	Deep neural networks
ELM	Extreme learning machine
GO	Gene ontology
hERG	Human ether-a-go-go-related gene
HT	High-throughput
Ikr	Rectifier potassium channel
Ils	Ionic liquids
LD_{50}	Lethal dose 50
MLR	Multiple linear regression

S. Ghosh, R. Dasgupta, *Machine Learning in Biological Sciences*,
https://doi.org/10.1007/978-981-16-8881-2_21

nB	Naive Bayesian
NN	Neural network
QSAR	Quantitative structure–activity relationships
R&D	Research and development
RF	Random forest
RFE	Recursive feature elimination
SVM	Support vector machines

21.1 Introduction

Toxicity can be induced by drugs and chemicals from the environment or administered as drug and can have morbid consequences on the human and animal lives. Therefore detection of toxicity and its effect on the body finds importance prior to its release as drug. The conventional model of toxicity study is through preclinical studies in the animal model system and it is extremely time-consuming, costly, requiring trained manpower, specific laboratory set ups with concerns on the health and ethics of the animal body.

Recently diverse applications of computational studies and machine learning based algorithms have been instrumental in detection and study of toxicity.

21.2 Machine Learning and Application in Detection of Toxicity

Ionic liquids (ILs) are salt-like consistency which are liquid at unusually low temperatures, and their wide spread synthesis is toxic to the environment. Quantitative structure–activity relationships (QSAR) model that specifies the function, the molecular system, structural and chemical property has been used to detect toxicity of ILs in leukemia rat cell line (ICP-81). Extreme learning machine (ELM) model revealed better than multiple linear regression (MLR) and SVM in toxicity detection (Cao et al. 2018). High-throughput (HT) analysis of large datasets with new machine learning (ML) tools has found applications in analysis of high-dimensional profiling data in drug discovery by image-based screening and profiling (Scheeder et al. 2018). Machine learning tools including support vector machine (SVM), random forest (RF), naive Bayesian (nB), back propagation neural network (NN) have been used to design drug toxicity and prediction of carcinogenicity, mutagenicity, and hepatotoxicity (Zhang et al. 2018). It has also been applied in lead optimization or in the process of design of drug candidate identified by qualitative classification and quantitative regression (QR) studies (Yang et al. 2018).

Pharmacovigilance is a primary step to monitoring, detection, and preventing adverse drug reactions (ADRs) and machine learning and deep learning has been instrumental in enabling in silico studies (Basile et al. 2019). Machine learning (ML) tools of artificial neural network (ANN) and random forest (RF) have been used to predict toxicity response for ecological parameters to understand environmental risk implications towards application of suitable remediation (Cipullo et al. 2019). Using regression model and recursive feature elimination (RFE) method together with the SVM algorithm it has been possible to measure the bio-concentration factors and median lethal concentrations (LD_{50}) to assess risks of organic pollutants to aquatic ecosystems (Ai et al. 2019).

Advancements in machine learning methods enable them now to handle increased big toxicity data of molecular descriptors, toxicogenomics, and high-throughput (ht) bioactivity in assessment of toxicity by in silico methods (Vo et al. 2020). Using the three parameters of acute toxicity, carcinogenicity, and inhibition of the human ether-a-go-go-related gene ion channel (hERG) which is a pore-forming subunit of delayed rectifier potassium channel (I_{Kr}), playing role in cardiac repolarization, it has been possible to design and develop new drug candidates with low toxicity (Lu et al. 2018).

Application of tools including natural language processing (NLP), speech recognition, image recognition, computational chemistry, bioinformatics, deep learning (DL),RF, k-nearest neighbors (kNN), and SVM are enabling toxicity prediction (Wu and Wang 2018; Idakwo et al. 2018). It is being tested if transcriptome data analysis is possible together with chemical structural description to improve the prediction accuracy (Wu and Wang 2018). Based on classification and regression analysis, models have been designed that enabled distinction between toxins and non-toxins (Sharma et al. 2017).

Considerable progresses have been made in machine learning classification algorithms and using a multi-level comparison of different performance metrics it has been possible to study acute and aquatic toxicities (Rácz et al. 2019) and predict toxicity of chemicals (Tang et al. 2018) by using deep neural networks (DNNs), hemolytic toxicity of saponins (Zheng et al. 2019).

TargeTox a machine learning tool has been applied to in identification of drug and their targets and off-targets.

Gene ontology (GO) annotations tools has been used to indicate functions and biological network data has been designed to predict toxicity of drugs (Lysenko et al. 2018). Apart from deep learning, shallow learning methods have been used for quantitative prediction of acute chemical toxicity (Liu et al. 2018). Machine learning tools have been instrumental in predicting and screening for mitochondrial toxicity from prospective compound libraries (Hemmerich et al. 2020). Artificial neural networks (ANN), SVM and supervised learning algorithm were applied successfully in high-throughput chemical screening (Zhang et al. 2016).

Machine learning algorithms of AdaBoost, C.45 Decision Tree, k-Nearest Neighbors (kNN), Multilayer Perceptron, RF used in a cascade model has been instrumental in predicting carcinogenicity in rats (Guan et al. 2018). A machine learning based tool has been designed to predict chemical toxicity in prospective drugs (Pu et al. 2019). Machine learning tools could enable prediction of reproductive toxicity in chemicals, (Jiang et al. 2019) development of drug (Dobchev et al. 2014), toxicity of xenobiotics (Maltarollo et al. 2015), oral toxicity of chemicals (Xu et al. 2017), toxicity from emissions (Marvuglia et al. 2015).

21.3 Discussion

Deep Learning has been used in toxicology and prediction of toxicity using different pipeline, together with the application of Big Data and graphical processing units (GPUs). The Tox21 (Attene-Ramos et al. 2013) is one such example in the toxicological research. It is thought that deep learning is a promising tool in predicting toxicity and future research may be based on the expansion and improvement in this domain.

References

Ai H, Wu X, Zhang L et al (2019) QSAR modelling study of the bioconcentration factor and toxicity of organic compounds to aquatic organisms using machine learning and ensemble methods. Ecotoxicol Environ Saf 179:71–78

Attene-Ramos MS, Miller N, Huang R et al (2013) The Tox21 robotic platform for the assessment of environmental chemicals--from vision to reality. Drug Discov Today 18(15–16):716–723

Basile AO, Yahi A, Tatonetti NP (2019) Artificial intelligence for drug toxicity and safety trends. Pharmacol Sci 40(9):624–635

Cao L, Zhu P, Zhao Y, Zhao J (2018) Using machine learning and quantum chemistry descriptors to predict the toxicity of ionic liquids. J Hazard Mater 352:17–26

Cipullo S, Snapir B, Prpich G, Campo P, Coulon F (2019) Prediction of bioavailability and toxicity of complex chemical mixtures through machine learning models. Chemosphere 215:388–395

Dobchev DA, Pillai GG, Karelson M (2014) In silico machine learning methods in drug development. Curr Top Med Chem 14(16):1913–1922

Guan D, Fan K, Spence I, Matthews S (2018) Combining machine learning models of in vitro and in vivo bioassays improves rat carcinogenicity prediction. Regul Toxicol Pharmacol 94:8–15

Hemmerich J, Troger F, Füzi B, F Ecker G (2020) Using machine learning methods and structural alerts for prediction of mitochondrial toxicity. Mol Inform 39(5):e2000005

Idakwo G, Luttrell J, Chen M et al (2018) A review on machine learning methods for in silico toxicity prediction. J Environ Sci Health C Environ Carcinog Ecotoxicol Rev 36(4):169–191

Jiang C, Yang H, Di P, Li W, Tang Y, Liu G (2019) In silico prediction of chemical reproductive toxicity using machine learning. J Appl Toxicol 39(6):844–854

Liu R, Madore M, Glover KP, Feasel MG, Wallqvist A (2018) Assessing deep and shallow learning methods for quantitative prediction of acute chemical toxicity. Toxicol Sci 164(2):512–526

Lu J, Lu D, Fu Z, Zheng M, Luo X (2018) Machine learning-based modeling of drug toxicity. Methods Mol Biol 1754:247–264

Lysenko A, Sharma A, Boroevich KA, Tsunoda T (2018) An integrative machine learning approach for prediction of toxicity-related drug safety. Life Sci Alliance 1(6):e201800098

Maltarollo VG, Gertrudes JC, Oliveira PR, Honorio KM (2015) Applying machine learning techniques for ADME-Tox prediction: a review. Expert Opin Drug Metab Toxicol 11(2): 259–271

Marvuglia A, Kanevski M, Benetto E (2015) Machine learning for toxicity characterization of organic chemical emissions using USEtox database: learning the structure of the input space. Environ Int 83:72–85

Pu L, Naderi M, Liu T, Wu HC, Mukhopadhyay S, Brylinski M (2019) eToxPred: a machine learning-based approach to estimate the toxicity of drug candidates. BMC Pharmacol Toxicol 20(1):2

Rácz A, Bajusz D, Héberger K (2019) Multi-level comparison of machine learning classifiers and their performance metrics. Molecules 24(15):2811

Scheeder C, Heigwer F, Boutros M (2018) Machine learning and image-based profiling in drug discovery. Curr Opin Syst Biol 10:43–52

Sharma AK, Srivastava GN, Roy A, Sharma VK (2017) ToxiM: a toxicity prediction tool for small molecules developed using machine learning and Chemoinformatics approaches. Front Pharmacol 8:880

Tang W, Chen J, Wang Z, Xie H, Hong H (2018) Deep learning for predicting toxicity of chemicals: a mini review. J Environ Sci Health C Environ Carcinog Ecotoxicol Rev 36(4):252–271

Vo AH, Van Vleet TR, Gupta RR, Liguori MJ, Rao MS (2020) An overview of machine learning and big data for drug toxicity evaluation. Chem Res Toxicol 33(1):20–37

Wu Y, Wang G (2018) Machine learning based toxicity prediction: from chemical structural description to transcriptome analysis. Int J Mol Sci 19(8):2358

Xu Y, Pei J, Lai L (2017) Deep learning based regression and multiclass models for acute oral toxicity prediction with automatic chemical feature extraction. J Chem Inf Model 57(11): 2672–2685

Yang H, Sun L, Li W, Liu G, Tang Y (2018) In silico prediction of chemical toxicity for drug design using machine learning methods and structural alerts. Front Chem 6:30

Zhang Y, Wong YS, Deng J et al (2016) Machine learning algorithms for mode-of-action classification in toxicity assessment. BioData Min 9:19

Zhang L, Zhang H, Ai H, Hu H, Li S, Zhao J, Liu H (2018) Applications of machine learning methods in drug toxicity prediction. Curr Top Med Chem 18(12):987–997

Zheng S, Wang Y, Liu H, Chang W, Xu Y, Lin F (2019) Prediction of hemolytic toxicity for Saponins by machine-learning methods. Chem Res Toxicol 32(6):1014–1026

Applications of Machine Learning in Study of Cell Biology

22

Abstract

The study of cell biology is extremely complex involving the diverse structural, functional features, development, differentiation, metabolism, and regulation. While it is intriguing to know the features, functions, and regulations in a normal cell and it is also important to understand the deviations leading to disease phenotype. The study of extremely complex mechanisms involved in cell fate, biology of lineage specific differentiation from the pluripotent stem cell capable of differentiating into different cell types, is far from complete and it is also complex to understand the biology of the cancer cell. While wet lab based techniques and omics based sequencing studies have helped in our understanding in the intricate biology of the cell in health and disease, machine learning tools and computational biology have immense applications in in-depth study of the biology of the cell. In this chapter we focus of the diverse application of computation and machine learning tool in the study of the biology of the cell.

Keywords

Virtual cell · I cell · Cell biology

Abbreviations

CSC	Cancer stem cell
ESC	Embryonic stem cells
iPSC	Induced pluripotent stem cells
OCLR	One-class logistic regression
PSC	Pluripotent stem cells

S. Ghosh, R. Dasgupta, *Machine Learning in Biological Sciences*,
https://doi.org/10.1007/978-981-16-8881-2_22

22.1 Introduction

Machine leaning is the recently developing advanced science that has immense application in diverse domains of biology including plant, animal, agriculture science, veterinary sciences, biology of toxicity, heath and disease, medicine, drug design, and epidemiological studies.

Cell is a basic unit of life and the study of the function and structure and biology of cell is termed as cytology. The complex biology of cellular mechanism encompassing its origin, development and fate, stem cells maintaining pluripotency and the differentiation into different lineages, diverse function of each cell in maintaining the biological homeostasis, cell cycle in normal cells and the affected cell leading to tumor or cancer, genes and network controlling the normal biology of a healthy cell and deviation in disease cell, cellular metabolism, communication in between cells, cell cycle, cellular transport and signaling, composition and biochemistry of cell all find importance to understand the overall cell functions and therefore understand the altered pathophysiology in disease.

Various wet lab techniques including culture of cell, cytology, sequencing, microscopy, cell fractionation, biochemical studies, molecular biology studies are used to study cells function and find utmost importance in research in the domain of unraveling complexity of diseases like cancer and in the domain of biomedical and clinical applications.

Recently with the advances of machine learning algorithms, it has been possible to study the cell biology and understand the health of a cell and its organelles and functions in greater depth and accuracy. We will discuss in this chapter the different applications of machine learning in the study of diverse domains of cell biology.

22.2 Study of Cell Biology

The study of cell biology is extremely intriguing and its understanding is a multidimensional problem, with different cellular organization, compartmentalization, regulation, cell-cell signaling, differentiation and cell fate, regeneration. With the advent and application of high-throughput sequencing data, a huge repertoire of information is available but it is important to analyze the data properly. Cellular metabolites and their levels are known to be indicators of health and disease and drivers of cell activities and therefore understanding their spatial distribution and the cellular mechanisms they govern can reveal metabolic disorders and deciphering their spatial distribution remains a major challenge in cell biology research.

22.3 Machine Learning in Study of Cell Biology

Machine learning algorithms, tools, emerging technology, and computation have been applied to address understanding of basics of cell biology to young students and also to understand the complex unsolved cellular mechanisms and is aiding in gaining insights into more in-depth and fundamental understanding of diverse mechanisms in cell biology.

22.3.1 Education and Research

Apps for touch-pad devices have enabled education along with other diverse applications. iCell has been designed with 3D computer graphics for study of cell organelles and their functions in animal, bacterial, and plant cells with features that enable study of each organelle by the 3D Cell Simulation and Stain Tool (iPad) app and finds importance for study by students at different levels of middle, high school, and college. The apps provide advantage of rotation of a stereotypical animal cell and understand the movements. Movie on live cells further aids in showing live cells in action. Features like Stain Your Own Cell enable users to stain cells virtually by clicking the options enabling a student's understanding of study of cell (Stark 2012). Computational modeling has enabled the design of the Virtual Cell that enable scientists and cell biologists to construct various cell biological models and generate simulations (Loew and Schaff 2001).

22.3.2 Cell Structure and Networks

Machine learning (ML) and supervised learning algorithms have enabled study of cell images obtained from light microscopy experiments with applications in biomedical sciences. It has been possible to study of segmenting and tracking of each cells, detection of cell variability and their modeling (Kan 2017). Machine learning has been used to study complex biological networks in cell (Camacho et al. 2018). Based on deep learning algorithm, U-Net has been designed and has been applied to detect cell and quantitative estimation of shape from biomedical image data (Falk et al. 2019).

Visible neural networks (VNNs) have been used to develop DCell (http://d-cell.ucsd.edu/) that has been trained with million genotypes and is designed to simulate cellular growth as correct as that of laboratory observations and enable understanding of underlying mechanisms of association between genotype and phenotype, understanding the genetics of disease, drug resistance (Ma et al. 2018).

Machine learning algorithms have been used to enable image-based cell sorting, from high-throughput cell microscopy, real-time images of micro-algal and blood cells from photosynthesis and atherothrombosis based studies in both plant and animal system, respectively (Nitta et al. 2018).

22.3.3 Cell Health

The study of mitochondria finds importance in understanding the health of a cell as it differs in normal and diseased states. Although automated analysis of images is used as the current method, it suffers from challenges of clarity and biasness. MitoMo integrated software together with machine learning algorithms is an advanced development in detection of mitochondrial morphology, texture, motion, morphogenesis, study of intra mitochondrial molecules, thus enabling study of normal mitochondria and its differences from mitochondria from diseased cell, studies of drug testing and molecular medicine (Zahedi et al. 2018).

22.3.4 Stem Cells

Induced pluripotent stem cells (iPSCs) are stem cells of pluripotent nature that develops from a somatic cell. In 2006 it was first discovered by introduction of four specific genes of transcription factors including Klf4, Myc, Oct 3, Oct 4, Sox2 that converted a somatic cell was reprogrammed into pluripotent stem cells (PSC), (Takahashi and Yamanaka 2006). Due to this property that they can be reprogrammed from somatic cell to pluripotent state, with properties of self-renewal and differentiation to different cell lineage they find important application in therapeutics and could replace embryonic stem cells (ESC), thus overcoming the ethical issues associated with use of embryos in research (Singh et al. 2015). Identifying (iPSc) progenitor cells, during initial stages of reprogramming, remains a major challenge as biomarkers appear after 6 days postinitiation of reprogramming. Machine learning algorithms have been instrumental for the identification of iPS progenitor cells from microscopic image analysis within 48 h using XGBoost model (Zhang et al. 2019).

iPSC colony images analysis finds importance in monitoring their health and growth but to check for large-scale cultures manually is a tedious process. Using Multiclass Support Vector Machines (SVM) and using a k-Nearest Neighbors (kNN) classifier it has been possible to monitor the iPSC large colonies with accuracy over and above conventional modes of studies (Joutsijoki et al. 2016).

22.3.5 Cell Segmentation and Trafficking

Single-cell data on microscopy analysis suffers from the challenges of accurate quantification due to human errors of oversight and curation. Machine learning has been advantageous in analyzing high-throughput image data from single cell. It has been possible to gain knowledge in cell segmentation and trafficking based on application of machine learning tools Fiji's Trainable Weka Segmentation and using open source Python scripts to single-cell data obtained from time-lapse microscopy images (Deter et al. 2019).

Deep learning based tools applied to studies on *Escherichia coli* (*E coli*) have enabled image analysis for segmentation, tracking, and lineage reconstruction (Lugagne et al. 2020).

22.3.6 Cancer Cells

Cancer therapy suffers from the challenges of detection and quantization and distribution of cancer metastatic cells, minimal residual disease (MRD), relapse and detection of distribution and targeting of cancer cells by therapeutic antibody. Machine learning tool DeepMACT has been designed for quantitative estimation of cancer metastasis and to detect targeted therapeutic antibody in transparent mice, animal model of breast, lung, and pancreatic cancer which is a boon for the biomedical field (Pan et al. 2019).

Detection of cancer cells from normal ones in early stages of breast cancer finds importance in targeted therapy and improves chances of survival and better health. Conventional methods of identification of breast cancer include performance of biopsy, studies from microscopic images, mammography to ultrasonography and magnetic resonance images (MRI). Extreme Learning Machine (ELM) classification as compared to other tools of Naive Bayes, Support Vector Machine (SVM), and Artificial Neural Network (ANN) finds importance in accurate and fast detection of malignant cells from image analysis of mammograms (Toprak 2018) thus aiding in early detection of breast cancer.

Stem cell like features in cancer also called as cancer stem cells (CSCs) have been known to play role in cancer initiation and progression and recently a new parameter in detection and progression of cancer has been attributed to stemness indices using one-class logistic regression (OCLR) machine learning algorithm form transcriptomic and epigenetic data from non-transformed pluripotent stem cells (PSC) and their differentiated progeny. This has also been instrumental in throwing insights into immune checkpoint, cancer cell heterogeneity, infiltrating immune cells and is a step towards design of novel therapeutics (Malta et al. 2018).

Machine learning applications of kNN and SVMs algorithm have enabled early detection of cervical cancer from pap-smear images analysis (William et al. 2018).

22.4 Discussion

A major focus of research of scientist working in the domain of cell biology is towards understanding a disease and prevent disease in life forms including plants and animals and to improve the quality of life through the entire lifespan. Despite advances in the study of cell biology deciphering its complexity, more-so in the post genomic era, major questions still remain unresolved in the domain of cell biology including movement of secreted proteins across Golgi, membrane sources of autophagosomes, exosomes, what factors govern cell, organelle, cytoskeletal structures, intriguing questions on factors governing cell fate (Munro 2013). Cancer cell biology still remains a major domain of study where machine learning tools can have more applications in solving problems and design of new and effective therapies Although machine learning has been instrumental in solving some major problems in this domain, it has to be studied and applied more in solving the problems hitherto unresolved in this domain of cell biology.

References

Camacho DM, Collins KM, Powers RK, Costello JC, Collins JJ (2018) Next-generation machine learning for biological networks. Cell 173(7):1581–1592

Deter HS, Dies M, Cameron CC, Butzin NC, Buceta J (2019) A cell segmentation/tracking tool based on machine learning. Methods Mol Biol 2040:399–422

Falk T, Mai D, Bensch R et al (2019) U-Net: deep learning for cell counting, detection, and morphometry [published correction appears in Nat Methods. 2019 Apr;16(4):351]. Nat Methods 16(1):67–70

Joutsijoki H, Haponen M, Rasku J, Aalto-Setälä K, Juhola M (2016) Machine learning approach to automated quality identification of human induced pluripotent stem cell colony images. Comput Math Methods Med 2016:3091039, 15p

Kan A (2017) Machine learning applications in cell image analysis. Immunol Cell Biol 95(6): 525–530

Loew LM, Schaff JC (2001) The Virtual Cell: a software environment for computational cell biology. Trends Biotechnol 19(10):401–406

Lugagne JB, Lin H, Dunlop MJ (2020) DeLTA: automated cell segmentation, tracking, and lineage reconstruction using deep learning. PLoS Comput Biol 16(4):e1007673

Ma J, Yu MK, Fong S et al (2018) Using deep learning to model the hierarchical structure and function of a cell. Nat Methods 15(4):290–298

Malta TM, Sokolov A, Gentles AJ et al (2018) Machine learning identifies stemness features associated with oncogenic dedifferentiation. Cell 173(2):338–354.e15

Munro S (2013) Open questions: what is there left for cell biologists to do? BMC Biol 11:16

Nitta N, Sugimura T, Isozaki A et al (2018) Intelligent image-activated cell sorting. Cell 175(1): 266–276.e13

Pan C, Schoppe O, Parra-Damas A et al (2019) Deep learning reveals cancer metastasis and therapeutic antibody targeting in the entire body. Cell 179(7):1661–1676.e19

Singh VK, Kalsan M, Kumar N, Saini A, Chandra R (2015) Induced pluripotent stem cells: applications in regenerative medicine, disease modeling, and drug discovery. Front Cell Dev Biol 3:2

Stark LA (2012) Cell biology apps for Apple devices. CBE Life Sci Educ 11(3):226–230

Takahashi K, Yamanaka S (2006) Induction of pluripotent stem cells from mouse embryonic and adult fibroblast cultures by defined factors. Cell 126(4):663–676

Toprak A (2018) Extreme learning machine (ELM)-based classification of benign and malignant cells in breast cancer. Med Sci Monit 24:6537–6543

William W, Ware A, Basaza-Ejiri AH, Obungoloch J (2018) A review of image analysis and machine learning techniques for automated cervical cancer screening from pap-smear images. Comput Methods Prog Biomed 164:15–22

Zahedi A, On V, Phandthong R et al (2018) Deep analysis of mitochondria and cell health using machine learning. Sci Rep 8(1):16354

Zhang H, Shao X, Peng Y et al (2019) A novel machine learning based approach for iPS progenitor cell identification. PLoS Comput Biol 15(12):e1007351

Genomics and Machine Learning 23

Abstract

Machine learning tools have wide scale application the science of biology and health and agricultural sciences and industry. In this chapter we discuss the applications of machine learning tools in the domain of genetic and genomics based studies that find importance in predicting phenotypes from genomics based studies and diagnosis and study of deadly disease like Cancer.

Keywords

GWAS · kNN · mGWAS

Abbreviations

B anthracis	*Bacillus anthracis*
CNV	Copy number variation
D melanogaster	*Drosophila melanogaster*
GEO	Gene Expression Omnibus
GWAS	Genome-wide association studies
ht.	High throughput
ICE	Integrative and conjugative elements
INDEL	Insertion–deletion
IS	Insertion sequences
KNN	*k-Nearest neighbors*
MDR	Multidimensional reduction
NGS	Next-generation sequencing
rDt	Recombinant DNA technology
RF	Random forest
rt.	Real time

S. Ghosh, R. Dasgupta, *Machine Learning in Biological Sciences*,
https://doi.org/10.1007/978-981-16-8881-2_23

S pneumonia	*Streptococcus pneumonia*
S pyrogenes	*Streptococcus pyogenes*
SEER	*Sequence element enrichment*
SNP	Single nucleotide polymorphism

23.1 Introduction

Genetics is a branch of biological sciences that involves the study of genes, heredity, genetic variation which is essentially a biological process by which genes are transferred from parents to offsprings. Along with the genes the offsprings inherit the parental characteristics. Heredity was studied for the first time by Gregor Johann Mendel, in the nineteenth century, with his observations in pea plants.

Although the earliest study in genetics revealed genes as units of inheritance, concepts of inheritance of traits and its mechanisms involved have considerably been studied in the twenty-first century. Modern day genetics focuses on gene structure, function, variation, and distribution in cell and population. Genetics has now been studied in diverse subdivisions of epigenetics, molecular and population genetics.

Genomics involves the study of whole genomes of organisms employing recombinant DNA technology (rDt), DNA sequencing, high throughput (ht) sequencing, real time (rt) and RNA sequencing, application of bioinformatics to sequence, assemble, and analysis of the structure and function of genomes. Genomics highlights interactions within genomes between loci, alleles, epistasis where one or more genes called as the epistatic gene(s) can alter the phenotypic expression of another gene called as the hypostatic gene, pleiotropy also termed as effect of a single gene on other phenotypic traits by coding for product and has signaling function on various targets and heterosis where due to selective breeding, the offspring reveals improved or better qualities with genetically superiority to their parents.

Genomics and classical genetics differ in that while the former focuses on entire organism's hereditary material in understanding the biology in health and disease, the latter focuses on one gene or one gene product at a time.

Considerable progress in the study of genomes has been made by the advancements of technology of next-generation sequencing (NGS). Genome-wide association studies (GWAS) are enabling the study of genetic variants on human traits, aiding in understanding of disease risks, function of specific genomic regions (Libbrecht and Noble 2015). However such data is huge and manual curation is both time-consuming, requires highly skilled specialized manpower and prone to human errors.

Machine learning tools offer advantage and in this chapter we discuss the diverse application of machine learning in the study of the genomics.

23.2 Machine Learning Applications in Genomics

Machine learning automated algorithms for extraction of information have enabled the design of a knowledge base of genetic associations GWASKB, including collection of information on studies based on genome wide association studies (GWAS) with greater precision (Libbrecht and Noble 2015).

GWAS on bacterial population has enabled detection and study of variations in genomes, indicate factors that affect host or pathogen, drug resistance, virulence, host specificity, features of prognosis. Genetic variation in microbes differs from human genetic variations in its complexity including a wide source of variations arising from causes of single nucleotide polymorphisms (SNPs), deletions, insertions, duplications, presence or absence of whole genes (San et al. 2020; Power et al. 2016; Dutilh et al. 2013; Read and Massey 2014). Thus one or more variations could affect the phenotypic expression and microbial GWAS or mGWAS is based on prior knowledge of such variations including SNPs and insertions and deletion mutations or INDELs, ranging from 1 to 10,000 base pairs long occurring during transcription for viruses and bacteria or duplication of cell (Weber et al. 2002; Mills et al. 2006). Microarray genotyping chips and variant calling pipelines, GATK, and SAMtools (DePristo et al. 2011; Li et al. 2009) find importance in detection of SNPs and INDELs and their association with phenotypes of interest. Presence or absence of genes occurs due to loss or gain of genes which may occur due to transposons, insertion sequences (IS) (Fitch 1970), infection by bacteriophages or viruses leading to plasmid acquisition or integrative and conjugative elements (ICEs) (Sobecky and Hazen 2009; Partridge et al. 2018; Langille et al. 2010) often helping the cells to survive harsh and adverse conditions or to confer pathogenicity (Schmidt and Hensel 2004). Copy number variations (CNVs) and sequence inversions (SIs) can arise due to transposons, genome deletions, or duplications and are known contribute to adaptation and microbial phenotypic variation (Kirkpatrick and Barton 2006) and also may lead to speciation.

With the advanced machine learning applied in the domain of genomics, and prediction of phenotypes from genotypes, the microbiology based GWAS tools, based on statistics and phylogenetic methods includes now categories from traditional studies including (1) phylogeny, (2) non-phylogeny (3) hybrid tools and (4) machine learning.

However performance of machine learning tools is proportional to the availability of more whole genomic sequences. CCTSWEEP and VENN (Habib et al. 2007) tools are reported to correlate and find the statistical significance between single nucleotide polymorphisms (SNPs) through phylogenetic and they are built to take into account missing data not used by other methods. They have been applied to study correlation of SNPs with that of resistance to *Bacillus anthracis* (*B anthracis*) causative agent of infectious disease Anthrax in inbred mouse strains (Habib et al. 2007).

GWAMAR a recently developed tool finds importance in detection of drug resistance-associated mutations in bacteria (Wozniak et al. 2014). Sequence element enrichment (SEER) analysis tool is based on a *k*-mer and together with string-mining algorithm (Lees et al. 2016) has been applied to understand resistance of

Streptococcus pneumonia (S. pneumonia) and *Streptococcus pyogenes (S. pyrogenes)* to antibiotics (Aun et al. 2018).

Bugwas is bacterial GWAS tool with end-to-end pipeline implemented in R, Python, and C++ (Earle et al. 2016), performing SNP, *k*-mer, and gene differential analysis.

PhenotypeSeeker is a machine learning tool based on k-mer generated statistical model that can predict host-phenotypes from bacterial genotype data (Aun et al. 2018). *Kover* based on k-mer approaches has been used to identify biomarkers (Earle et al. 2016; Drouin et al. 2015; Drouin et al. 2016). PySEER involves application of Python in SEER with several enhancements (Lees et al. 2018) and is applied in testing association of SNPs and indels against a reference genome followed by prediction by use of machine learning. MAGNAMWAR is based on R package and has been applied to understand genotype to phenotype relationships in bacterial orthologous genes (Sexton et al. 2018) and identification of bacterial genes in relation with variation in fruitfly *Drosophila melanogaster* (*D melanogaster*) phenotypes (White et al. 2018).

23.3 k-Nearest-Neighbors (kNN) and Application in Genomic Based Studies

Generation of large volumes of data is a major challenge in analyzing epistatic interactions in large studies of GWAS. k-nearest neighbors (kNN), method, being a supervised machine learning algorithm, has been used for classification and regression and multidimensional Reduction (MDR) approaches of reduction of dimensionality of data from the high-dimensional to fewer dimensions, with reduced number of variables, together forming the (kNN-MDR) method. Using this method, it has been possible to study genetic interactions and analysis of complex characters (Abo Alchamlat and Farnir 2017).

A kNN based approach with noise-reduction filters has been used to analyze microarray expression data, to find genes involved in osteogenic pathway from studies in immortalized mouse cell line derived from mouse skeletal myoblasts C2C12 (Theilhaber et al. 2002).

Pancreatic cancer suffers from the challenges of lack of effective biomarkers in disease diagnosis, early diagnosis and prognosis. Therefore search for main genes in pancreatic cancer finds importance as biomarkers in cancer diagnosis and therapy. Using kNN and random forest (RF) algorithms based analysis of GSE15471 expression profiles from Gene Expression Omnibus (GEO) database, two hub genes, MMP7 and ITGA2, has been identified that finds importance in diagnosis and prognosis of pancreatic cancer (Li et al. 2018).

Breast cancer is the most commonly occurring cancer in women, affecting 2.1 million women globally, leading to highest recorded number of cancer-related deaths among women, claiming 627,000 lives of women globally in 2018 and the increase in the number of incidences reported across the globe mostly from the more developed regions is alarming (https://www.who.int/cancer/prevention/

diagnosis-screening/breast-cancer/en/). Thus the study of diagnosis, prognosis, and years of survival in breast cancer patients finds importance. Using kNN and dimensional reduction methods (DRM), it has been possible to predict 5-year survival in patients with breast cancer (Zhao et al. 2018).

23.4 Discussions

Machine leaning application to genomics is undoubtedly aiding to the understanding of the scientists in the domain of genomics, metagenomics, and disease biology. It is particularly a boon in search for key genes involved in diagnosis and prognosis of deadly disease like cancer from genomics based approaches.

Although there have been considerable advances in the study of microbial GWAS, tools that are user-friendly and freely available and easy to use still are required that calls in for more research in this domain of biology (San et al. 2020).

References

Abo Alchamlat S, Farnir F (2017) KNN-MDR: a learning approach for improving interactions mapping performances in genome wide association studies. BMC Bioinformatics 18(1):184

Aun E, Brauer A, Kisand V, Tenson T, Remm M (2018) A k-mer-based method for the identification of phenotype-associated genomic biomarkers and predicting phenotypes of sequenced bacteria. PLoS Comput Biol 14:e1006434

DePristo MA, Banks E, Poplin R, Garimella KV, Maguire JR, Hartl C et al (2011) A framework for variation discovery and genotyping using next-generation DNA sequencing data. Nat Genet 43: 491–498

Drouin A, Gigù Ere S, Déraspe M, Marchand M, Tyers M, Loo VG et al (2015) Predictive computational phenotyping and biomarker discovery using reference-free genome comparisons. BMC Genomics 17:754. https://doi.org/10.1101/045153

Drouin A, Raymond F, St-Pierre GL, Marchand M, Corbeil J, Laviolette F (2016) Large scale modeling of antimicrobial resistance with interpretable classifiers. arXiv [Preprint]. http://arxiv.org/abs/1612.01030

Dutilh BE, Backus L, Edwards RA, Wels M, Bayjanov JR, van Hijum SAFT (2013) Explaining microbial phenotypes on a genomic scale: GWAS for microbes. Brief Funct Genomics 12:366–380

Earle SG, Wu CH, Charlesworth J, Stoesser N, Gordon NC, Walker TM et al (2016) Identifying lineage effects when controlling for population structure improves power in bacterial association studies. Nat Microbiol 1:1–21

Fitch WM (1970) Distinguishing homologous from analogous proteins. Syst Zool 19:99–113

Habib F, Johnson AD, Bundschuh R, Janies D (2007) Large scale genotype phenotype correlation analysis based on phylogenetic trees. Bioinformatics 23:785–788

Kirkpatrick M, Barton N (2006) Chromosome inversions, local adaptation and speciation. Genetics 173:419–434

Langille MGI, Hsiao WWL, Brinkman FSL (2010) Detecting genomic islands using bioinformatics approaches. Nat Rev Microbiol 8:373–382

Lees JA, Vehkala M, Välimäki N, Harris SR, Chewapreecha C, Croucher NJ et al (2016) Sequence element enrichment analysis to determine the genetic basis of bacterial phenotypes. Nat Commun 7:12797

Lees JA, Galardini M, Bentley SD, Weiser JN, Corander J (2018) Pyseer: a comprehensive tool for microbial pangenome-wide association studies. Bioinformatics 34:4310–4312

Li C, Zeng X, Yu H, Gu Y, Zhang W (2018) Identification of hub genes with diagnostic values in pancreatic cancer by bioinformatics analyses and supervised learning methods. World J Surg Oncol 16(1):223

Li H, Handsaker B, Wysoker A, Fennell T, Ruan J, Homer N et al (2009) The sequence alignment/map format and SAMtools. Bioinformatics 25:2078–2079

Libbrecht MW, Noble WS (2015) Machine learning applications in genetics and genomics. Nat Rev Genet 16(6):321–332

Mills RE, Luttig CT, Larkins CE, Beauchamp A, Tsui C, Pittard WS et al (2006) An initial map of insertion and deletion (INDEL) variation in the human genome. Genome Res 16:1182–1190

Partridge SR, Kwong SM, Firth N, Jensen SO (2018) Mobile genetic elements associated with antimicrobial resistance. Clin Microbiol Rev 31:1–61

Power RA, Parkhill J, De Oliveira T (2016) Microbial genome-wide association studies: lessons from human GWAS. Nat Rev Genet 18:41–50

Read TD, Massey RC (2014) Characterizing the genetic basis of bacterial phenotypes using genome-wide association studies: a new direction for bacteriology. Genome Med 6:109

San JE, Baichoo S, Kanzi A, Moosa Y, Lessells R, Fonseca V, Mogaka J, Power R, de Oliveira T (2020) Current affairs of microbial genome-wide association studies: approaches, bottlenecks and analytical pitfalls. Front Microbiol 10:3119

Schmidt H, Hensel M (2004) Pathogenicity islands in bacterial pathogenesis. Clin Microbiol Rev 17:14–56

Sexton CE, Smith HZ, Newell PD, Douglas AE, Chaston JM (2018) MAGNAMWAR: an R package for genome-wide association studies of bacterial orthologs. Bioinformatics 34:1951–1952

Sobecky PA, Hazen TH (2009) Horizontal gene transfer and mobile genetic elements in marine systems. Methods Mol Biol 532:435–453

Theilhaber J, Connolly T, Roman-Roman S et al (2002) Finding genes in the C2C12 osteogenic pathway by k-nearest-neighbor classification of expression data. Genome Res 12(1):165–176

Weber JL, David D, Heil J, Fan Y, Zhao C, Marth G (2002) Human diallelic insertion/deletion polymorphisms. Am J Hum Genet 71:854–862

White KM, Matthews MK, Hughes RC et al (2018) A metagenome-wide association study and arrayed mutant library confirm *Acetobacter* lipopolysaccharide genes are necessary for association with *Drosophila melanogaster*. G3 (Bethesda) 8(4):1119–1127

Wozniak M, Tiuryn J, Wong L (2014) GWAMAR: genome-wide assessment of mutations associated with drug resistance in bacteria. BMC Genomics 15:S10

Zhao M, Tang Y, Kim H, Hasegawa K (2018) Machine learning with K-means dimensional reduction for predicting survival outcomes in patients with breast cancer. Cancer Inform 17: 1176935118810215

Cell Fate Analysis and Machine Learning 24

Abstract

One of the most intriguing and fundamental questions in developmental biology is in the domain of understanding of cell fate, molecular mechanisms, regulatory factors, signaling molecules, key molecules and networks involved in the process of decision-making, and cell fate determination and differentiation. It is also intriguing to know the phenomena of differentiation and dedifferentiation and the molecular mechanisms involved. Recently high-throughput based technologies, omics based approaches and advances in line cell imaging and computational tools and machine learning algorithms which is a developing domain with applications in developmental biology have enabled study of cells at a single-cell level and could be a step towards understanding of cell fate choices and also in understanding the aberrant phenotype and behavior of cell at the initiation of biology of disease like cancer. We discuss in this chapter the application of machine learning in the study of cell fate.

Keywords

Cell fate · Developmental biology · Drosophila

Abbreviation

C elegans	*Caenorhabditis elegans*
D melanogaster	*Drosophila melanogaster*
D rerio	*Danio rerio*
HSC	Hematopoietic stem cells
KO	Knock-outs

S. Ghosh, R. Dasgupta, *Machine Learning in Biological Sciences*,
https://doi.org/10.1007/978-981-16-8881-2_24

M. musculus	*Mus musculus*
mESC	Mouse embryonic stem cell
SVM	Support vector machine

24.1 Introduction

In the field of developmental biology, an intriguing question to the human mind and that is unresolved in biology is the development of a cell from the parent cell and is studied under the domain of cell fate determination.

Development of an embryo, to a complete organism includes complex biological events of direction and number of cell division, cell cycle, cell growth and differentiation, signaling, cell shape changes, migration, changes in the composition of cell membrane or secreted products and programmed cell death or apoptosis. Signaling reactions are vital to such development and each cell receives signals by proteins, RNA, surface interactions from neighboring cells.

The process of embryo development is conserved and called as embryogenesis. Embryogenesis involves six vital processes:

1. Fertilization, involving fusion of the mature egg and sperm cell or gametes, leading to initiation of development, the fusion of the gamete nuclei gives rise to the embryo genome that provides instructions to the embryo to develop into the adult.
2. Cleavage is a post-fertilization event with extremely fast mitotic divisions that causes the large zygote cytoplasm to divide into smaller cells called as blastomeres. As cleavage comes to end, blastomeres form a sphere called as blastula.
3. Gastrulation: Post-cleavage, the cell division rate lowers, blastomeres rearrangement occurs followed by series of extensive cell rearrangements called as gastrulation leading to the formation of gastrula and the embryo now contains three germ layers including ectoderm which leads to generation of outer layer of embryo, producing epidermis, surface layer of skin, brain and nervous system, endoderm giving rise to innermost layer of embryo, forming epithelium of gastrointestinal tract, lungs and mesoderm lying between the ectoderm and endoderm, giving rise to blood, heart, kidney, gonads, muscles, and connective tissues.
4. Organogenesis occurs postformation of germ layers and involves the interaction and rearrangement of cells to give rise to tissues and organs. Chemical signals between cells of germ layer lead to the formation of specific organs at designated sites. Cells including precursors of blood cells, lymph cell, gametes, pigments migrate long distances from their site of origin to their destination location.

5. Metamorphosis occurs in many organisms who develop from the egg to become sexually mature adult. The young stage or larva may have different morphology as compared to the adult.
6. Gametogenesis occurs whereby in many species, a group of cells called the precursors of the gametes are set aside for their future role in reproduction and production of the next generation. Together with the gametes, these precursors are called as germ cells. The separation of germ cells from somatic cells or cells that give rise to individual's body is the first differentiation occurring during animal development. The germ cells migrate to the gonads where they differentiate and develop into gametes, by a process called as gametogenesis. On maturity these cells play role in fertilization process to form a new embryo.
   Considerable progress has been made in the last two and half decades revealing molecules, proteins, and mRNAs involved in embryogenesis.

   The events of embryogenesis and developmental biology are considerable studied in model organisms including fruitfly (*Drosophila melanogaster*, *D melanogaster*), mouse (*Mus musculus*), *Caenorhabditis elegans* (*C elegans*), zebrafish (*Danio rerio*) that finds importance in our understanding of development in other animals including humans.

   It is intriguing to know the differential expression of RNAs and proteins by different cells, temporally and spatially, and their roles in cell fate determination leading to the diversity in organisms.

24.2 Cell Fate and Fate Maps

The process of development is so dynamic that tracing of the cell lineages is an intriguing problem in developmental biology research. It is very difficult to trace the development of a cell from its embryonic stage to the adult stage. Staining cells in experimental setup has enabled to understand the later area of development. This is called the fate maps. The fate of a cell determines its future, or its daughter cells, prior to its phenotypic detection after differentiation or division. Endogenous developmental factors, interaction with adjacent cells, or signals by morphogens and hormones control cell fate. Wet lab techniques involving application of green fluorescent protein (GFP), with advances in fluorescence microscopy, injection of fluorescein-conjugated dextran in single cell of an early embryo and tracing the embryo under UV light, cell lineage mapping has been possible that enable study of cells while they differentiate into different lineages with specific function. Application of molecular biology tools of gene and protein knock-outs (KO) or their overexpression enables detection of mechanisms involved in cell fate determination. Live cell imaging, confocal and super-resolution microscopy have enabled study of molecular changes in experimental setups. Transplantation of embryonic tissues from one species of newt to other newt species leading to the formation of chimeric embryos was done as early as in 1920 by German embryologist Hilde Mangold and Hans Spemann (Gilbert 2000; Spemann 2001). Later techniques ofgenetic manipulation, lineage tracing, inducible, Cre/lox transgenic mice where the Cre/lox system

(Kratochwil and Rijli 2014; Kim et al. 2018) is a genetic engineering tool involving the site-specific recombinase technology, has been applied to perform deletions, insertions, translocations, and inversions at specific sites in the DNA of cells is used to make knockout (KO) alleles, used to activate gene expression and has enabled study of cell fate and linage commitment. Cre is the recombinase, and loxP is the recognition site and these have been adapted from the P1 bacteriophage.

A cell fate of a cell can be seen in two states: (1) a cell is committed or specified and (2) a cell is determined. In the former state, the cell fate can be reversed to another fate by manipulation but in the latter state the cell's fate is irreversible. Thus a cell determined to differentiate into a nerve cell cannot be reversed to a blood cell. Post determination, differentiation events occur leading to altered biochemistry, structure, and function resulting in cell types with specificity. Under normal conditions, a cell with a specified fate undergoing cellular differentiation in specific direction is irreversible to undifferentiated or less specified states. However recently in aberrant malignant states of cancer, dedifferentiation has been reported.

Understanding of complexity and plasticity of cell fate and the molecular biochemical signaling events that play role in regulating cells with same genetic information with different cell is an intriguing unresolved problem in biology. Cell differentiation, into different lineages with unique and specialized functions, is associated with altered gene expression, chromatin characteristics, and nuclear architecture.

Recent reports on reprogramming of cell fate are further intriguing to understand factors controlling cell fate decisions, multiple different fates, degree of mutability and reversibility of cell fates.

It is further interesting to know that misregulation of cell fate has been associated with the pathophysiology of origin of different diseases including cancer (Buttitta 2016).

24.3 Cell Fate and Machine Learning

Live cell imaging and cell tracing and multifactorial classification by machine learning based tools have been used to identify, locate a cell, and aid in understanding of organ regeneration in studies conducted on neuromasts of larval zebrafish, controlled by a self-regulatory mechanism for regeneration (Viader-Llargués et al. 2018). Machine learning and imaging based methods have been instrumental in understanding of single-cell dynamics and interactions and competition and such application finds importance in understanding of complex process in decision-making during choices of cell fate by cells (Gradeci et al. 2020).

Machine learning applications on analysis of data from array-based chromatin immunoprecipitation (ChIP) and transcription factor (TF) sequence motifs have been used to identify the two distinct types of cells from Drosophila heart including the contractile cardiac cells (CCs) and the surrounding non-muscle pericardial cells (PCs, Ahmad et al. 2014).

Developments in the field developmental biology have improved in study of single cell based on its identity: phenotype, lineage, and state and finds importance in understanding and manipulation of cell fate (Morris 2019). Using a Bayesian network machine learning approach on high-throughput protein function, it has been possible to predict network involved in molecular programming of mouse embryonic stem cell (mESC, Dowell et al. 2013).

Studies on hematopoietic stem cells (HSC) have revealed that application of machine learning algorithms has enabled our understanding of variability and stem cell fate and the differentiation potential of stem cells (Quesenberry et al. 2019). Application of support vector machine (SVM) classifier on live cell imaging data has enabled prediction of cell fates of individual lymphocytes together with understanding of the cells spatiotemporal dynamics and lineage information (Chakravorty et al. 2014).

24.4 Discussion

Computational tools, bioinformatics, live cell imaging, high-throughput sequencing approaches, omics based studies of transcriptomics, metabolomics, genomics, epigenetics and advances of machine learning, tools like feature extraction, algorithms, segmentation, image enhancement to the domain of developmental biology have enabled the study of cell at a single level. This has also enabled our better understanding of single cell dynamics, molecular mechanism behind complex phenomena of cell fate and factors involved in their regulation and also the biology behind aberrant cell phenotypes (Yalcin et al. 2016).

However major challenges in understanding the stem cell fates and differentiation are complex events of decision-making, phenomena and regulation involved in steps of differentiation and dedifferentiation. It remains for the ongoing and future researches and researchers working in this domain to develop the applications of machine learning in the domain of cell biology and developmental biology in understanding the fundamental question of cell fate and unraveling the initiation of disease biology like in cancer.

References

Ahmad SM, Busser BW, Huang D, Cozart EJ, Michaud S, Zhu X, Jeffries N, Aboukhalil A, Bulyk ML, Ovcharenko I, Michelson AM (2014) Machine learning classification of cell-specific cardiac enhancers uncovers developmental subnetworks regulating progenitor cell division and cell fate specification. Development 141(4):878–888

Buttitta L (2016) Editorial: cell fate. Front Genet 6:363

Chakravorty R, Rawlinson D, Zhang A, Markham J, Mark R, Dowling MR, Cameron Wellard C, Jie HS, Zhou J HS, Hodgkin PD (2014) Labour-efficient in vitro lymphocyte population tracking and fate prediction using automation and manual review. PLoS One 9(1):e83251

Dowell KG, Simons AK, Wang ZZ, Yun K, Hibbs MA (2013) Cell-type-specific predictive network yields novel insights into mouse embryonic stem cell self-renewal and cell fate. PLoS One 8(2):e56810

Gilbert SF (2000) Developmental biology, 6th edn. Sinauer Associates, Sunderland. Available from: https://www.ncbi.nlm.nih.gov/books/NBK9983/

Gradeci D, Bove A, Charras G, Lowe AR, Banerjee S (2020) Single-cell approaches to cell competition: High-throughput imaging, machine learning and simulations. Semin Cancer Biol 63:60–68

Kim H, Kim M, Im SK, Fang S (2018) Mouse Cre-LoxP system: general principles to determine tissue-specific roles of target genes. Lab Anim Res 34(4):147–159

Kratochwil CF, Rijli FM (2014) The Cre/lox system to assess the development of the mouse brain. Methods Mol Biol 1082:295–313

Morris SA (2019) The evolving concept of cell identity in the single cell era. Development 146(12): dev169748

Quesenberry P, Borgovan T, Nwizu C, Mark Dooner M, Goldberg L (2019) Heuristic bias in stem cell biology. Stem Cell Res Ther 10(1):241

Spemann H (2001) Induction of embryonic primordia by implantation of organizers from a different species. Int J Dev Biol 45(1):13–38

Viader-Llargués O, Lupperger V, Pola-Morell L, Marr C, López-Schier H (2018) Live cell-lineage tracing and machine learning reveal patterns of organ regeneration. elife 29(7):e30823

Yalcin D, Hakguder ZM, Otu HH (2016) Bioinformatics approaches to single-cell analysis in developmental biology. Mol Hum Reprod 22(3):182–192. https://doi.org/10.1093/molehr/gav050. Epub 2015 Sep 10

Study of Biomarker and Machine Learning 25

Abstract

Diagnostic and prognostic markers for identification of disease and the effect of therapy on the disease still remain a major bottleneck in disease biology and hence due to lack of suitable markers, in early diagnosis and prognosis of the disease, diagnosis is late or at advanced stage of the disease or it is difficult to understand a recurrent or relapse, delays immediate therapy and many lives are lost from myriads of diseases. Thus search for suitable and effective biomarkers for a disease is a major focus of research in the biology of disease. Recently machine learning has helped in the search of disease specific biomarkers. In this chapter we highlight the important applications of machine learning in study of biomarkers, although at it its infancy, it is promising tool in future day biomarker research.

Keywords

CKD · Sleep disorders · Biomarkers · Cancer

Abbreviations

AD	Alzheimer's disease
CKD	Chronic kidney disease
CT	Computed tomography
IE	Infective endocarditis
miRNA	microRNA
MRI	Magnetic resonance imaging
ncRNA	Noncoding RNA
PSG	Polysomnography
SDB	Sleep disordered breathing

S. Ghosh, R. Dasgupta, *Machine Learning in Biological Sciences*,
https://doi.org/10.1007/978-981-16-8881-2_25

sncRNAs	Small noncoding RNAs
UPLC-HDMS	Ultraperformance liquid chromatography together with high-definition mass spectrometry

25.1 Introduction

Biomarkers aid in early diagnosis, screening, and prognosis of disease. During disease condition, alterations in cellular, biochemical, molecular, or physiological changes sometimes lead to aberrant expression of molecules in the body components tissue or body fluid. These are known as disease markers that aid in identification, classification of a disease and associated risk factors, and enable our understanding of disease biology and pathogenesis. Same or different biomarkers can reveal disease condition from early to the end-terminal stages. Biomarkers find importance in identifying the stage of the disease and help in clinical investigation. Some biomarkers find importance as drug targets. Thus accessing the expression of biomarkers and its monitoring of its status in pharmacological responses to therapy is vital to disease diagnosis and prognosis. Although there are several wet lab genomic, metabolomic, and proteomic techniques in the detection of biomarkers, there are several challenges to the study of biomarkers, like validation of these markers which is costly and challenging, and thereby their successful application to clinical practice is a lengthy, time-consuming process. Besides in addition to clinical validation, understanding the role of the biomarker in the pathophysiology of the disease and the molecular pathways that they activate or downregulate or affect is most important to understand the molecular basis of the disease. In this chapter we discuss the (1) biomarkers and their importance in disease biology and (2) recent advances and application of machine learning in biomarker study.

25.2 Biomarkers and their Importance in Disease Biology

The National Cancer Institute (NCI) has defined a biomarker as "*a biological molecule found in blood, other body fluids, or tissues that is a sign of a normal or abnormal process, or of a condition or disease*."

The biomarker is a factor or biomolecule with huge variety ranging from proteins like enzyme like proteinases or receptor, nucleic acids like microRNA (miRNA), noncoding RNA (nc-RNA), antibodies, peptides, sugars like glycoconjugates, sialic acids and their modifications, lipids, small metabolites, cytogenetic, cytokinetic parameters collection of total alterations in gene expression, proteomic, and metabolomic profiles that could have been generated from germline, somatic mutations, transcriptional changes, and post-translational modifications and they enable to differentiate between the healthy from a diseased individual (Mayeux 2004; Henry and Hayes 2012; Saalberg and Wolff 2016; Bhatt et al. 2010;

Srivastava and Creek 2019; Hayes et al. 2014; Ghosh 2015; Ghosh et al. 2007; Ghosh et al. 2005a, b, c; Pal et al. 2004a, b; Hakomori 1986; Li et al. 2017).

These biomarkers can be detected noninvasively like expressed on cell surface, altered T cell markers as diagnostic and prognostic markers in T cell leukemia, expressed in the body fluids like saliva, whole blood, serum, or plasma, volatile organic compounds in breath in lung cancer, excretions or secretions like their presence in stool, urine, sputum, or nipple discharge, tissue-derived, derived from bone marrow, detected biopsy or special imaging techniques like CT-scan, radiology (Mayeux 2004; Henry and Hayes 2012; Saalberg and Wolff 2016; Bhatt et al. 2010; Srivastava and Creek 2019; Hayes et al. 2014; Ghosh 2015; Ghosh et al. 2007; Ghosh et al. 2005a, b, c; Pal et al. 2004a, b; Hakomori 1986; Li et al. 2017).

Over the last two decades, together with the application of conventional techniques of immunology, histopathology, clinical biochemistry, molecular biology, advances in the omics based high-throughput technologies, considerable progress has been made in the methods used in disease diagnosis leading to the discovery of novel disease biomarkers in a plethora of diseases including different cancer, autoimmune, gastrointestinal, cardiovascular, endocrine, genetic and neural disorders.

25.3 Machine Learning Applications in Biomarkers

Radiomics is an emerging field of research and study in the domain of medical study. Machine learning applied to the domain of radiology involving high-throughput automated extraction of features from medical and clinical images by algorithms enables study of features in tumor. Machine learning has been able to extract information from computed tomography (CT) scan images of lung cancer patients and finds importance in search of clinical radiomic biomarkers, by a noninvasive method, thereby helping in characterizing and quantifying monitoring tumor-phenotype (Parmar et al. 2015).

Machine learning applications have been used to image biomarkers in diagnosis, prognosis, and monitoring of therapeutic responses in neuro-oncological disorders including high-grade glioma from magnetic resonance imaging (MRI, Booth et al. 2020).

Attempts are being made by scientists to utilize artificial intelligence and machines to measure pain by detecting pain biomarkers like electroencephalography (EEG) for pain prediction (Levitt and Saab 2019).

Alzheimer's disease (AD) is a progressive neurodegenerative disease in which brain cells undergo degeneration and die and beta-amyloid plaques form in the brain. It causes dementia or continuous loss in thinking ability, behavioral and social skills and prevents an individual's independently function. The disease suffers from the challenges of detection, and conventional modes of detection remain in neuropathologic examination of the brain. Therefore alternate blood based biomarkers are being searched for. Blood-borne small noncoding RNAs (sncRNAs) find importance as biomarkers in different biochemical, pathological diagnostic tests. In neurodegenerative diseases like Alzheimer's disease (AD), attempts are being made to identify biomarkers. microRNA (miRNA) biomarkers and application of machine learning based application are finding application as disease biomarker in AD correlating with neurodegeneration (Ludwig et al. 2019). Machine learning based approaches of supervised, unsupervised, and probabilistic techniques are being applied on imaging data from AD patients to find out biomarkers in AD (Mirzaei et al. 2016).

Chronic kidney disease (CKD) is a disease involving loss of kidney function and eventual renal failure leading to building up of waste in the body. The chief challenges in this disease lie in their early prediction and monitoring of progression from early to end-stage renal disease and therefore search of biomarkers find utmost important domain of study in this domain. Machine learning tools applied to data from experiments on ultraperformance liquid chromatography together with high-definition mass spectrometry (UPLC-HDMS)-based metabolomics performed on serum samples of CKD patients at different stages of the disease have enabled establishing metabolite biomarkers in diagnosis of the disease as compared to normal individuals (Guo et al. 2019).

Sleep disorders involve altered sleep patterns or habits affecting negatively the health. Patients with sleep disordered breathing (SDB) reveal altered breathing sounds during sleep. Characterizing these altered sounds by different acoustic specificities, detected by snoring or beating sound patterns by individuals during sleep by polysomnography (PSG) recorded in a microphone together with application of machine learning tools and feature extraction tools have led to the identification of acoustic biomarkers indicating of the severity of SDB (Kim et al. 2018).

Infective endocarditis (IE) is the cardiac disease, involving inflammation of the inner lining of the heart. Infection in heart by bacteria can cause bacterial endocarditis, claiming lives across the globe. The detection of the disease suffers from the challenges of appropriate search of biomarkers in early diagnosis and prognosis and associated risk of mortality of the disease. Machine learning tools have been able to designate cytokines, chemokines, and growth factors in diagnosis of IE and predict mortality (Ris et al. 2019).

Breast cancer occurs in women and is the large killer of human lives. Machine learning applications when applied on microRNAs (miRNA) expression data in breast cancer have led to identification of differentially expressed miRNAs as biomarkers in this cancer (Sherafatian 2018).

25.4 Discussions

Several diseases lack appropriate and early detection and treatment and prognosis of treatment due to the lack of suitable markers. Although attempts are being made by scientist by wet lab, radiological, scanning methods, several diseases like AD, CKD, cancer suffers, sleep disorders suffer from early diagnosis and thereby affect the lives of humans negatively. Machine learning is a big boon in this domain enabling the study of markers in disease detection, diagnosis, and prognosis. More applications of machine learning in diverse and deadly diseases like cancer, infectious viral diseases, inflammation need to be carried out and remain future scope of the study.

References

Bhatt AN, Mathur R, Farooque A, Verma A, Dwarakanath BS (2010) Cancer biomarkers - current perspectives. Indian J Med Res 132:129–149

Booth TC, Williams M, Luis A, Cardoso J, Ashkan K, Shuaib H (2020) Machine learning and glioma imaging biomarkers. Clin Radiol 75(1):20–32

Ghosh S (2015) Sialic acids: biomarkers in endocrinal cancers. Glycoconj J 32(3-4):79–85

Ghosh S, Bandyopadhyay S, Mallick A, Pal S, Vlasak R, Bhattacharya DK, Mandal C (2005a) Interferon gamma promotes survival of lymphoblasts overexpressing 9-O-acetylated sialoglycoconjugates in childhood acute lymphoblastic leukaemia (ALL). J Cell Biochem 95(1):206–216

Ghosh S, Bandyopadhyay S, Pal S, Das B, Bhattacharya DK, Mandal C (2005b) Increased interferon gamma production by peripheral blood mononuclear cells in response to stimulation of overexpressed disease-specific 9-O-acetylated sialoglycoconjugates in children suffering from acute lymphoblastic leukaemia. Br J Haematol 128(1):35–41

Ghosh S, Bandyopadhyay S, Bhattacharya DK, Mandal C (2005c) Altered erythrocyte membrane characteristics during anemia in childhood acute lymphoblastic leukemia. Ann Hematol 84(2): 76–84

Ghosh S, Bandyopadhyay S, Mukherjee K et al (2007) O-acetylation of sialic acids is required for the survival of lymphoblasts in childhood acute lymphoblastic leukemia (ALL). Glycoconj J 24(1):17–24

Guo Y, Yu H, Chen D, Zhao YY (2019) Machine learning distilled metabolite biomarkers for early stage renal injury. Metabolomics 16(1):4

Hakomori S (1986) Tumor-associated glycolipid markers: possible targets for drug and Immunotoxin delivery. In: Gregoriadis G, Senior J, Poste G (eds) Targeting of drugs with synthetic systems, NATO ASI series (series a: life sciences), vol 113. Springer, Boston

Hayes J, Peruzzi PP, Lawler S (2014) MicroRNAs in cancer: biomarkers, functions and therapy. Trends Mol Med 20(8):460–469

Henry NL, Hayes DF (2012) Cancer biomarkers. Mol Oncol 6. https://doi.org/10.1016/j.molonc.2012.01.010

Kim T, Kim JW, Lee K (2018) Detection of sleep disordered breathing severity using acoustic biomarker and machine learning techniques. Biomed Eng Online 17(1):16

Levitt J, Saab CY (2019) What does a pain 'biomarker' mean and can a machine be taught to measure pain? Neurosci Lett 702:40–43

Li G, Li L, Joo EJ et al (2017) Glycosaminoglycans and glycolipids as potential biomarkers in lung cancer. Glycoconj J 34:661–669

Ludwig N, Fehlmann T, Kern F et al (2019) Machine learning to detect Alzheimer's disease from circulating non-coding RNAs. Genomics Proteomics Bioinformatics 17(4):430–440

Mayeux R (2004) Biomarkers: potential uses and limitations. NeuroRx 1(2):182–188

Mirzaei G, Adeli A, Adeli H (2016) Imaging and machine learning techniques for diagnosis of Alzheimer's disease. Rev Neurosci 27(8):857–870

Pal S, Ghosh S, Bandyopadhyay S, Mandal C, Bandhyopadhyay S, Kumar Bhattacharya D, Mandal C (2004a) Differential expression of 9-O-acetylated sialoglycoconjugates on leukemic blasts: a potential tool for long-term monitoring of children with acute lymphoblastic leukemia. Int J Cancer 111(2):270–277

Pal S, Ghosh S, Mandal C, Kohla G, Brossmer R, Isecke R, Merling A, Schauer R, Schwartz-Albiez R, Bhattacharya DK, Mandal C (2004b) Purification and characterization of 9-O-acetylated sialoglycoproteins from leukemic cells and their potential as immunological tool for monitoring childhood acute lymphoblastic leukemia. Glycobiology 14(10):859–870

Parmar C, Grossmann P, Bussink J, Lambin P, Aerts HJWL (2015) Machine learning methods for quantitative Radiomic biomarkers. Sci Rep 5:13087. Published 2015 Aug 17. https://doi.org/10.1038/srep13087

Ris T, Teixeira-Carvalho A, Coelho RMP et al (2019) Inflammatory biomarkers in infective endocarditis: machine learning to predict mortality. Clin Exp Immunol 196(3):374–382

Saalberg Y, Wolff M (2016) VOC breath biomarkers in lung cancer. Clin Chim Acta 459:5–9

Sherafatian M (2018) Tree-based machine learning algorithms identified minimal set of miRNA biomarkers for breast cancer diagnosis and molecular subtyping. Gene 677:111–118

Srivastava A, Creek DJ (2019) Discovery and validation of clinical biomarkers of cancer: a review combining metabolomics and proteomics. Proteomics 19(10):e1700448

26 Animal Behavior: An Introduction

Abstract

Different life forms exhibit differences in their behavior ranging from the feeding, mating, locomotion, singing, social behavior. While factors including circadian rhythms, role of nervous system, process of learning, hormones, together with the triggering role played by environmental changes regulate and control animal behavior. While the most conventional mode of study of animal behavior lies in the recording of observations by the scientists, experiments in closed environment have been instrumental in the study and understanding of the animal behavior. However conventional modes of experimentation in controlled laboratory setup may reveal differences with that of observation in the animal behavior in nature. Therefore computational methods of imaging and analysis offer advantage in study and analysis of animal behavior. We highlight in this chapter the different animal behavior and factors involved, conventional modes of study and bottlenecks and importance of computational studies of animal behavior.

Keywords

Animal behavior · Ethology · Courtship

Abbreviations

ADHD	Attention deficit hyperactivity disorder
C pallida	*Centris pallida*
D melanogaster	*Drosophila melanogaster*

S. Ghosh, R. Dasgupta, *Machine Learning in Biological Sciences*,
https://doi.org/10.1007/978-981-16-8881-2_26

26.1 Introduction

A behavior involves a sequence of movements exhibited by an animal or person towards a specified goal like navigating to a burrow, escaping a predator, courtship behavior, mating strategies winning over rival, reproductive benefits from bird song, feeding behavior, habitat selection, migration, territoriality, parental care, eu-social behavior. Behaviors may be stereotyped or different, may vary in duration or magnitude, and can be composed of smaller behaviors. The study of the hugh repertoire expressions of animal behavior, ranging in its variety and diversity, was not only intriguing observations to the curious naturalists, but finds importance in understanding the underlying reason behind a particular behavior exhibited by an animal.

We discuss in this chapter the range and diversity of different animal behavior, traditional ways of study of animal behavior.

26.2 Diversity of Animal Behavior and the Factors Involved

26.2.1 Movement/Locomotion

Animal movement is intricately associated with animal behavior. Time and space are essential parameters in animal ecology and are used to study behavior patterns (Torres et al. 2017).

26.2.2 Courtship

Courtship behavior is a complex behavioral pattern in animals by which different species select their mates or partners for reproduction. Many animal species reveal diverse mate selection courtship behavior like dancing or touching, songs, or beauty display or fighting prowess. In the bee *Centris pallida* (*C pallida*) is found to be regulated by the neurons. Studies in fruitfly *Drosophila melanogaster* (*D. melanogaster)* have revealed the evolution in courtship behaviors by a new chimeric gene (Dai et al. 2008).

26.2.3 Chemical Mimicry

Signaling by species could be deceptive, as manifested in Alcon blue butterfly where the caterpillar emits smell like the ant species which can attract the worker ants, who will carry the caterpillar to the nest and feed and protect the young butterfly like an immature ant. Such a behavior is also observed among the European cuckoo and the North American cowbird where the egg laying females lay eggs in the other birds nest. The young hatchings then using visual and acoustic signals begs for food from the foster parents. Studies in wild on bolas spiders reveal to attract the male moth prey, mimicking the female moth sex hormone odor (Stowe et al. 1987).

26.2.4 Social Behavior

Sequences of movements by a male fruitfly such as orienting, tapping, and singing to a female can be grouped together as courtship behavior of a mate. *Drosophila* has been studied for neuronal circuits involved sexual behavior (Pavlou and Goodwin 2013).

26.2.5 Bat flight

Bats emit ultrasonic sounds to fly at nights. Flying moths when exposed to ultrasounds emitted by flying predatory bats in controlled experimental conditions reveal activated auditory receptors leading to onset of neural events, helping the insect respond to acoustic stimuli (Hedenström and Johansson 2015; Surlykke et al. 2009).

26.2.6 Ethology

Dutch biologist Nikolaas Tinbergen (1907–1988) and Austrian biologists Konrad Lorenz were the proponents of ethology and the work by Tinbergen, Lorenz, and Karl von Frisch (1886–1982), starting in 1930s in this field led to the award of the Nobel Prize in Physiology and Medicine in the year 1973 (The Nobel Prize in Physiology or Medicine 1973). The science of ethology involves observation, description characterization, and estimated measurement of the acts and postures constituting animal behavior (Dixon et al. 1989).

26.2.7 Hormones

Hormones can regulate animal behavior. A male ring courts a female dove by a chasing aggressive behavior which is influenced by the testosterone hormone. Whereas the estrogen hormone facilitates nesting behavior. Three aspects including (1) neuronal behavior enabling communication, (2) biological clocks scheduling the activity, (3) hormonal pathways that track changing physical and social environment. These behavioral priorities have been attributed to the command centers comprising of neurons at a brain center and interconnected such centers which also plays role in decision-making in the praying mantis like search for mate, sunbathing, copulation, flying, attacking a prey with powerful grasping movements. Ovarian hormones have been reported to regulate female sociosexual behavior in female rodents (Takahashi 1990).

26.2.8 Behavioral Schedules and Biological Clock

Biological clocks are essential as they regulate life functioning and organization and coordination of behavior like active and inactive periods, behavior related to day and night cycle in order to increase productivity and minimize risk based on internal clock also called the biological clock. Internal clocks can be set by exogenous or external stimuli or environmental factors or by endogenous or internal rhythm. Some animal activities are done in specific times of the day (Kronfeld-Schor et al. 2013). Like the female *Teleogryllus* crickets are hidden in burrows or under leaf litter during the day times and come out after dusk in search of mates in response to call by male crickets after dusk, indicative of a biological 24 h clock in operation. The biological clocks have been known to play role in controlling the songs of birds, adjusting activities by the monarch butterfly during day time, changing proprieties of activities during day time. This phenomena has been attributed to two probable mechanisms of inbuilt system and/or based on feedback relationships of the command center in response to changes in the environment (Alcock 2013).

In some groups of crickets, juvenile hormones and their cyclical release regulates behavioral activity. A master clock has been summarized to be composed of sensory receptors that are activated in response to changes in sunlight or other environmental conditions that set the clock mechanism and revealed by observed rhythms of locomotion patterns, feeding behaviors, hormone pattern release, and other rhythms (Alcock 2013). The white crowned sparrows reveal a clock mechanism of sensitivity to light between 17–19 h each day (Alcock 2013).

26.2.9 Bird Song

It has been observed by the birders and observers that birds can be identified by their appearance and by their distinct vocalizations and songs like the common yellow-throat and the ovenbird which sings songs different from each other. These species specific songs help in the adaption and attraction of the male by the female of the right species. Like the song of the male ovenbird attracts the female ovenbird in his reproductive interest and helps the female track the right male avoiding males of other species, thereby reducing the risk of producing infertile hybrid offsprings. Limnologist Peter Marler, in the1950s, observed that songs of chaffinches varied with the lakes and inferred that songs within the same species could vary with the different environment like different geographical location of lakes. However, these could also be due to genetic differences between populations like the white crowned sparrows.

However to understand the underlying cause for the observed different variations of songs within a species or dialect and its association with development, experimental analysis was set up. Experiments of Marler of hatching of eggs of white crown sparrows and hand rearing them away from the adult birds revealed that although they could make some noise of twittering after 150 days, they could not sing like the male in the wild. These processes when repeated allowing the birds to listen to the recorded songs of the males, of different dialects, revealed that they could mimic the dialect of the song of the wild bird song in which the laboratory bird was exposed to indicating that the environment had a major role to play in regulating the dialects of the white-crown sparrows. Experiments also revealed that deaf birds could not sing and therefore emphasized that hearing one's own voice enables complete development of singing behavior in development. Experiments also revealed that social experience played a dominant role on the singing behavior in white-crown birds by the study that the young's in course of development could learn the dialect of song listening to an adult whom they could only hear and not see.

Studies in the molecular biology of singing in birds have revealed the activation of the gene ZENK in some part of the brain while listening and in other parts of the brain while singing of the song by birds. A detailed understanding of the process involved includes differences in gene-environment interactions, hormonal differences, leading to differences in song system construction and its operation. This in turn leads to song differences within individuals including song differences within species, different sex, and different dialect which may affect the reproductive success. These characters are then transmitted to the next-generation leading to the differences in gene-environment interactions (Alcock 2013; Slater 2003; Nagui and Riebel 2006; Williams 2004; Waldvogel 2000).

26.2.10 The Behavior of Worker Bee

While the mother queen carries the duty of laying eggs in her lifetime, the worker bees reveal different occupations at different time intervals of their lifespan including cleaning the comb cells, nurse and feed the honey to the larvae, and provides food to the other workers and lastly by foraging for pollen and nectar gathered from outside the hive. This kind of behavior has also been attributed to the gene–environment interaction. The transition from nursing to foraging behavior has recently been implicated to switching on of genes by the molecular biology analysis (Alcock 2013; Darwin 1871).

26.2.11 Predator Animal Behavior

Predator behavior in animals includes chase or pursuit, capture, and killing of the prey or animals for food. Animals can show predatory behaviors solo involving only self-attack or solitary hunting like leopard, or they may exhibit hunting in a group like pack of wolves. The senses of predators are adapted in different ways to enable hunting process. Predation finds importance as a fundamental cause driving animal evolution (Darwin 1871; Ricklefs 1969; Dawkins and Krebs 1979; Quinn and Cresswell 2004). How prey species of animals escape or manage such predation risk is an intriguing domain of study behavioral ecology (Quinn and Cresswell 2004).

26.3 Traditional Ways of Study of Animal Behavior

Traditional ways of study of animal behavior include observation based recording of behavior like focal sampling or simple assays designed to understand the complex behavior (Alcock 2013; Altmann 1974; Arrington 1943) like tube test in mice (Alcock 2013; Lindzey et al. 1961). Recording behavior manually is, however, time-consuming, laborious, and varies under laboratory conditions and therefore human observations are converted into a numeric index attempting to capture some aspect of behavior (Alcock 2013; Michael et al. 1967) but this method also suffers from obscuring the mechanistic detail of underlying structure of a behavior (Anderson and Perona 2014). Human observations also suffer from the limitation of being subjective and can vary within the observers and some important behavior manifestations may be missed by the errors of the human eye. Behavioral assays, however, are easy to implement, fast, and more easily standardized across labs but suffers from obscuring of useful mechanistic details.

26.4 Computational Studies of Animal Behavior

Computational analysis methods have been employed in the study of quantitative estimation of animal behavior. The experimental design has accordingly been done to extract biologically important animal behavioral data with the use of computation and automation. However, for the purpose of the study, while it is easier to study behavior by restricting the environment of the animal for study purpose, but it always suffers from the challenges of missing important behavioral trait or observation as the animal would have exhibited in wild within unrestricted and its natural environment. To solve this problem, suitable algorithms encompassing tracking, feature extraction, automated behavior classification, and automated behavior analysis are being designed to understand and analyze and infer at fruitful animal behavior from experiments (Egnor and Branson 2016).

Attention deficit hyperactivity disorder (ADHD) like behavior is exhibited by dogs revealing altered aggressive behavior, impulsive nature, inattentiveness, affecting dog and the owner sometimes requiring treatment based on reports of the owner. However there is a lack of suitable tools to detect the disorder. Thus computational tool has found important application in detection of dog movement automatically from video footage (Bleuer-Elsner et al. 2019).

Ethology involving the study of casual, phenomenological, ontogenetic, and evolutionary aspects concerned with animal behavior (Waldvogel 2000; Tinbergen 1951; Tinbergen 1963). However the conventional modes of study and analysis of animal behavior based on human observation, suffers from challenges on data acquisition and analysis as it is slow, labor intensive, low in throughput, can include limited number of different experiments performed, restrictions in sample size, statistical power, reliability of results (Button et al. 2013) which may be imprecise and subjective, results may vary with the preciseness of the observer which is subjective and fails to be standardize, inconsistent between different observers thus posing challenges to data consistency and reproducibility, inter and intra laboratories working on the same behavior study. It is also difficult to manually study and record single behavior with different spatial and temporal scales together with human limitations of vision to different between two closely related and similar animals but revealing different behavior. Computational ethology is a promising tool overcoming the bottlenecks in manual study providing scope for automated neurophysiological recordings or optical imaging based study of neuronal activity, unbiased high-throughput study of genes, and neural networks in relation to animal behavior (Brown et al. 2013; Vogelstein et al. 2014).

26.5 Discussions

Machine learning is finding important applications in study of animal behavior, which have enabled to circumvent such aforementioned problems in behavior study and enables observation and quantification of behavior of interest directly and enabling high-quality, high-throughput analysis of behavior, rather than studies based on correlated behavioral test or laborious manual annotations. We discuss in the next chapter in greater detail.

References

Alcock J (2013) Animal behavior, an evolutionary approach, 9th edn. Sinauer Associates, Inc, Publishers, Sunderland

Altmann J (1974) Observational study of behavior: Sampling methods. Behaviour 49(3–4): 227–267

Anderson DJ, Perona P (2014) Toward a science of computational ethology. Neuron 84:1

Arrington RE (1943) Time sampling in studies of social behavior: a critical review of techniques and results with research suggestions. Psychol Bull 40:81–124

Bleuer-Elsner S, Zamansky A, Fux A, Kaplun D, Romanov S, Sinitca A, Masson S, van der Linden D (2019) Computational analysis of movement patterns of dogs with ADHD-like behavior. Animals 9:1140

Brown AE, Yemini EI, Grundy LJ, Jucikas T, Schafer WR (2013) A dictionary of behavioral motifs reveals clusters of genes affecting Caenorhabditis elegans locomotion. Proc Natl Acad Sci U S A 110:791–796

Button KS, Ioannidis JP, Mokrysz C, Nosek BA, Flint J, Robinson ES, Munafò MR (2013) Power failure: why small sample size undermines the reliability of neuroscience. Nat Rev Neurosci 14: 365–376

Dai H, Chen Y, Chen S et al (2008) The evolution of courtship behaviors through the origination of a new gene in Drosophila. Proc Natl Acad Sci U S A 105(21):7478–7483

Darwin C (1871) On the origin of species. Murray, London

Dawkins R, Krebs JR (1979) Arms races between and within species. Proc R Soc Lond Ser B 205: 489–511

Dixon AK, Fisch HU, Huber C, Walser A (1989) Ethological studies in animals and man, their use in psychiatry. Pharmacopsychiatry 22(Suppl 1):44–50

Egnor SER, Branson K (2016) Computational analysis of behavior. Annu Rev Neurosci 39(1): 217–236

Hedenström A, Johansson LC (2015) Bat flight: aerodynamics, kinematics and flight morphology. J Exp Biol 218(Pt 5):653–663

Kronfeld-Schor N, Bloch G, Schwartz WJ (2013) Animal clocks: when science meets nature. Proc Biol Sci 280(1765):20131354

Lindzey G, Winston H, Manosevitz M (1961) Social dominance in inbred mouse strains. Nature 191:474–476

Michael RP, Saayman GS, Zumpe D (1967) Sexual Attractiveness and Receptivity in Rhesus Monkeys. Nature 215:554

Nagui M, Riebel K (2006) Birdsong: a key model in animal communication. In: Encyclopedia of language & linguistics, 2nd edn. Elsevier, Amsterdam, pp 40–53

Pavlou HJ, Goodwin SF (2013) Courtship behavior in Drosophila melanogaster: towards a 'courtship connectome'. Curr Opin Neurobiol 23(1):76–83

Quinn JL, Cresswell W (2004) Predator hunting behaviour and prey vulnerability. J Anim Ecol 73: 143–154

Ricklefs RE (1969) An analysis of nesting mortality in birds. Smiths Cont Zoology 9:1–48

Slater PJB (2003) Fifty years of bird song research: a case study in animal behaviour. Anim Behav 65(4):633–639

Torres LG, Orben RA, Tolkova I, Thompson DR (2017) Classification of animal movement behavior through residence in space and time. PLoS One 12(1):e0168513

Stowe MK, Tumlinson JH, Heath RR (1987) Chemical mimicry: bolas spiders emit components of moth prey species sex pheromones. Science 236(4804):964–967

Surlykke A, Boel Pedersen S, Jakobsen L (2009) Echolocating bats emit a highly directional sonar sound beam in the field. Proc Biol Sci 276(1658):853–860

The Nobel Prize in Physiology or Medicine 1973. NobelPrize.org. Nobel Media AB 2020. Sun. 13 Sep 2020. https://www.nobelprize.org/prizes/medicine/1973/summary/

Takahashi LK (1990) Hormonal regulation of sociosexual behavior in female mammals. Neurosci Biobehav Rev 14(4):403–413

Williams H (2004) Birdsong and singing behavior. Ann N Y Acad Sci 1016:1–30

Waldvogel JA (2000) Birdsong playback as a tool for teaching animal behavior. In: Karcher SJ (ed) Tested studies for laboratory teaching, Proceedings of the 22nd Workshop/Conference of the Association for Biology Laboratory Education (ABLE), vol 22, pp 247–260. 489 pages

Tinbergen N (1951) The Study of Instinct. Clarendon Press/Oxford University Press, Oxford

Tinbergen N (1963) On aims and methods of ethology. Z Tierpsychol 20:410–433

Vogelstein JT, Park Y, Ohyama T, Kerr RA, Truman JW, Priebe CE, Zlatic M (2014) Discovery of brainwide neural-behavioral maps via multiscale unsupervised structure learning. Science 344: 386–392

27 Study of Animal Behavior and Machine Learning

Abstract

Application of machine learning has enabled better understanding of observation and analysis of animal behavior as compared to conventional modes of manual observation revealing limitations in accurate observation and analysis. Automated videos and audio recording are enabling close understanding of animal behavior, like movement, courtship, guard of territory, mating, and others. It is also a boon to the understanding of taxonomy and biodiversity based on analysis of auditory signals by insects. We discuss in this chapter the different applications of machine learning in understanding of animal behavior.

Keywords

Bioacoustics · Auditory signals · Video

Abbreviation

UML Unsupervised machine learning

27.1 Introduction

In the preceding chapter we have discussed the diverse domains of animal behavior, which is intriguing to the human mind. The studies have been chiefly through the observations in nature and in isolated laboratory environment which lacked the flavor of the actual behavior of the animal in its natural habitat ecology, niche, and settings.

S. Ghosh, R. Dasgupta, *Machine Learning in Biological Sciences*,
https://doi.org/10.1007/978-981-16-8881-2_27

Besides, study of animal behavior by human observations, manually alone suffers from subjectivity as observations can vary between ant two observers, human error of vision and capability of rightly recording very similar yet different and subtle behavior by animals, or observing similar very looking animals, thus results may vary within observers in the same laboratory as well as in between different laboratories working on similar problems. Problems also arise from handling a large number of animals and their behavior. Thus alternative automated computer based studies with designed algorithms to analyze the behavior is a welcome step.

Recently imaging based studies in natural environment and analysis of computational methods of imaging hold importance in the study of the animal behavior which is over and above the studies by conventional methods. We discuss in this chapter some of the recent applications of machine learning in study of animal behavior.

27.2 Machine Learning and Study of Animal Behavior

27.2.1 Unsupervised Machine Learning

Unsupervised machine learning has been applied in the study of animal behavior. It has been applied both in the automated process of detection, and also in understanding behavior and its for neurobiological control. UML has been used to mine the structure in behavior from large video or trajectory datasets by clustering the video sequences into behavior categories so that members of different clusters are more different than members of the same clusters. Euclidean distance between the video segments in one of the low-dimensional behavior representations. Then the data is modeled based on assumptions of the algorithm (Ikizler-Cinbis and Sclaroff 2010; Laptev et al. 2008; Stikic et al. 2011).

27.2.2 Supervised Learning

Supervised learning, application in study of animal behavior has been used to obtain more trustworthy signals (Ikizler-Cinbis and Sclaroff 2010; Laptev et al. 2008; Stikic et al. 2011).

27.2.3 Semi-Supervised Learning

Semi-supervised learning is a branch of machine learning lying between supervised and unsupervised learning and has been used to combine small set of labeled data with the structure implicit in a large set of unlabeled data (Ikizler-Cinbis and Sclaroff 2010; Laptev et al. 2008; Stikic et al. 2011).

27.3 Recording Behavior

After the identification of the behavior it has to be recorded which is based on properties of the behavior. Behaviors range from short and simple like a rapid movement of the eye between fixation points to an auditory target (Knudsen et al. 1979) and long and complex behavior like courtship (Spieth 1974) behavior observed in *Drosophila*.

Diverse sensors, microphone arrays, laser Doppler vibrometry, accelerometers, touch screens, search coils for monitoring eye or head position have been used to capture animal behavior (Knudsen et al. 1979; Friedman et al. 1997; Niebles et al. 2008; Elias et al. 2003; Spence et al. 2010; Bermejo et al. 1994; Mendes et al. 2013) which is relatively noninvasive, inexpensive. It has been possible to extract features from video. But the key challenges remain in defining of the behavior, its recording, extraction of descriptive features, and combined features to identify behaviors.

27.4 Tracking Animal Position and Pose

Behavioral studies have been designed based on analyses of the position and pose of the animals over time. Animal tracking from video by manual methods of flight movements of *Drosophila melanogaster* (Buelthoff et al. 1980), wolf (Moran et al. 1981), and human social behavior (Kendon 1975) suffered from the disadvantage of time-consumption which was improvised by automatic tracking programs (Dell et al. 2014). Tracking has the main challenge of data-association (Forsyth et al. 2005; Poppe 2007), of each pixel to each frame of a video with either an animal or the background, tracking of multiple animals, required each pixel within an animal with its animal identity to be labeled, and separate labeling was required for each pixel with the body part identity to track body parts. Most tracking algorithms divide the main problem into smaller or sub-problems (Branson et al. 2009; Noldus et al. 2002; Ohayon et al. 2013; Swierczek et al. 2011) and remain a difficult domain to assign for pixels in entire video. Machine learning is aiming at tackling optimal assignment for pixels in the entire video, thus enabling a clarity in understanding of the whole behavior of the animal.

27.5 Automation and Behavioral Analysis

Behavior and movement of walking, grooming, or chasing are considered as behavior and time-series studies enable measuring changes in the manner and frequency of such animal behavior. This can be detected by taking video footage and analyzing the video sequence input to the machine learning classification algorithm based on trajectory-based features with features extracted from tracked changes in positions of the animals or their parts with time (Bulling et al. 2014; Burgos-Artizzu et al. 2012; Eyjolfsdottir et al. 2014; Jhuang et al. 2010; Kabra et al. 2013; Mirat et al. 2013; Ramanan and Forsyth 2003; Rao et al. 2002; Rohrbach et al. 2012) and help in study of continuous properties of behavior. This feature is also helping in analyzing group movement behavior like chasing of prey by predator group.

Pixel-based features enable extraction of pixels from local patches of video, thereby retaining relevant information for object detection and discarding nonrelevant information like lighting-dependent variations.

27.6 Automatic Classification of Behavior

It is done by three different ways. It includes rule-based classifiers (Dankert et al. 2009; de Chaumont et al. 2012) involving setting of rules to designate if the animal is performing a behavior or not, use of supervised machine learning applications for training classifiers from manually labeled video sequences (Burgos-Artizzu et al. 2012; Kabra et al. 2013) and use of unsupervised learning for clustering groups of video sequences into classes (Berman et al. 2014; Vogelstein et al. 2014), however, the machine learning algorithms are based on several assumptions and suffers challenges on data interpretation on animal behavioral analysis properly, training behavioral classifiers and suffers from challenges of noise in the input, and therefore requires training.

27.7 Behavioral Analysis from Audio Based Methods

While video based observations find importance in study of movement related behavior like foraging, migration, chasing, etc., vocal behavior, like songs of birds by which male birds attract females for mating and guard territories, sounds of insects can be traced by audio based recordings that finds importance in analyzing the social and foraging behavior in the animal kingdom (Konishi 1989) which is a far more advanced form of observation and analysis of manual observation of bird songs initially by Thorpe (1954).

Further advancement in machine learning methods has led to far more advancement by the addition of feature extraction including Wiener entropy, spectral continuity, pitch, and frequency modulation, (Tchernichovski et al. 2000) with the help of the software http://soundanalysispro.com and has been applied to understanding of birdsong, in absence of a tutor (Feher et al. 2009), song and language learning involves separate steps (Lipkind et al. 2013). Studies are being conducted on songs by zebra finches and role of basal ganglia (Kojima et al. 2013).

Studies have revealed the application of automated sound parameterization technique to identify and classify acoustic signals of crickets under order Orthoptera, family Gryllidae, katydids, related to grasshoppers and crickets under order Orthoptera, and cicadas under order Hemiptera and is finding importance in the study of their biodiversity and is used as a tool for species level identification and classification in insects (Noda et al. 2019).

Using bioacoustics and application of machine learning tools to acoustic sounds of many insects belonging to orders Diptera, Orthoptera, and Hemiptera, under Class

Insecta under Phylum Arthropoda has been used in taxonomy, classification, species identification together with morphologic identification and DNA analysis (Zamanian et al. 2011; Chesmore 2004).

27.8 Discussion

Researchers in computer science and biology are working on features extracted from video efficient for training behavioral classifiers or for the identification of intrinsic behavioral structure to understand and decipher complex behaviors. The development in automated bioacoustics and application of machine learning in the study of animal behavior is throwing new light in species identification, and a classification and is overall new dimension in animal taxonomy.

References

Berman GJ, Choi DM, Bialek W, Shaevitz JW (2014) Mapping the stereotyped behaviour of freely moving fruit flies. J R Soc Interface 11:20140672

Bermejo R, Houben D, Zeigler HP (1994) Dissecting the conditioned pecking response: an integrated system for the analysis of pecking response parameters. J Exp Anal Behav 61:517–527

Branson K, Robie AA, Bender J, Perona P, Dickinson MH (2009) High-throughput ethomics in large groups of *drosophila*. Nat Methods 6:451–457

Buelthoff H, Poggio T, Wehrhahn C (1980) 3-D analysis of the flight trajectories of flies (*Drosophila melanogaster*). Z Naturforschung C 35:811–815

Bulling A, Blanke U, Schiele B (2014) A tutorial on human activity recognition using body-worn inertial sensors. ACM Comput Surv 46:33

Burgos-Artizzu XP, Dollár P, Lin D, Anderson DJ, Perona P (2012) Social behavior recognition in continuous video. In: Presented at 2012 IEEE conf. Comput. Vis. Pattern Recognit. (CVPR), Providence, RI

de Chaumont F, Coura RD, Serreau P, Cressant A, Chabout J et al (2012) Computerized video analysis of social interactions in mice. Nat Methods 9:410–417

Chesmore D (2004) Automated bioacoustic identification of species. An Acad Bras Ciênc, Rio de Janeiro 76(2):436–440

Dankert H, Wang L, Hoopfer ED, Anderson DJ, Perona P (2009) Automated monitoring and analysis of social behavior in *drosophila*. Nat Methods 6:297–303

Dell AI, Bender JA, Branson K, Couzin ID, de Polavieja GG et al (2014) Automated image-based tracking and its application in ecology. Trends Ecol Evol 29:417–428

Elias DO, Mason AC, Maddison WP, Hoy RR (2003) Seismic signals in a courting male jumping spider (Araneae: Salticidae). J Exp Biol 206:4029–4039

Eyjolfsdottir E, Branson S, Burgos-Artizzu XP, Hoopfer ED, Schor J et al (2014) Detecting social actions of fruit flies. In: Presented at 13th Eur. Conf. Comput. Vis. (ECCV), Zurich

Feher O, Wang H, Saar S, Mitra PP, Tchernichovski O (2009) De novo establishment of wild-type song culture in the zebra finch. Nature 459:564–568

Forsyth DA, Arikan O, Ikemoto L, O'Brien J, Ramanan D (2005) Computational studies of human motion: part 1, tracking and motion synthesis. Found Trends Comput Graph Vis 1:77–254

Friedman D, Haim A, Zisapel N (1997) Temporal segregation in coexisting spiny mice (genus *Acomys*): role of photoperiod and heterospecific odor. Physiol Behav 1997(62):407–411

Ikizler-Cinbis N, Sclaroff S (2010) Object, scene and actions: combining multiple features for human action recognition. In: Presented at 11th Eur. Conf. Comput. Vis. (ECCV), Crete, Greece

Jhuang H, Garrote E, Yu XL, Khilnani V, Poggio T et al (2010) Automated home-cage behavioural phenotyping of mice. Nat Commun 1:68

Kabra M, Robie AA, Rivera-Alba M, Branson S, Branson K (2013) JAABA: interactive machine learning for automatic annotation of animal behavior. Nat Methods 10:64–67

Kendon A (1975) Some functions of the face in a kissing round. Semiotica 15:99–334

Knudsen EI, Blasdel GG, Konishi M (1979) Sound localization by the barn owl (*Tyto alba*) measured with the search coil technique. J Comp Physiol A 133:1–11

Kojima S, Kao MH, Doupe AJ (2013) Task-related "cortical" bursting depends critically on basal ganglia input and is linked to vocal plasticity. PNAS 110:4756–4761

Konishi M (1989) Birdsong for neurobiologists. Neuron 1989(3):541–549

Laptev I, Marszalek M, Schmid C, Rozenfeld B (2008) Learning realistic human actions from movies. In: IEEE Conference on Computer Vision and Pattern Recognition. IEEE, Piscataway, pp 1–8

Lipkind D, Marcus GF, Bemis DK, Sasahara K, Jacoby N et al (2013) Stepwise acquisition of vocal combinatorial capacity in songbirds and human infants. Nature 498:104–108

Mendes CS, Bartos I, Akay T, Marka S, Mann RS (2013) Quantification of gait parameters in freely walking wild type and sensory deprived *Drosophila melanogaster*. elife 2:e00231

Mirat O, Sternberg JR, Severi KE, Wyart C (2013) ZebraZoom: an automated program for high-throughput behavioral analysis and categorization. Front Neural Circuits 2013(7):107

Moran G, Fentress JC, Golani I (1981) A description of relational patterns of movement during ritualized fighting in wolves. Anim Behav 29:1146–1165

Niebles JC, Wang H, Fei-Fei L (2008) Unsupervised learning of human action categories using spatialtemporal words. Int J Comput Vis 79:299–318

Noda JJ, Travieso-González CM, Sánchez-Rodríguez D, Alonso-Hernández JB (2019) Acoustic classification of singing insects based on MFCC/LFCC fusion. Appl Sci 9:4097. https://doi.org/10.3390/app9194097

Noldus LPJJ, Spink AJ, Tegelenbosch RAJ (2002) Computerised video tracking, movement analysis and behaviour recognition in insects. Comput Electron Agric 35:201–227

Ohayon S, Avni O, Taylor AL, Perona P, Egnor SER (2013) Automated multi-day tracking of marked mice for the analysis of social behaviour. J Neurosci Methods 219:10–19

Poppe R (2007) Vision-based human motion analysis: an overview. Comput Vis Image Underst 108:4–18

Ramanan D, Forsyth DA (2003) Automatic annotation of everyday movements. In: Presented at 24th Annu. Conf. Neural inf. Process. Syst. (NIPS), Vancouver, Canada

Rao C, Yilmaz A, ShahM. (2002) View-invariant representation and recognition of actions. Int J Comput Vis 50:203–226

Rohrbach M, Amin S, Andriluka M, Schiele B (2012) A database for fine grained activity detection of cooking activities. In: Presented at. 2012 IEEE conf. Comput. Vis. Pattern Recognit. (CVPR), Providence, RI

Spence AJ, Revzen S, Seipel J, Mullens C, Full RJ (2010) Insects running on elastic surfaces. J Exp Biol 213:1907–1920

Spieth HT (1974) Courtship behavior in *drosophila*. Annu Rev Entomol 19:385–405

Stikic M, Larlus D, Ebert S (2011) Weakly supervised recognition of daily life activities with wearable sensors. IEEE Trans Pattern Anal Mach Intell 3:2521–2537

Swierczek NA, Giles AC, Rankin CH, Kerr RA (2011) High-throughput behavioral analysis in *C. elegans*. Nat Methods 8:592–598

Tchernichovski O, Nottebohm F, Ho CE, Pesaran B, Mitra PP (2000) A procedure for an automated measurement of song similarity. Anim Behav 59:1167–1176
Thorpe WH (1954) The process of song-learning in the chaffinch as studied by means of the sound spectrograph. Nature 173:465–469
Vogelstein JT, Park Y, Ohyama T, Kerr RA, Truman JW et al (2014) Discovery of brain wide neural-behavioral maps via multiscale unsupervised structure learning. Science 344:386–392
Zamanian H, Mehdipour M, Riede K (2011, July) Insect sound application in identification, classification using ARMA model based on machine learning and pest control. In: Conference: invertebrate sound and vibration, at: Missouri, volume: 13th meeting

Machine Learning and Precision Farming 28

Abstract

With the increase in the world population, the demands on the food supply are rising globally. Thus there is a need for increased supply of food from both plant and animal sources which has a direct need for increased agricultural productivity globally. Challenges to farming and increased productivity are manifold including the monitoring of individual plant health, environmental, soil, pathogens disease conditions for crops and monitoring of animal health, growth and well-being, animal behavior with relation to its alteration in response to disease onset or changed environmental conditions. This is done by taking pictures, videos, from satellites, drones and modeling the data by application of algorithms and software. This particularly becomes challenging when maintaining big animal herds by farmers. Precision agriculture/animal farming is an interdisciplinary science incorporating the inputs from informatics, algorithms, engineering, animal and veterinary sciences, physiologists in order to monitor the health, behavior, growth and production of animals in real time so that immediate decisions in tackling problems can be taken up by the farmers. We discuss in this chapter briefly applications of precision agriculture in (1) plant production, (2) animal production, (3) the big data and livestock application, (4) precision animal breeding, (5) precision livestock management.

Keywords

Precision agriculture · Precision mariculture and aquaculture · Precision live stock farming · Precision animal breeding (PAB) · Precision livestock management

S. Ghosh, R. Dasgupta, *Machine Learning in Biological Sciences*,
https://doi.org/10.1007/978-981-16-8881-2_28

Abbreviations

BV	Breeding values
EU	European Union
GPS	Global positioning system
PAB	Precision animal breeding
PLF	Precision livestock farming
RFID	Radiofrequency identification
Rt	Real time

28.1 Introduction

The growing world's population puts an increased demand on the agricultural productivity and long-term food supply. An estimated increase by 70% in demands for animal products of eggs, meat, and milk by 2050, together with the decreased number of farmers has led to farmers keeping bigger herds, but which suffers from the problems of difficulty to monitor individual animal health, productivity, and environmental impact. The challenges on increased agricultural productivity are manifold because of threats from climate changes, environmental pollution, diseases in plants and animals. Monitoring of animal health is a major concern from the point of view of animal welfare as well as prevention of transmission of zoonotic diseases to humans. Besides the maintenance of safety and quality of the food is a requisite. Environmental impact of the livestock is a major concern since it contributes to above 90% NH_3, 37% CH_4, and 65% N_2O in the atmosphere and uses about 30% of all land and 8–15% of water resources and their high density could adversely impact the environment.

Thus to tackle the manifold problems associated with increased production in agriculture, digital technologies have found importance. Precision livestock farming (PLF) is an interdisciplinary science involving inputs from animal biologists, ethologists, engineers, computational studies, informatics, physiologists, veterinarians that enables, real-time monitoring of animal health and productivity during different stages of rearing by providing warning messages and signaling problems for immediate steps to be taken by the farmer to resolve the issue and therefore enforces better managing system for farmers (Fig. 28.1).

28.1.1 Precision Agriculture

Precision agriculture impacts better management of plant productivity. We discuss briefly, different application of precision agriculture on plant health and productivity in this section. Images taken from satellites, drones reveal important information on plant health like images of healthy plants with higher biomass appear green, while

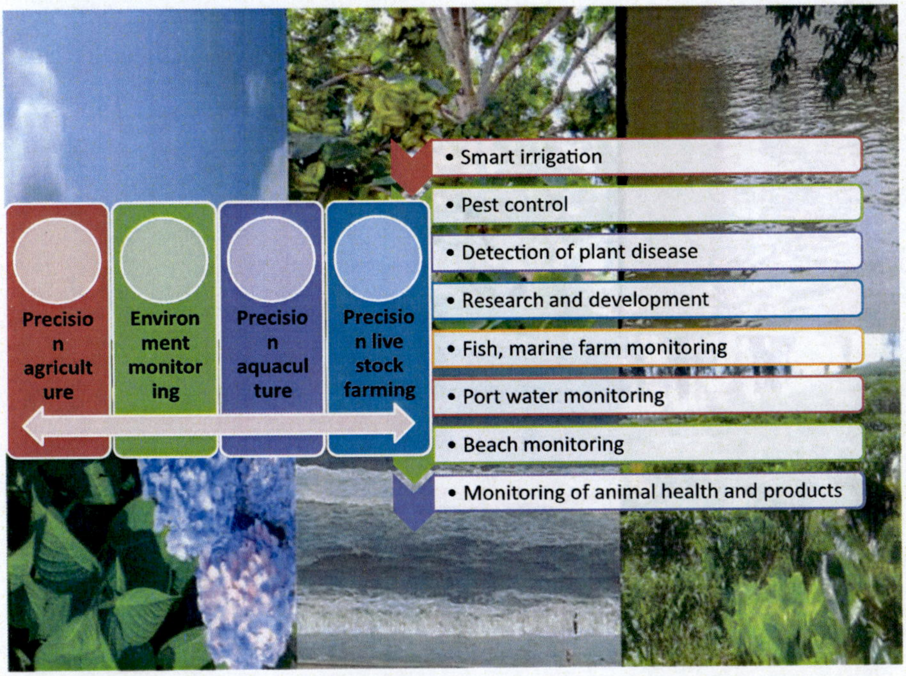

Fig. 28.1 Precision agriculture and important areas of applications

plants under stress of drought, infections by pathogens, and pests appear red or yellow in color. Sensors on devices including tractors, harvesters, irrigation systems reveal soil types and condition, water, and nutrient availability.

Internet of things (IOT), together with application of remote sensing, wireless communications, cloud-based computing, use of computers, smart phones, laptops, satellite based images, drone based images, has applications in food production and precision agriculture and has enabled smart irrigation, detection of plant disease, smart pest control which are the key challenges to productivity in agriculture (Bhakta et al. 2019; Moscoso et al. 2020). Integrated efforts of farmers and scientist have enabled study of weather parameters including temperature, pressure, humidity, soil humidity, leaf humidity, etc. in the control of agricultural produce (Fig. 28.2).

Precision agriculture (PA), with aim to attain optimized productivity and sustainability has advanced in the application in agriculture science domain by the application of automation and wireless sensor network (WSN) sensing soil parameters and condition and parameters of weather by filed based sensors, satellite or manned or unmanned air-borne devices to detect crop health at the earliest and take immediate decisions on unhealthy crops (Shafi et al. 2019). Energy efficient WSN and harvesting techniques have been used to effective control of weeds, pests, by pesticides and plant diseases in control of plant health (Jawad et al. 2017). Precision agriculture has been used to application nitrogen fertilizer to field which can potentially improve production and retaining nutrient and is a step towards

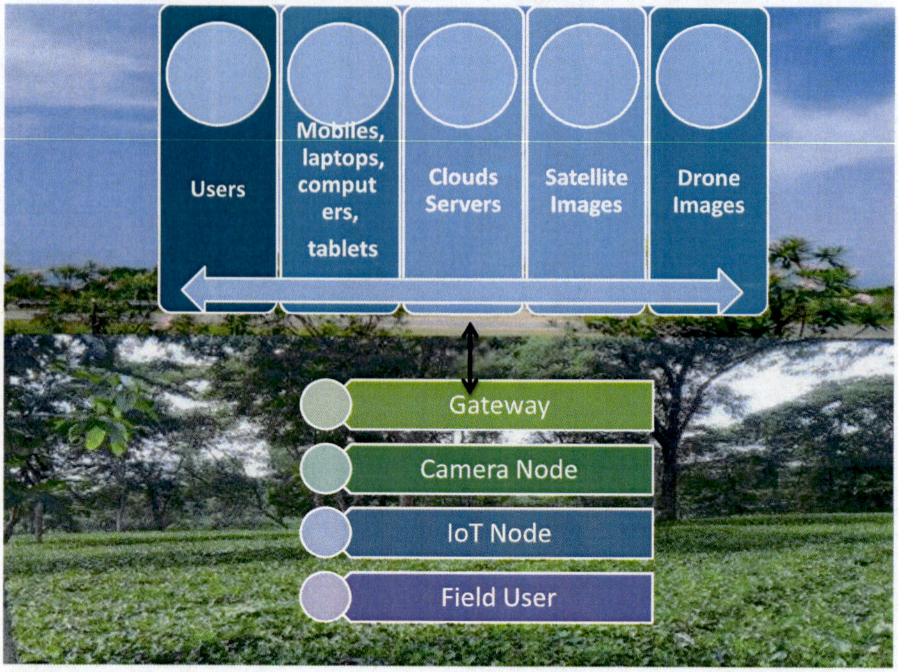

Fig. 28.2 Conceptual view of precision agriculture

efficient nutrient management in agriculture farms (Hedley 2015). Integration and modeling of this data obtained from satellites, drones, sensors in the field, irrigation systems, drones, and other input sources indicative of soil conditions, water, nutrients, climate changes, stress and challenges faced by plants, enabled by application of software, are being employed nowadays by the farmers to take individual decisions in the crop season till harvesting towards an overall increased agricultural productivity and get specific tailor made solution to specific problems, ranging from selection of the right crop variety for that individual field based on the soil, climate, and region, selecting the ideal time of cropping, harvesting, selecting the active ingredient, time to time assessing crop health and growth, required crop protection, prevent the waste of resources, reduce negative impacts of agriculture on the environment and to get a sustainable, abundant harvest.

Quantitative estimation, fast automated, real-time analysis by electrochemical sensing of soil macronutrients including nitrogen (N), phosphorus (P), and potassium (K) finds importance in increasing efficiency in agricultural production and site-specific crop management (SSCM) by adjusting the application of external fertilizers (Kim et al. 2009). Precision agriculture techniques of remote sensing and GIS or geographic information system which is a framework for collecting, managing, and analyzing geographical data have helped in prediction of potato crop yield (Al-Gaadi et al. 2016).

28.1.2 Ecological Monitoring

Internet of things (IoT) together with different sensors has enabled the applications of ecological monitoring by collecting and monitoring weather data, including environmental parameters of air quality, humidity, temperature, pressure, wind velocity, and water quality and has been useful to diverse range of users including common men, scientist, agriculturist, farmers, ports, government offices, athlete, swimmer, beach goers. Climate FieldView™ (https://climate.com/), is one such computational application tool enabling farmers by providing concise image of their operation. Precision agriculture with its tool of sensing, information technologies, and mechanical systems forms an advanced step towards better crop management.

28.1.3 Precision Mariculture and Aquaculture

Precision aquaculture with diverse use of sensors finds application in marine environment for monitoring, analyzing, understanding, making decision and providing support to farmers. Internet of things (IoT) applied to aquaculture and mariculture based studies have enabled farmers equipped with sensors and servers for storing data and connect with a fog and cloud-based technologies. IOT cloud-based solutions to aquaculture related problems are being designed (O'Donncha and Grant 2019; Acar et al. 2019).

Smart aquaculture using automated infrastructure of sensors and machine learning tools has been developed that enable real-time monitoring of aquatic environment in which the organism fish, oysters, lobsters, prawns, etc., are being cultivated, predict, warn, risk factors and enable prompt decision-making and control and condition and recognition of behavior of the aquatic cultured organism in response to change of the aquatic environmental conditions. This has highlighted the impact of altered aquatic environment on the altered behavior of fish and marine cultured organism and is a boon to the aquaculture industry (Hu et al. 2020) (Fig. 28.3).

28.1.4 Precision Livestock Farming (PLF)

Tools in computation have been able to target the challenges related to aquaculture like animal care, feeding, measuring, and monitoring of health systems and growth.

Animal production also faces challenges of growth, development, reproduction, and health, infection by pathogens, environmental conditions, individual animal feed consumption, feedlot movement, temperature, lameness, milk production, meat composition and quality and weight gain.

An animal facing non-ideal conditions reveal alterations in behavior which could be monitored by noninvasive and contactless way detected by PLF sensing, by image and/or sound analysis or body sensors. In 2003, the first European Conference

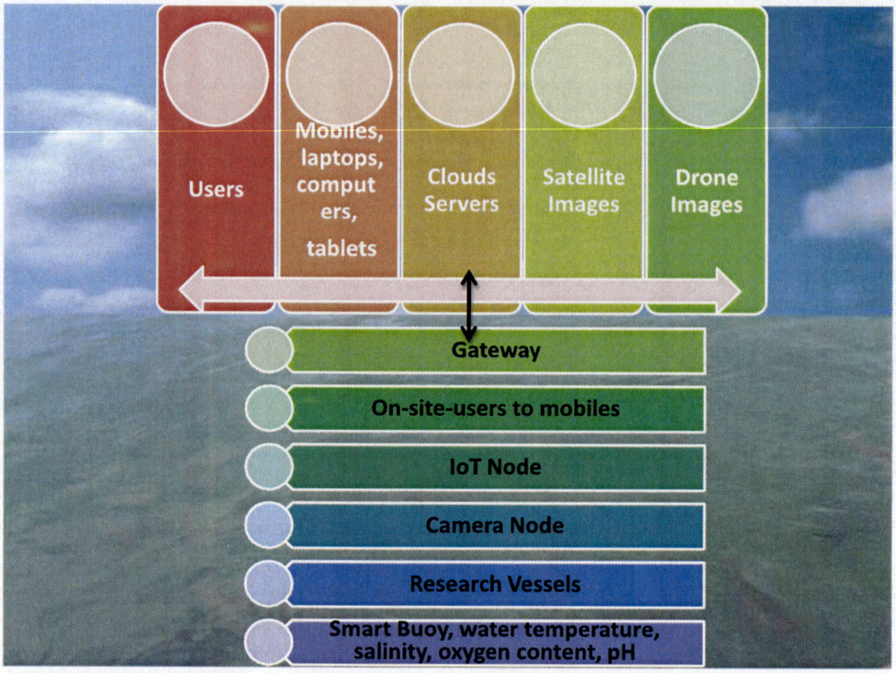

Fig. 28.3 Schematic representation of mariculture and ecological monitoring

PLF in Berlin, a living organism was referred to as CITD system or complex, individually different, time-varying, and dynamic system finding importance in design of algorithms to monitor these individuals (Berckmans 2017) at varying time.

Hydrologic modeling has enabled, frequent or continuous air and water monitoring by data collection and its analysis. Animal tagging has been in operation since a decade. It is now possible to attach electronic tag measuring and recording animal health by tracing the temperature or heart rate. While an increase in temperature can be indicative of estrus onset or a disease condition, electronic tags have been applied for identification of the individual animal and thereby facilitate their marketing. A PLF based eYeNamic system comprising of 3–4 ceiling cameras taking pictures of birds and their distribution, from Fancom BV, Netherlands has algorithms that has enabled detection of the activity, distribution, day to day problems in broiler houses based on broilers' behavior analysis, thus enabling the farmer to take immediate decisions Fig. 28.4.

Use of sensors including accelerometers, an instrument used to acceleration of a moving animas, or vibrating body surfaces, cameras to record images and microphones to record sounds together with use of machine learning algorithms have been used to noninvasively monitor animals including swines and finds importance in increasing their productivity and their welfare and can be monitored from a distance by the farm owners (Benjamin and Yik 2019).

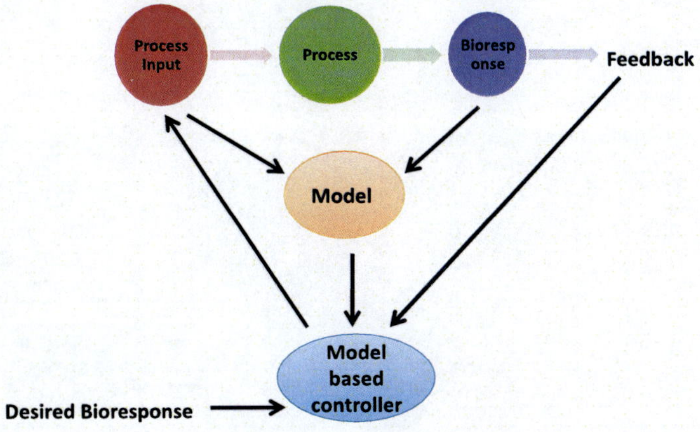

Fig. 28.4 Model and management of biological processes (Image adapted with modification from (Berckmans 2017) with permission)

PLF together with the use of environmental and wearable sensors and cameras has been applied to the poultry farming and finds application in improving the productivity and animal welfare. However more studies need to be done with technologies focused on animal welfare prior to their large-scale commercial application (Rowe et al. 2019). The earliest developments of PLF including automatic milking systems (AMS) is a boon to the dairy industry globally (John et al. 2016).

28.2 IOT Platforms and Precision Agriculture

Information and communication technology (ICT) has led to the development of Internet of things (IoT) integrating physical objects with computer based technologies enabling sensing of distant objects of interest and their remote control offers advantages over conventional methods of monitoring in terms of accuracy, efficiency, less cost, better quality and coveys promises in application in science and life.

Cloud computing service models including infrastructure as a service (IAAS), platform as a service (PAAS), and software as a service (SAAS), deployed as public, private, or hybrid have been used in precision agriculture and ecological monitoring requirements with applications in sustainable agriculture, crop health monitoring, monitoring of forest and water ecosystems and land quality, steps to control and reduce pollution, food products standardization. An architecture has been proposed including IoT nodes, communicating to the platform or through a gateway with data import. The IoT platform integrates and manages sensor data, configuration portal for administrative functions (Bajceta et al. 2016) (Fig. 28.5).

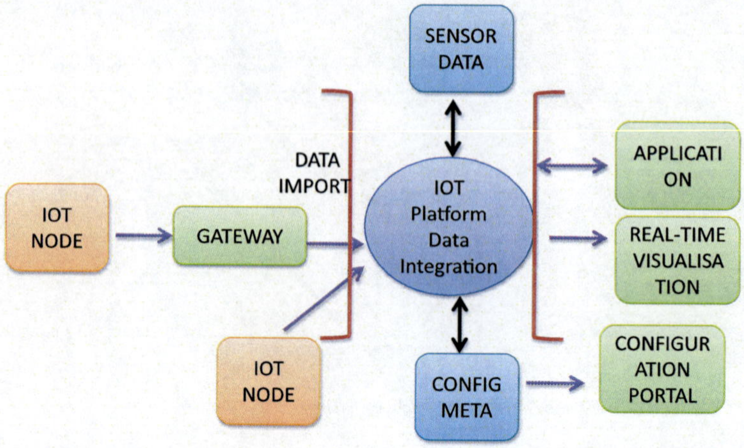

Fig. 28.5 IOT platforms. Adapted with modification with permission from *Bajceta M, Sekulić P, Krstajic B, Popovic T A Private IoT Cloud Platform for Precision Agriculture and Ecological Monitoring Conference: 13th IcETRAN At: Zlatibor, Serbia*

28.3 The Big Data and Livestock Application

Precision animal agriculture enables better management, effective production, better animal welfare, sustainability, farm animal health surveillance, and environmental factors in livestock industry. Diverse tools and applications of machine learning have been used to monitor and collect information in a routine, disciplined continuous yet less labor intensive and automated manner recording generating throughput data by digital images, sensors based and auditory data, real-time noninvasive computer mediated vision and videos and is a big welcome step towards solving problems in animal based industry and animal welfare (Morota et al. 2018).

Precision livestock farming or PLF approach is based on real time (rt) measurements of the animals responses and bio-signals from temperature measurement, *global positioning system* (GPS) position, accelerometer data, real-time image or sound analysis which has the advantages of no risk of physical contact, transmission of infection or disease spread, no risk of influencing the animal response, reduced costs, continuous monitoring like stress monitoring in every second or weight monitoring once a day and appears to be better than the big data based monitoring from the point of view of real-time analysis (Fig. 28.6).

Big data has found importance in the dairy industry and studies with diverse classes including dairy farm, animal feed, animal breed, animal health, farm and network, together with studies on volume, velocity, variety, data analytics, and use of machine learning tools of supervised, unsupervised, semi-supervised and reinforcement learning used on data collected with characteristics of time-series, streaming, sequence, graph, spatial, multimedia, etc., are being studied and requires further

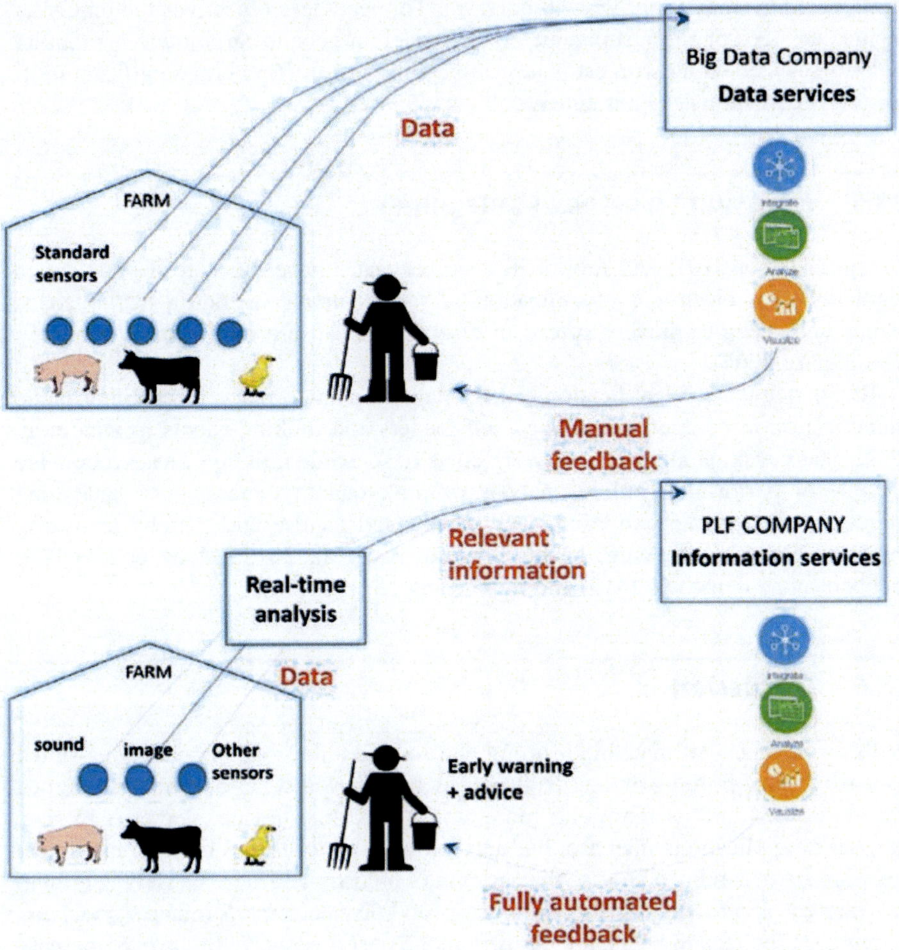

Fig. 28.6 Big data vs precision livestock farming (PLF) processing (Adapted with permission from (Berckmans 2017)

research to fully utilize the huge role of big data on farm industry (Lokhorst et al. 2019).

28.4 Precision Animal Breeding (PAB)

PAB is used when animals are bred with a purpose and has the major goals encompassing (1) predictions of the outcomes of breeding decisions, (2) avoid the introduction and adaptation of features negative to animal well-being or, (3) maintain the species well-being and (4) maintenance of genetic resources and diversity between and within populations following the guidelines of Convention on

Biological Diversity http://www.biodiv.org. To meet these objectives the requisites include the need of a robust parameterized model for genetic variation in heritability of traits selected along with estimates of breeding values (BVs) for significant traits for individual animals (Flint and Woolliams 2008).

28.5 Precision Livestock Management

European Union (EU) and few of its member states have introduced practices in regulations for electronic identification of farm animals including cattle, goats, swine, and sheep to prevent spread of diseases and ensure food safety (European Community 2007).

Radio frequency identification (RFID) tags find importance in enabling cattle identification in computer-controlled self-feeders and milking robots (Ipema et al. 2009; Stankovski et al. 2012) allowing cows to schedule milking. These robots are capable of performing online analysis of milk quantity and quality, analyzing indicators of hygienic condition, fat, protein, and lactose and thereby analyzing individual animal's feeding, health (Stankovski et al. 2012; Maltz et al. 2009; Gebbers and Adamchuk 2010) and monitoring of behavior.

28.6 Discussion

With the advances of machine learning and several tailor made applications suited for each area of farming and agricultures, including plants, animals being designed and researched and as more and more smart electronic devices decrease in size, number of applications increase, big data and internet of things or IOT, intelligent software are becoming a boon to the agriculture industry and is immensely helpful to the farmers in growing, maintaining crop and livestock performance. Precision farming is predicted to be a huge industry in the next 5 years. With the ever increase of requirement of food globally, such a boon to the farming industry is a step to increase farming in meeting demands to increasing hunger globally.

References

Acar U et al (2019) Designing an IoT cloud solution for aquaculture. In: *2019 global IoT summit (GIoTS)*, Aarhus, Denmark, pp 1–6

Al-Gaadi KA, Hassaballa AA, Tola E et al (2016) Prediction of potato crop yield using precision agriculture techniques. PLoS One 11(9):e0162219

Bajceta M, Sekulić P, Krstajic B, Popovic T A (2016) Private IoT Cloud Platform for Precision Agriculture and Ecological Monitoring Conference: 13th IcETRAN At: Zlatibor, Serbia

Benjamin M, Yik S (2019) Precision livestock farming in swine welfare: a review for swine practitioners. Animals (Basel). 9(4):133

Berckmans D (2017) General introduction to precision livestock farming. Anim Front 7(1). https://doi.org/10.2527/af.2017.0102

Bhakta I, Phadikar S, Majumder K (2019) State-of-the-art technologies in precision agriculture: a systematic review. J Sci Food Agric 99(11):4878–4888

European Community, Council Regulation (EC)No 1560/2007(European Commission, Brussels, 2007)

Flint APF, Woolliams JA (2008) Precision animal breeding. Philos Trans R Soc Lond Ser B Biol Sci 363(1491):573–590

Gebbers R, Adamchuk VI (2010) Precision agriculture and food security. Science 327(5967): 828–831

Hedley C (2015) The role of precision agriculture for improved nutrient management on farms. J Sci Food Agric 95(1):12–19

Hu Z, Li R, Xia X, Yu C, Fan X, Zhao Y (2020) A method overview in smart aquaculture. Environ Monit Assess 192(8):493

Ipema et al (2009) In: Lokhorst PWG, Koerkamp G (eds) Precision Livestock Farming'09,C. Wageningen Academic Publishers, Wageningen, pp 135–142

Jawad HM, Nordin R, Gharghan SK, Jawad AM, Ismail M (2017) Energy-efficient wireless sensor networks for precision agriculture: a review. Sensors (Basel). 17(8):1781

John AJ, Clark CE, Freeman MJ, Kerrisk KL, Garcia SC, Halachmi I (2016) Review: milking robot utilization, a successful precision livestock farming evolution. Animal 10(9):1484–1492

Kim HJ, Sudduth KA, Hummel JW (2009) Soil macronutrient sensing for precision agriculture. J Environ Monit 11(10):1810–1824

Lokhorst C, de Mol RM, Kamphuis C (2019) Invited review: big data in precision dairy farming. Animal 13(7):1519–1528

Maltz E, Antler A, Halachmi I, Z. (2009) In: Lokhorst C, Groot Koerkamp PWG (eds) Schmilovtich, in precision livestock Farming'09. Wageningen Academic Publishers, Wageningen, pp 17–23

Morota G, Ventura RV, Silva FF, Koyama M, Fernando SC (2018) Big data analytics and precision animal agriculture symposium: Machine learning and data mining advance predictive big data analysis in precision animal agriculture. J Anim Sci 96(4):1540–1550

Moscoso CG, Potz KR, Tan S, Jacobson PA, Berger KM, Steer CJ (2020) Precision medicine, agriculture, and genome editing: science and ethics. Ann N Y Acad Sci 1465(1):59–75

O'Donncha F, Grant J (2019) Precision Aquaculture. IEEE Internet of Things Magazine 2(4):26–30

Rowe E, Dawkins MS, Gebhardt-Henrich SG (2019) A systematic review of precision livestock farming in the poultry sector: is technology Focussed on improving bird welfare? Animals (Basel) 9(9):614

Shafi U, Mumtaz R, García-Nieto J, Hassan SA, Zaidi SAR, Iqbal N (2019) Precision agriculture techniques and practices: from considerations to applications. Sensors (Basel) 19(17):3796

Stankovski S et al (2012) Dairy cow monitoring by RFID. Sci agric (Piracicaba, Braz), Piracicaba 69(1):75–80

29 Machine Learning in the Study of Animal Health and Veterinary Sciences

Abstract

Machine learning is being used in the prediction of health and disease status in the animals and has wide scale application in the animal health care and well-being and veterinary sciences.

In this chapter we have highlighted some of the applications of machine learning in the health and disease prediction in animals. Although the applications are at its infancy, the future scope of such application science in this domain of biology is immense and needs to be exploited more in the welfare of animal life.

Keywords

Cow · Cats · Dogs · IBD · ALA · Epilepsy

29.1 Introduction

Artificial intelligence (AI) and machine learning have touched diverse applications of life and life science research is greatly benefitted by its advanced applications. Veterinary medicine, practitioners, and pet owners are applying AI-based technologies, in animal welfare and maintaining of health and detection of disease and taking proper care by proper automated diagnostic approaches mediated by AI which is much better, effective, and reliable as compared to the conventional approaches of monitoring of health of animals and pets and treatment by practitioners based on only the pet owners observation.

Machine learning, with its ability to analyze multidimensional and complex data has been applied in the domain of veterinary sciences and finds importance in early prediction of diseases, identifying risk factors of disease, thereby enabling early treatment and animal welfare.

S. Ghosh, R. Dasgupta, *Machine Learning in Biological Sciences*,
https://doi.org/10.1007/978-981-16-8881-2_29

Scientists are working towards generating algorithms to detect and diagnose complicated diseases like Additions Disease, leptospirosis which can later develop kidney problems and other complicated diseases. AI has been applied to radiography and algorithms are being designed to detect and recognize patterns from imaging data (https://www.avma.org/javma-news/2020-07-15/artificial-intelligence-veterinary-medicine, https://vhive.buzz/artificial-intelligence-animal-veterinary-sciences/).

29.2 Machine Learning Applications in Disease

Machine learning, associated with data mining tools, artificial intelligence-based tools, statistical tools, probability theory, pattern recognition, algorithms, and computer sciences has been important in enabling computer enabled automated disease diagnosis, understanding the relation of risk factors with diseases and is vital for both human and animal health studies. We discuss in this chapter the applications of machine learning in different diseases of animals.

29.2.1 Infection

29.2.1.1 Cattle

Mastitis disease in dairy cattle is caused by gram-negative bacteria, *Escherichia coli* (*E. coli*) and early diagnosis finds importance in preventing the rapid spread of the disease, in the farm, selection of efficient treatment strategies and prevents unnecessary use of antibiotics. Omics based, microarray based, transcriptomic profiling studies, and availability of databases for easy access and retrieval of data together with applications of data mining tools, meta-analysis and machine learning revealed hypoxia, drug response, upregulated genes, role of chemokines including C-X-C Motif Chemokine Ligand 8 (CXCL8), Chemokine (C-X-C motif) ligand 2 (CXCL2), GRO1(CXCL1), Chemokine (C-C motif) ligand 20 CC L20, transcriptional regulator NFKB Inhibitor Zeta (NFKBIZ), homeoprotein (HP), RNase ZC3H12A, phosphodiesterase 4B (PDE4B), Caspase 4 (CASP4), Complement factor B (CFB), S100 calcium-binding protein A9 (S100A9), and S100A8, a calcium- and zinc-binding protein in immune response, inflammation in the disease of which ZC3H12A, CXCL2, GRO, CFB were identified as disease biomarkers (Sharifi et al. 2018). Sub-clinical mastitis in cow can cause reduction of milk quality and its production and suffers from the challenges of early detection and the conventional method of somatic cell count (SCC) day-test is not sufficient. Lack of early diagnosis leads to treatment by antibiotics leading to the risk of rise of antibiotic-resistant bacteria. Machine learning tools of deep learning (DL), Naïve Bayes, generalized liner model, logistic regression, decision tree, gradient-boosted tree (GBT), and RF used to data on monitoring of parameters of milk production and quality of milk produced including: milk volume production, lactose concentration in milk, electrical conductivity (EC), protein concentration in milk, peak flow and milking time for each cow and predict incidence of mastitis at early stages (Ebrahimi et al. 2019). Support Vector Machines (SVM) was used to analyze and predict the zoonotic potential of bovine E. coli strain *O157* which could be also infect humans and finds importance in future design of vaccination strategies (Lupolova et al. 2016).

29.2.1.2 Canines

Chiari-like malformation (CM) disease in dogs reveals pathophysiology of complex neuromorphological alterations of the skull and cranial cervical vertebrae, causing pain also called as Chiari-like malformation-associated pain (CM-P) and secondary syringomyelia (SM) which suffers from the challenges of improper and insufficient diagnosis. Machine learning models and support vector machines (SVM) have been used to identify and characterize such neuromorphological alterations not possible to be detected by human eyes by image-based studies and therefore identification of image-based biomarkers for the detection of the disease (Spiteri et al. 2019). Canine hypoadrenocorticism (CHA) is a fatal disease in dogs, killing a large number of dogs annually, suffers from lack of proper diagnostic features although reveals manifestations of kidney and gastrointestinal diseases which is common to other disorders as well and thus search for appropriate markers for early diagnosis of the disease finds importance in proper treatment of the sick dog. Machine learning methods of boosted tree algorithm (AdaBoost) were used to analyze diagnostic tools of complete blood count (CBC) and serum parameters and has been effective in screening of diseased and suspected dogs (Reagan et al. 2020).

29.2.1.3 Felines

Inflammatory bowel disease (IBD) in cats is a specific reaction to chronic irritation of the stomach or intestines caused by injury, food related allergy, foreign agents like bacterial infection such as *Salmonella* or *E. coli*, parasitic infection by *Giardia, Tritrichomonas*, worm infection leading to inflammation. Alimentary lymphoma (ALA) in cats is a lymphoma or malignant tumor developed in the alimentary form with a pathophysiology of lesions observed in the stomach, small intestine, liver, mesenteric lymph nodes (Paulin et al. 2018; Karakuş et al. 2018).

Both IBD and ALA in cats are common gastrointestinal (gi) disorders and reveal similar and overlapping symptoms of chronic vomiting, chronic diarrhea, weight loss, poor appetite, and similar histopathology such that the detection of each is sometimes confusing and challenging, hence there is a need for suitable diagnostic tools to clearly diagnose each disease. Supervised machine learning algorithms of naive Bayes, decision trees, and artificial neural networks (ANN) have been applied on noninvasive diagnostic test reports from CBC and serum parameters to identify each disease, IBD or ALA with normal cats as controls (Awaysheh et al. 2018) and proved to help in the diagnosis and categorization of the two disorders and could identify the normal cats (Awaysheh et al. 2018; Awaysheh et al. 2016). Application of semantic deep learning approach which is an integration of semantic web technologies and deep learning on gastrointestinal (GI) biopsies data of IBD and alimentary ALA in cats and has enabled improved accurate diagnosis (Awaysheh et al. 2018; Awaysheh et al. 2016).

Machine learning tools and game theory have been used to understand infection process by predicting risk of infection by canine distemper virus (CDV) and feline parvovirus pathogens in African lions (Fountain-Jones et al. 2019).

29.2.1.4 Birds

Machine learning tools and supervised machine learning have been applied to epidemiology studies, for classification, diagnosis, and predicting risk of infection. SVM has been applied to study hock burn, dermal infection in skin of broilers, indicative of their overall health from data collected from commercial broiler farms and is a promising step towards control and management of broiler in farms (Hepworth et al. 2012).

29.2.2 Health

29.2.2.1 Cattle

For ruminants, grazing and ruminating are important behavioral patterns and thus, monitoring their eating behavior can be informative of their health, productivity, and overall welfare. However, it is extremely difficult, subjective, time-consuming, and error prone to observe such behavior manually by the human eye, especially for large herd of cows. Automated recording of data by the use of sensors like accelerometer or gyroscope sensor with suitable software and machine learning algorithms of random forest (RF), support vector machine (SVM), *k*-nearest neighbor (kNN) and adaptive boosting (AdaBoost), has been used to classify and identify grazing and ruminating behavior (Mansbridge et al. 2018).

Supervised machine learning tools have been used to monitor the vaginal temperature (VT) and conductivity (VC) in cattle by the use of vaginal sensor and finds application in efficient and accurate detection of estrous in cattle (Higaki et al. 2019).

Plasma levels of metabolites and metabolic hormones including β-hydroxybutyrate (BHB), free fatty acids (FFA), glucose, insulin, and insulin-like growth factor 1 (IGF-1) are used to study the metabolic status of dairy cows in early lactation. But detection of such parameters for a large herd is time-consuming and difficult and therefore parameters like milk production traits find importance in easy prediction of metabolic status of cows in farm. Machine learning algorithms of Random Forest (RF) and SVM has been applied in prediction of metabolic status of cow revealing three categories of poor, good, and average metabolic status revealed by highest level of expression of BHB and FFA in the blood plasma, lowest glucose level in plasma, insulin hormone, and IGF-1 indicative of poor metabolic status while the reverse indicative of good metabolic status and in between levels of these parameters were indicative of average metabolic status (Xu et al. 2019).

Machine learning techniques of RF, linear discriminant analysis, and neural network were used in automated monitoring of lying and rumination behaviors in dairy cattle to understand prepartum behavior indicative of calving and thus alerts based on such behavior find importance in drawing attention of the farm owner and providing appropriate care to the animal (Borchers et al. 2017). Machine learning approaches of sequential minimal optimization (SMO), RF, alternating decision tree (ADTree), and naïve Bayes (NB) classifiers applied to metabolomics data of serum components have been used to identify association of metabotypes in over-conditioned cows, in lactation, blood metabolites, and hormones (Ghaffari et al. 2020). An automated SVM classifier trained to on histologic images of bovine liver, lung, spleen, and kidney, revealing normal, acute necrosis and inflammation both acute and chronic tissues could accurately identify normal tissue and tissue lesions (Hattel et al. 2013).

29.2.2.2 Felines

Machine learning methods of feature extraction and recurrent neural network (RNN) have been applied parameters including age, blood urea nitrogen, creatinine, and urine specific gravity, from electronic health records (EHRs) of routine veterinary practice and have been instrumental in design of a model with high sensitivity and specificity for prediction of risk of cats developing chronic kidney disease (CKD, Bradley et al. 2019) and is a step towards better prediction and help the practitioner for better decision-making.

29.2.2.3 Zebrafish

Sex determination in fresh water living zebrafish *Danio rerio*, under family Cyprinidae, order Cypriniformes, manually is extremely labor and time-consuming and subjective, varying with the human observer. Machine learning methods of deep convolutional neural networks (DCNNs) and SVM have enabled sex identification based on color of caudal fin. The study also revealed impact of altered environmental parameters like increased water temperature during embryogenesis on altered sex-related color differences in adult zebrafish with lower color intensity in male animals (Hosseini et al. 2019).

29.2.2.4 Giardia Duodenalis

Data from I-TASSER suite or Iterative threading assembly refinement is a package that is designed for prediction of protein structure and function (Yang et al. 2015; Yang and Zhang 2015) and Protein Data Bank (PDB, (Burley et al. 2019)) together with application of RF classifier has been used to predict *Giardia duodenalis (G duodenalis)* proteome (Ansell et al. 2019), a microscopic flagellated parasite, causing, diarrheal disease of giardiasis.

29.2.2.5 Sea Lions

California sea lions (*Zalophus californianus*) under Order Carnivora, family Otariidae, predominant in to western North America, are known for their intelligence and playfulness. Data from microarray analysis of blood samples and altered gene expression has been reported to be an indicator of altered health status of wildlife sea lions (Mancia et al. 2012). This would enable monitoring and maintaining their health.

29.2.2.6 Sharks

Accelerometers are used to detect acceleration and find importance in detection of behavior. Machine learning (ML) classifiers, logistic regression, artificial neural network, random forest (RF) models, gradient boosting model, and voting ensemble (VE) model on data acquired from accelerometer applied to commonly called juvenile lemon sharks (*Negaprion brevirostris*) under Class Chondrichthyes, Order Carcharhiniformes and family Carcharhinidae revealed different behavior of chafing, burst swimming, headshaking, resting, swimming and found importance in predicting prey capture (Brewster et al. 2018).

29.2.2.7 Horses

Horses suffering from colic or acute abdomen colic disorder reveal extreme abdominal pain in horses due to a diseases of the abdomen and acute gastrointestinal disease and the multiple risk factors include pasture or drylot, access to water, intake of carbohydrates, parasite infection, breed of horses (Curtis et al. 2019). Machine learning tools of decision trees, multilayer perceptron, Bayes network, and Naïve Bayes were used to predict need of surgery and chances of surviving of horses suffering from the disorder (Fraiwan and Abutarbush 2020).

29.2.2.8 Canines

Epilepsy in dogs is a most commonly observed neurological disorder, affecting around 0.75% of the canine population (Berendt et al. 2015). The disease is characterized by recurrent, unprovoked seizures, leading to brain abnormality. Description of seizures finds importance in diagnosis of epilepsies in dogs which includes two types of seizures, namely categorized as generalized and focal. Generalized seizures at the onset of the disease occur at both the sides of the brain and clinical signs appear on both sides of the body revealing sudden loss or increased muscle tone, occurrence of salivation, urination and/or defecation. Focal seizures however initiate in discrete brain area affecting one side or specific body site, revealing facial twitches, chewing movements, paddling of limb, signs of fear, attention seeking, dilation of pupil, salivation and vomiting. Epilepsy has been categorized into stages including the preictal state that occurs at the initiation of seizure, ictal state, initiating at seizure onset and ending in attack, the postictal state that starts postictal state. Thus identifying the disease at the preictal stage is challenging and finds importance in early diagnosis of the disease. Specific patterns in the *electroencephalography* or EEG is used to detect and diagnose seizure type in dogs (Berendt et al. 2015; Ekenstedt and Oberbauer 2013; Heske et al. 2014). Therefore early forecasting of seizures and early treatment by antiepileptic drugs might help in the making better the lives of the dogs. A semi-supervised algorithm was used to forecast seizure in chronic EEG recordings studies in dogs with epilepsy (Nasseri et al. 2020).

29.2.3 Disease Surveillance and Health

Disease surveillance finds importance in predicting potential disease outbreaks from zoonotic infections and bioterrorism threats. Deep learning tools applied on free-text veterinary necropsy reports from veterinary diagnostic center with gastrointestinal (gi), respiratory, or urinary pathology records could devise a model in understanding epidemiological trends focused on gastrointestinal disease (Bollig et al. 2020).

Cryptosporidium and *Giardia* are parasites that can cause zoonotic infectious diseases like diarrheal cryptosporidiosis and giardiasis, respectively, in animals and humans through exposure to contaminated water and is a public health concern. Thus monitoring of oocytes in the water finds importance in risk assessment and could prevent infection and disease outbreaks models. Machine learning tools together with biological, meteorological, and physicochemical parameters have been employed to predict contamination intensity by both organisms (Ligda et al. 2020).

29.3 Discussions

There is an increased trend to monitor health of farm animals by the use of sensors that can be worn, video footages, smart cameras, accelerometers, and other smart devices using AI, using machine learning algorithms and their real-time analysis. A lot of data is being generated, analyzed, visualized by automated means, using classifiers and feature extraction techniques and diverse machine learning algorithms to diverse complex problems related to health and welfare of animals, predicting disease, understanding trend of infection, sex determination, optimize milk production.

Certainly AI and machine leaning are important tools for the animal industry, farm owners, and vets. Industrial giants like IBM may be the forerunners in this domain, but at this juncture there is need to understand the animal health, disease pathology, and infection biology and normal behavior in order to apply the machine learning tools to resolve the exact complex problems in this domain of life sciences.

References

Ansell BRE, Pope BJ, Georgeson P, Emery-Corbin SJ, Jex AR (2019) Annotation of the Giardia proteome through structure-based homology and machine learning. *Gigascience*. 8(1):giy150

Awaysheh A, Wilcke J, Elvinger F, Rees L, Fan W, Zimmerman K (2018) Identifying free-text features to improve automated classification of structured histopathology reports for feline small intestinal disease. *J Vet Diagn Invest*. 30(2):211–217

Awaysheh A, Wilcke J, Elvinger F, Rees L, Fan W, Zimmerman KL (2016) Evaluation of supervised machine learning algorithms to distinguish between inflammatory bowel disease and alimentary lymphoma in cats. J Vet Diagn Invest. 28(6):679–687

Berendt M, Farquhar RG, Mandigers PJ, Pakzody A, Bhatti SF, De Risio L, Fischer A, Long S, Matsiasek K, Munana K, Patterson EE, Penderis J, Platt S, Podell M, Potschka H, Pumarola

MB, Rusbridge C, Stein VM, Tipold A, Volk HA (2015) International Veterinary Epilepsy Task Force consensus report on epilepsy definition, classification and terminology in companion animals. BMC Vet Res 11:182

Bollig N, Clarke L, Elsmo E, Craven M (2020) Machine learning for syndromic surveillance using veterinary necropsy reports. *PLoS One* 15(2):e0228105

Borchers MR, Chang YM, Proudfoot KL, Wadsworth BA, Stone AE, Bewley JM (2017) Machine-learning-based calving prediction from activity, lying, and ruminating behaviors in dairy cattle. J Dairy Sci. 100(7):5664–5674

Bradley R, Tagkopoulos I, Kim M et al (2019) Predicting early risk of chronic kidney disease in cats using routine clinical laboratory tests and machine learning. J Vet Intern Med 33(6):2644–2656

Brewster LR, Dale JJ, Guttridge TL et al (2018) Development and application of a machine learning algorithm for classification of elasmobranch behavior from accelerometry data. *Mar Biol.* 165(4):62

Burley SK, Berman HM, Bhikadiya C et al (2019) RCSB Protein Data Bank: biological macromolecular structures enabling research and education in fundamental biology, biomedicine, biotechnology and energy. Nucleic Acids Res. 47(D1):D464–D474

Curtis L, Burford JH, England GCW, Freeman SL (2019) Risk factors for acute abdominal pain (colic) in the adult horse: A scoping review of risk factors, and a systematic review of the effect of management-related changes. PLoS One 14(7):e0219307

Ebrahimi M, Mohammadi-Dehcheshmeh M, Ebrahimie E, Petrovski KR (2019) Comprehensive analysis of machine learning models for prediction of sub-clinical mastitis: Deep Learning and Gradient-Boosted Trees outperform other models. *Comput Biol Med.* 114:103456

Ekenstedt KJ, Oberbauer AM (2013) Inherited Epilepsy in Dogs. Topics in Compan An Med 28: 51–58

Fountain-Jones NM, Machado G, Carver S, Packer C, Recamonde-Mendoza M, Craft ME (2019) How to make more from exposure data? An integrated machine learning pipeline to predict pathogen exposure. *J Anim Ecol.* 88(10):1447–1461

Fraiwan MA, Abutarbush SM (2020) Using Artificial Intelligence to Predict Survivability Likelihood and Need for Surgery in Horses Presented With Acute Abdomen (Colic). *J Equine Vet Sci.* 90:102973

Ghaffari MH, Jahanbekam A, Post C et al (2020) Discovery of different metabotypes in overconditioned dairy cows by means of machine learning. J Dairy Sci. S0022-0302(20):30592

Hattel A, Monga V, Srinivas U et al (2013) Development and evaluation of an automated histology classification system for veterinary pathology. J Vet Diagn Invest. 25(6):765–769

Hepworth PJ, Nefedov AV, Muchnik IB, Morgan KL (2012) Broiler chickens can benefit from machine learning: support vector machine analysis of observational epidemiological data. J R Soc Interface. 9(73):1934–1942

Heske L, Nodvedt A, Jaderlund KH, Berendt M, Egenvall A (2014) A cohort study of epilepsy among 665,000 insured dogs: incidence, mortality and survival. Vet J 202:471–476

Higaki S, Miura R, Suda T et al (2019) Estrous detection by continuous measurements of vaginal temperature and conductivity with supervised machine learning in cattle. *Theriogenology.* 123: 90–99

Hosseini S, Simianer H, Tetens J, Brenig B, Herzog S, Sharifi AR (2019) Efficient phenotypic sex classification of zebrafish using machine learning methods. Ecol Evol. 9(23):13332–13343

Karakuş A, Kasap S, Mutlu A, Akbala Tavus M, Salcı H, Sönmez G (2018) Alimentary lymphoma in a cat. Uludağ Üniversitesi Veteriner Fakültesi Dergisi 37(2):159–162

Ligda P, Claerebout E, Kostopoulou D et al (2020) Cryptosporidium and Giardia in surface water and drinking water: Animal sources and towards the use of a machine-learning approach as a tool for predicting contamination. *Environ Pollut.* 264:114766

Lupolova N, Dallman TJ, Matthews L, Bono JL, Gally DL (2016) Support vector machine applied to predict the zoonotic potential of E. coli O157 cattle isolates. *Proc Natl Acad Sci U S A.* 113(40):11312–11317

Mancia A, Ryan JC, Chapman RW et al (2012) Health status, infection and disease in California sea lions (Zalophus californianus) studied using a canine microarray platform and machine-learning approaches. *Dev Comp Immunol.* 36(4):629–637

Mansbridge N, Mitsch J, Bollard N et al (2018) Feature Selection and Comparison of Machine Learning Algorithms in Classification of Grazing and Rumination Behaviour in Sheep. *Sensors (Basel).* 18(10):3532

Nasseri M, Kremen V, Nejedly P et al (2020) Semi-supervised Training Data Selection Improves Seizure Forecasting in Canines with Epilepsy. *Biomed Signal Process Control* 57:101743

Paulin MV, Couronné L, Beguin J et al (2018) Feline low-grade alimentary lymphoma: an emerging entity and a potential animal model for human disease. BMC Vet Res 14:306

Reagan KL, Reagan BA, Gilor C (2020) Machine learning algorithm as a diagnostic tool for hypoadrenocorticism in dogs. *Domest Anim Endocrinol.* 72:106396

Sharifi S, Pakdel A, Ebrahimi M, Reecy JM, Fazeli Farsani S, Ebrahimie E (2018) Integration of machine learning and meta-analysis identifies the transcriptomic bio-signature of mastitis disease in cattle. *PLoS One.* 13(2):e0191227

Spiteri M, Knowler SP, Rusbridge C, Wells K (2019) Using machine learning to understand neuromorphological change and image-based biomarker identification in Cavalier King Charles Spaniels with Chiari-like malformation-associated pain and syringomyelia. J Vet Intern Med. 33(6):2665–2674

Xu W, van Knegsel ATM, Vervoort JJM et al (2019) Prediction of metabolic status of dairy cows in early lactation with on-farm cow data and machine learning algorithms. J Dairy Sci. 102(11): 10186–10201

Yang J, Yan R, Roy A, Xu D, Poisson J, Zhang Y (2015) The I-TASSER Suite: Protein structure and function prediction. Nat Methods 12:7–8

Yang J, Zhang Y (2015) I-TASSER server: new development for protein structure and function predictions. Nucl Acids Res 43:W174–W181

Machine Learning and Animal Reservoirs 30

Abstract

Pathogens have been threatening to the life of wild animals and human beings since time immemorial and recently the globe has witnessed the havoc the pathogens have been causing in the form of epidemics and pandemics affecting animal and human lives adversely. Identifying strains of pathogens causing infection to humans, their intermediate hosts and animal reservoirs is a key to predict and prevent such future epidemics. Machine learning is being applied to understand the animal reservoirs, species involved in transmittance of deadly diseases. At this juncture, we discuss in this chapter the different applications of machine learning in this domain of biology.

Keywords

Animal reservoir · Disease · Epidemics · Zoonosis

30.1 Introduction: Zoonotic Infection

Zoonosis or zoonotic disease is an infectious disease in human caused by an infectious agent or pathogen including bacterium, fungus, parasite, prion, and virus that has crossed species and jumped from a non-human animal to a human. Few examples of bacterial zoonosis include diseases including anthrax, brucellosis, bovine tuberculosis, tuberculosis, cat scratch disease, campylobacteriosis, listeriosis, leptospirosis, salmonellosis, and psittacosis, parasitic zoonoses include balantidiosis, cryptosporidiosis, cysticercosis, hydatidosis, paragonimiasis, taeniasis, and toxoplasmosis, viral zoonosis includes avian flu, Ebola fever, Lassa fever, severe acute respiratory syndrome (SARS), mycotic zoonoses include cryptococcosis, dermatophytoses, histoplasmosis and sporotrichosis (Atawodi et al. 2013), prion mediated zoonosis including scrapie in sheep, chronic wasting disease in deer and

S. Ghosh, R. Dasgupta, *Machine Learning in Biological Sciences*,
https://doi.org/10.1007/978-981-16-8881-2_30

elk, bovine spongiform encephalopathy (BSE), in cattle and spongiform encephalopathies (TSEs) and Creutzfeldt–Jakob disease (CJD, Lee et al. 2013) in humans.

The close human animal association could be a key for type of disease transmittance like coming in contact with infected animals saliva, blood, urine, mucous, feces, or other body fluids, touches like petting or bites, scratches, contact with animal living and roaming areas, exposure to contaminated objects or surfaces called as fomites like aquarium tank water, pet habitats, chicken coops, barns, plants, and soil, pet food, water dishes. Disease could be transmitted by insect vector while bitting like tick, insect like mosquito or flea or it could be foodborne i.e., transmitted by eating contaminated food or drinking contaminated fluids, unpasteurized milk, raw or not properly cooked meat or eggs, or contaminated raw fruits, vegetables drinking or coming in contact with contaminated water by infected animal (https://www.who.int/topics/zoonoses/en/), (Burroughs et al. 2002). These are major public health problems across the globe, by adversely affecting the health of animals they prevent efficient food production from animals and is a challenge and loss to the animal farming and agriculture industry (Weber and Rutala 1999). The last few decades have witnessed an increase in the trend in the emergence of zoonotic virus. Studies have highlighted a One Health initiative approach (www.onehealthinitiative.com) which entails an interdisciplinary collaborative study at local, national and global levels in order to attain optimal health for life forms including animals, human, plants and our environment. Studies in zoonoses would have applications in human and veterinary medicine and in study, control and prevention of sudden disease outbreaks (Wang and Crameri 2014).

30.2 Animals as Reserviors

Animal reservoirs include bats, cows, dogs, pigs, rabbit, raccoons, rodents, sheep, swine, and other mammals. Bats and birds due to their high mobility are at a greater risk of zoonotic transmission.

Zoonotic diseases including Psittacosis or parrot fever in birds is caused by a infection by bacterium *Chlamydia psittaci*. The bacteria can infect human when transmitted through contact of infected birds including parrots, macaws, cockatiels, and budgerigars, pigeons, sparrows, ducks, hens, gulls. Newcastle Disease (ND), caused by a negative stranded RNA virus under family paramyxoviruses is a highly contagious, severe disease affecting birds, domestic poultry, pigeons, parrots, and can be transmitted to humans by aerosol, contaminated food, water and Campylobacter infections causing campylobacteriosis, a disease in pet and laboratory animal species. It can be transmitted to humans from feces of infected animals by the feces entering through mouth causing acute enteritis.

Avian influenza virus (AIV) and West Nile virus (WNV) have also crossed species and infected humans from animal origin.

Bats act as reservoirs for a number of virus including Coronavirus, Ebola hemorrhagic fever, Hendra virus, Lyssaviruses, and Nipah virus (Calisher et al. 2006).

Marine birds and fresh water fish act as reservoir for capillaria species, causing internal capillariasis and is transmitted to humans from infected fish or marine birds.

Diphyllobothriasis is caused by tapeworm infection in fish caused by tapeworm of the family Diphyllobothriidae, genus Diphyllobothrium, including *D. latum, D. latus,* and *D. pacificum* and can infect humans, dogs, bears, and other fish eating animals, transmitted by raw or partially cooked infected fish.

Rat-bite fever is caused by *Streptobacillus moniliformis* or *Spirillum minor* found in respiratory tracts and mouths of rodents and rats and is transmitted by rats. Dogs can transmit diseases including brucellosis, campylobacteriosis, cryptosporidiosis, dermatophytosis, leptospirosis, rabies, sporotrichosis, salmonellosis, to human by bites and scratches, licking or the fecal–oral route.

Cattle and other ruminants including sheep, goats, and camels can cause many human diseases including anthrax, brucellosis, Borna virus infection, bovine tuberculosis, campylobacteriosis, cowpox, Creutzfeldt–Jakob disease (CJD), cryptosporidiosis, etc.

Anthrax is caused in humans by Bacillus anthracis in contact with anthrax-infected animals or animal products like sheep wool or goat hair.

Bovine spongiform encephalopathy (BSE) or variant Creutzfeldt–Jacob Disease (vCJD) or mad-cow disease fatal in cows caused by prions can be transmitted to humans by eating contaminated food (Myrna 2008).

30.3 Machine Learning and Animal Reservoirs

30.3.1 Influenza

Zoonotic influenza is caused influenza virus by transmission from animal–human to infect people. Human infection can be caused by avian influenza virus (AIV) subtypes A(H5N1) and A(H9N2) and swine influenza virus (SIV) subtypes A (H1N1) and (H3N2). Horses (equine influenza viruses) and dogs also reveal different varieties of influenza viruses. Human infections of zoonotic influenza occur by direct contact with infected animals or contaminated environments (https://www.who.int/news-room/spotlight/influenza-are-we-ready/zoonotic-influenza).

Influenza viruses including types A, B, C, and D infect human, domestic, and wild animal life. Human influenza A and B viruses cause seasonal flu epidemics known to cause pandemics. Influenza type C infections cause mild illness, influenza D viruses on the other primarily infect cattle and are not known to infect human.

Influenza A viruses are composed of surface proteins hemagglutinin (H) with 18 different subtypes of H1-H18 and neuraminidase (N) of 11 different subtypes including N1-N11. A(H1N1) and A(H3N2) subtypes influenza A viruses routinely infect humans. Influenza A subtypes are composed of genetic clades and sub-clades or groups and subgroups, respectively (https://www.cdc.gov/flu/index.htm).

Aquatic birds are known to be reservoirs of influenza A virus (IAV) and influenza outbreaks have been reported from poultry, mammalian species, including humans, horses, pigs, mink, seals, domestic dogs (Parrish et al. 2015), ducks, chickens, and whales. IAV subtypes causing outbreaks in humans include H1N1 and H3N2 infections in pigs, H7N7 and H3N8 virus in horses (https://www.cdc.gov/flu/avianflu/virus-transmission.htm).

Wild birds are the natural reservoirs for these viruses, and active surveillance within wild bird populations is important to understand viral evolution and risk assessments, however, suffers from the disadvantage of being resource-intensive. Thus machine learning is a powerful tool in disease surveillance systems and has been used in estimation of probability of isolating AIV from wild bird samples from USA during a period of 2006–2011 focusing on features of age, sex, type of bird, geographic location, and matrix gene analysis from rRT-PCR experiments of which geographic location and rRT-PCR revealed as most important predictors for AIV isolation (Walsh et al. 2019).

30.3.2 Nipah

Nipah virus (NiV) is a zoonotic virus, first recorded in 1999, in pigs farmers in Malaysia transmitted from animals to humans, contaminated food or fruits like raw date palm juice, contaminated with urine or saliva from infected fruit bats in Bangladesh and India, causing illnesses from subclinical features or asymptomatic infection or acute respiratory illness and fatal neurological encephalitis with symptoms of fever, headaches, muscle pain, nausea, vomiting, sore throat dizziness, drowsiness, altered consciousness, atypical pneumonia, severe respiratory problems, acute respiratory distress (ARD).

Fruit bat (*Pteropus* sp) under family *Pteropodidae* and *Pteropus* genus is the natural reservoir. Human to human transmission could occur by secretions and excretions, contact of health care givers with infected individuals.

The main diagnostic tests include real-time polymerase chain reaction (RT-PCR) from body fluids and antibody detection via enzyme-linked immunosorbent assay (ELISA), polymerase chain reaction (PCR) assay, and virus isolation by cell culture. This infection lacks treatment by specific drugs or specific vaccines. Nipah virus has been reported to cause outbreaks in pigs, domestic animals including cats, dogs, horses, goats, sheep, and is most contagious in pigs with incubation period ranging from 4 to 14 days.

Prevention strategies lie in controlling Nipah virus in pigs, restriction of animal to human transmission, restriction reservoir-bat to human transmission, animal health and wildlife surveillance system, reducing or preventing human–human infection in people by spreading awareness of risk factors and mass education, and follow procedures to reduce infection transmission (https://www.who.int/news-room/fact-sheets/detail/nipah-virus).

The 2018, the Nipah virus outbreak in Kerala, India, highlighted the importance of global surveillance of RNA viruses, henipaviruses, under family

Paramyxoviridae, order *Mononegavirales* with 5 species, natural reservoirs being pteropid fruit bats and microbats of several species causing illness and death in domestic animals and humans (Plowright et al. 2019).

30.3.3 Rift Valley Fever (RVF)

RVF is a vector-borne viral zoonosis belonging to the genus *Phlebovirus and was* first identified in 1931 in an epidemic in sheep from Rift Valley of Kenya. It affects adversely the health of public, livestock, and production of domestic animals including buffalo, cattle, camels, goats, and sheep and thereby affecting the food security across the African continent, sub-Saharan Africa, and the Arabian Peninsula. It can be transmitted to human by blood, body fluids, or tissues of infected animals, or by bites from infected *Aedes* and *Culex* mosquitoes.

Machine learning methods have enabled prediction of RVF risk in Africa and the Arabian Peninsula in the different landscapes (Walsh et al. 2017).

30.3.4 Studies on Hosts and Reservoirs

Bats are reservoirs for pathogens, including henipaviruses and filoviruses (Crowley et al. 2020). Machine learning was used in surveillance of insectivorous bats (*Megaderma spasma*), fruit bats (*Pteropus medius, P. medius*) revealing Indian species that was Nipah virus seropositive, or which are reported as reservoirs of Nipah virus and could assess the risk of Nipah virus outbreaks in India (Plowright et al. 2019). Sero-survelliance on filovirus and henipavirus was performed using machine learning algorithm, phylofactorization, to understand the bat phylogeny and the clade structures in filovirus and henipavirus infection. However based on bat phyllogenetic tree more studies on surveillance for these pathogens need to be done (Crowley et al. 2020).

Machine learning has been used to identify arthropod vectors from viral genome sequences as animal origins of RNA viruses causing infectious diseases and finds importance in disease surveillance (Babayan et al. 2018). Machine learning has been used to find out drivers of viral diversity in bats like viral genome type and predict undetected viral carriers revealing that long living bat species in large groups distributed broadly in eastern hemisphere carry more viruses (Guy et al. 2020).

Avian strains of influenza can overcome species barrier and infect the humans causing pandemics. Wild birds being the IAV (influenza A virus) reservoir, new strains like H7N9 can emerge and therefore understanding host tropism is essential in understanding the underlying mechanism of virus strains crossing species barrier, thereby spreading the infection in humans. Machine learning tools of random forest have been employed in prediction of host tropism based on training on influenza protein sequences from both avian and human samples indicating the distinctness (Eng et al. 2014). Machine learning tools have been used to build computational model to predict zoonotic strains of influenza A virus and estimate zoonotic risk from protein sequences. This type of study finds importance in prediction on influenza outbreak and therefore prompt action can be taken to save the human population (Eng et al. 2017).

Random forests algorithm has been used to predict viral ecology and environmental niche of avian influenza virus in wild birds which is helpful in prediction of disease outbreak. The primary niche was found to be at areas of low annual rainfall and low temperatures and model revealed low level pathogenicity of avian influenza virus in wild birds (Herrick et al. 2013).

30.4 Discussion

Machine learning tools and algorithms enable identification of animals that are transmitting deadly pathogens to humans, but the model needs more refinement to understand disease reservoirs at species level. Attempts are being made to understand the genetic information in Ebola to predict, the virus and their reservoirs. Understanding the reservoir could prevent future emerging disease outbreaks in humans.

However certain bottlenecks in this domain remain in our lack of knowledge of the exact sequences of pathogens that are harbored by the animal, not causing it sick but if infected in humans can transmit the disease, better understanding if viral transmittance and strains of virus and their families infecting humans with known hosts as well as unknown hosts, data on infectious disease surveillance is required for in-depth analysis.

Thus although machine learning offers promising and vital application, in this domain, far more studies are needed in this area of biological sciences in human benefit.

References

Atawodi JC, Dzikwi AA, Odoba MB, Dagai DD (2013) Animals as reservoir of some human diseases. Electron J Biol 9(2):24–28

Babayan SA, Orton RJ, Streicker DG (2018) Predicting reservoir hosts and arthropod vectors from evolutionary signatures in RNA virus genomes. Science 362(6414):577–580

Burroughs T, Knobler S, Lederberg J, Institute of Medicine (US) Forum on Emerging Infections (eds) (2002) The emergence of zoonotic diseases: understanding the impact on animal and human health: workshop summary. National Academies Press, Washington. 2, The Importance of Zoonotic Diseases. Available from: https://www.ncbi.nlm.nih.gov/books/NBK98094/

Calisher CH, Childs JE, Field HE, Holmes KV, Schountz T (2006) Bats: important reservoir hosts of emerging viruses. Clin Microbiol Rev 19(3):531–545

Crowley D, Becker D, Washburne A, Plowright R (2020) Identifying suspect bat reservoirs of emerging infections. Vaccines (Basel) 8(2):228

Eng CL, Tong JC, Tan TW (2014) Predicting host tropism of influenza a virus proteins using random forest. BMC Med Genet 7(Suppl 3):S1

Eng CLP, Tong JC, Tan TW (2017) Predicting zoonotic risk of influenza a viruses from host tropism protein signature using random Forest. Int J Mol Sci 18(6):1135

Guy C, Ratcliffe JM, Mideo N (2020) The influence of bat ecology on viral diversity and reservoir status. Ecol Evol 10(12):5748–5758

Herrick KA, Huettmann F, Lindgren MA (2013) A global model of avian influenza prediction in wild birds: the importance of northern regions. Vet Res 44(1):42

Lee J, Kim SY, Hwang KJ, Ju YR, Woo HJ (2013) Prion diseases as transmissible zoonotic diseases. Osong Public Health Res Perspect 4(1):57–66

Myrna E (2008) Watanabe, animal reservoirs: Harboring the next pandemic. Bioscience 58(8): 680–684

Parrish CR, Murcia PR, Holmes EC (2015) Influenza virus reservoirs and intermediate hosts: dogs, horses, and new possibilities for influenza virus exposure of humans. J Virol 89(6):2990–2994

Plowright RK, Becker DJ, Crowley DE et al (2019) Prioritizing surveillance of Nipah virus in India. PLoS Negl Trop Dis 13(6):e0007393

Walsh DP, Ma TF, Ip HS, Zhu J (2019) Artificial intelligence and avian influenza: using machine learning to enhance active surveillance for avian influenza viruses. Trans bound Emerg Dis 66(6):2537–2545

Walsh MG, Willem de Smalen A, Mor SM (2017) Wetlands, wild Bovidae species richness and sheep density delineate risk of Rift Valley fever outbreaks in the African continent and Arabian peninsula. PLoS Negl Trop Dis 11(7):e0005756

Wang LF, Crameri G (2014) Emerging zoonotic viral diseases. Rev Sci Tech Off Int Epiz 33(2): 569–581

Weber DJ, Rutala WA (1999) Zoonotic infections. Occup Med 14(2):247–284

Challenging Problems in Plant Biology 31

Abstract

The immensely complex biology of the plant system is far from being completely understood and there are intriguing domains and research problems that remain to be understood in the plant sciences. With the advent and increased applications of computer science, bioinformatics, and lately machine learning tools and applications, the mysteries in plant sciences in diverse domain are beginning to unfold with importance in better plant growth, increased production and better agriculture and food resource with higher yield. In this chapter we discuss some of the intriguing, immensely complex domains in plant biology. In subsequent chapters we highlight the different applications of machine learning in diverse domains of plant biology research.

Keywords

Plant · Biology · Genome · Ecology

31.1 Introduction

Plants find importance in maintaining and improving environment and play role in controlling soil erosion, pollution, reducing air pollution. The human and industrial activity, population growth, increased pollution, cut down of forest and use of land in human benefit have impacted the global climate change, affecting adversely human health and well-being. Thus the significance of plants to human and human society is immense.

Plants find importance in meeting the ever-expanding growing population and its hunger and therefore focus on practices in agricultural sciences towards improved and increased yield finds importance. Significant improvements are necessitated in the cultivation and growth of basic crops rice (*Oryza sativa*), wheat (*Triticum*

S. Ghosh, R. Dasgupta, *Machine Learning in Biological Sciences*,
https://doi.org/10.1007/978-981-16-8881-2_31

aestivum), maize (*Zea mays*), the oil crops including Canola (*Brassica napus*), mustard (*Brassica campestris*), soy (*Glycine max*), sunflower (*Helianthus* spp.) oil palm (*Elaeis guineensis*), and commodity crops including cotton (*Gossypium* spp.), tea (*Camellia sinensis*), and coffee (*Coffea* spp.) for human consumption and animal feed. More research needs to be done to generate variants by genetic manipulation of disease resistant, stress resistant varieties and that do not lose to weeds, improve the efficiency of fertilizer (Grierson et al. 2011; Kossmann 2012).

Our basic needs of food, shelter, clothes, and energy are all derived from plants and the society needs to be made aware by mass education on the importance of plants. Attempts in the lines of use of plant science in prevent malnutrition, improving human health, improving the urban environment, understanding seed longevity and dormancy, understanding and control flowering time, understanding signaling and hormonal cross-talk, generation of salt, heavy metal, drought-tolerant varieties, practices of "clever farming," understanding plant succession of any habitat, understanding impact of climate change on crop varieties, control of weeds, understanding plant-plant communication, generating pathogen resistant strains, understanding differentiation and dedifferentiation, plant stem cell biology, importance of short RNA in plants are few of the important intriguing questions related to plant biology (Grierson et al. 2011; Kossmann 2012).

The science of plant biology focuses on the different vital domains of understanding complexity involved in biological systems, pathways, regulations, genomes, networks in control of biological processes, understanding the diversity within species, understanding plant diseases, pathogens, plant–pathogen interactions, role of soil, weather factors in plant growth, monitoring of plant health, improved plant growth, produce and yield.

31.2 Some of the Interesting Areas and Challenges in Study in Plant Sciences

Understanding and monitoring of the ecology and biology of the environment and the pathogen like microbial population is fundamental for healthy growth of crops and prevent loss of crop yield. Thus knowledge of plant pathogen ecology, factors encompassing pathogen survival, diversification, pathogenic potential of a strain, plant–pathogenic interactions finds importance more so in the context of altered global climate change and would help in predicting risk or emergency of a pathogen attack and plant disease (Morris et al. 2017) and calls in for more research. In the subsequent chapters we have highlighted the applications of machine learning on understanding of plant pathogen interaction. Although the applications have just started to be investigated, it remains a major scope of research in the ongoing world.

One major intriguing domain in plant biology research is understanding the growth regulations which is immensely complex involving several networks. Multiscale models have been designed to understand the regulatory circuits in plant biology including parameters of hormone response, metabolism, circadian clock, growth biomechanics, and decipher the complexity of growth and development of new organ (Hodgman and Ajmera 2015).

Another interesting domain in plant biology research is the study of genetic variation in species and plants at molecular level genomics. Although studies from genomics, comparative genomic analysis, phylogenomics, evolutionary analysis and genome-wide association studies (GWAS), have generated a high data, there are need of tools to handle, manage and analyze such high-throughput data. In the current times effective analysis of the genomics data and its application in the increase in plant productivity (Ong et al. 2016; Bevan and Walsh 2004) remains a major challenge. Thus computational biology, tools of machine learning find importance in resolving this challenge of the plant biology research. In the subsequent chapters we have highlighted the importance of machine learning in different domains of plant biology research.

Protein phosphorylation is known to regulate vital mechanisms of plant life; however, studies are limited due to lack of suitable methods to detect the dynamic nature of the proteins. Significant challenges remain in methods related to enrichment, separation, detection, quantification, and localization of phosphorylated proteins and peptides and is a predominant area of research in the plant signal transduction aiding in the immensely complex vital processes of growth and development (Li et al. 2015).

Protists form essential component of the soil microbiome and understanding the diversity and ecology of the soil protists is the key to understand dynamics of plant pathogens populations and communities that play role in affecting plant growth produce. Predatory soil protists play role in plant growth by releasing nutrients in soil that enhance plant growth and also find importance as soil quality bioindicators of soil quality, understanding of ecotoxicology, potential biofertilizers, and biocontrol agents. A major domain of Biology research should be focused on understanding, characterization, studies on morphology, taxonomy and diversity of soil protists and their role in plant growth produce and development (Geisen et al. 2018).

Citizen science is new nomenclature involving research as a collaborative project with professional scientists highlighting branches of physics and biology in solving problems of the public like challenges in agriculture or food research (Ryan et al. 1891).

Understanding gene function, redundancy, lethality in plant biology is finding importance and finds importance in pharmaceutical and drug discovery. However more research on chemical genetics and direct target identification are necessitated (Dejonghe and Russinova 2014).

Several genetic engineering techniques and tools including application of sequence-specific nucleases like zinc-finger nucleases (ZFNs), transcription activator-like effector nucleases (TALENs), and CRISPR/Cas-9 has been applied in design of genetic variations in genomes of diverse organisms including plants. Generating additions, insertion, deletion, and/or substitution in the genetic sequence and reverse-engineering techniques enables understanding of basic and fundamental mechanisms in plant biology research and is known to play role in food security research (Satheesh et al. 2019; Zhang et al. 2019; Sedeek et al. 2019).

Big data and applications in plant biology have enabled analysis of complex data from plant genomes and have been considerable developed in the last decade together with the advanced applications of next-generation sequencing (NGS), whole-genome sequencing (WGS), and mass-spectroscopic studies in plant models systems and crop plants. In plant science an open source cyberinfrastructure platform called iPlant has been developed (Popescu et al. 2016) to promote collaborative research with inputs from research community globally together with high-performance computational services.

Gene expression data across different species and strains is vital towards our understanding of Plant Biology, processes, but it suffers from the challenges from the lack of suitable technological innervations to compute and display expression networks from more than one species and thus plant comparative transcriptomics is a major domain of research in the recent years (Movahedi et al. 2012).

A plant cell atlas (PCA) is being designed to decipher the immensely complex functioning of a plant cell, mapping molecular mechanisms in cellular and subcellular domains (Rhee et al. 2019) and could be a step towards understanding and increasing plant growth and produce.

Together with genomics and proteomics, metabolomics studies in plants have been instrumental in understanding mechanisms of plant growth, development, and stress responses, which has immense applications towards increasing efficient plant growth, yield, produce, improve crop quality and food safety and this domain of biology is developing further and is an active area of research (Hong et al. 2016; Price et al. 2010). Environmental factors are a key challenge to agriculture and need sustainable environment for proper growth and yield of plants. European farmers have highlighted stress factors in cultivation of crops including barley, beet, grapevine, maize, oilseed rape, olive, potato, sunflower and wheat including challenges from pathogens like fungi, insects, viral, bacterial, parasitic infections that affect yield and harvest quality. Apart from wet lab techniques, computational tools like data mining, databases, etc., find importance in this domain of research (Ricroch et al. 2016; Allahverdiyeva et al. 2015).

Arabidopsis thaliana was the first plant to have a sequenced genome in 2000. Since then applications of new algorithms, technologies to generate genome, transcriptome, and epigenome data for plant model species and crop species have developed considerably in plant biology. Challenges in plant biology remain in sequencing of any genome including ploidy, heterozygosity, and paralogy. de novo transcriptome assemblies enable bypass of the complex genomes, together with developments in sequencing platforms, development of software and algorithm to handle large and complex genomes are improving results (Hamilton and Buell 2012). AgriSeqDB (https://expression.latrobe.edu.au/agriseqdb) a database to view, analyze and interpret data on developmental and tissue or cell-specific transcriptome from crops including wheat, rice, maize, barley, tomato and *Arabidopsis* is not only informative but a nice tool to allow a platform for collaborative research (Robinson et al. 2018).

Plant viruses interaction dynamics is not well studied and would be a key to the control of infections in plants that causes hugh loss to the agriculture industry. Thus it is important to understand ecology of plant viruses which also finds importance from the evolutionary point of view (Malmstrom et al. 2011).

Genetic Resources in the Face of New Environmental, Economic and Social Challenges held in Montpellier (France) during 20–22 September 2011 highlighted research on genetic diversity of plant, animal, fungal, and microbial species aimed at understanding of changes in agricultural industry, climate change adaptation, food security, food in meeting the demands of ever growing world population and altered dietary habits (Pautasso 2012).

31.3 Discussion

Plant science addresses vital questions faced by humanity including food security and quality addressing the increasing demands on food. Plants as food are also important in nutritional value and as herbal medicines. Thus improving crops is a major important research domain.

In this chapter we have discussed some of the major problems, challenges, ongoing research, and future scope in the ever expanding domain of plant sciences. In the subsequent chapters we have highlighted the different domains of plant Sciences and the applications of machine learning tools and computation in the study and resolving of the challenging domains in plant biology with the aim of increasing our in-depth understanding of underlying mechanisms and the immensely complex biology of plants, understanding the ecology and evolution, plant–pathogen interactions, solving the food security issues, improving growth, production, yield of plant and to solve the global food demand.

References

Allahverdiyeva Y, Battchikova N, Brosché M et al (2015) Integration of photosynthesis, development and stress as an opportunity for plant biology. New Phytol 208(3):647–655

Bevan M, Walsh S (2004) Positioning Arabidopsis in plant biology. A key step toward unification of plant research. Plant Physiol 135(2):602–606

Dejonghe W, Russinova E (2014) Target identification strategies in plant chemical biology. Front Plant Sci 5:352

Geisen S, Mitchell EAD, Adl S, Bonkowski M, Dunthorn M, Ekelund F, Fernández LD, Jousset A, Krashevska V, Singer D, Spiegel FW, Walochnik J, Lara E (2018) Soil protists: a fertile frontier in soil biology research. FEMS Microbiol Rev 42(3):293–323

Grierson CS, Barnes SR, Chase MW, Clarke M, Grierson D, Edwards KJ, Jellis GJ, Jones JD, Knapp S, Oldroyd G, Poppy G, Temple P, Williams R, Bastow R (2011) One hundred important questions facing plant science research. New Phytol 192:6–12. https://doi.org/10.1111/j.1469-8137.2011.03859.x

Hamilton JP, Buell CR (2012) Advances in plant genome sequencing. Plant J 70(1):177–190

Hodgman TC, Ajmera I (2015) The successful application of systems approaches in plant biology. Prog Biophys Mol Biol 117(1):59–68

Hong J, Yang L, Zhang D, Shi J (2016) Plant metabolomics: an indispensable system biology tool for plant science. Int J Mol Sci 17(6):767. Published 2016 Jun 1

Kossmann J (2012) Grand challenges in plant biotechnology. Front Plant Sci 3:61. https://doi.org/10.3389/fpls.2012.00061

Li J, Silva-Sanchez C, Zhang T, Chen S, Li H (2015) Phosphoproteomics technologies and applications in plant biology research. Front Plant Sci 16(6):430

Malmstrom CM, Melcher U, Bosque-Pérez NA (2011) The expanding field of plant virus ecology: historical foundations, knowledge gaps, and research directions. Virus Res 159(2):84–94

Morris CE, Barny MA, Odile Berge O, Kinkel LL, Lacroix C (2017) Frontiers for research on the ecology of plant-pathogenic bacteria: fundamentals for sustainability: Challenges in Bacterial Molecular Plant Pathology. Mol Plant Pathol 18(2):308–319

Movahedi S, Van Bel M, Heyndrickx KS, Vandepoele K (2012) Comparative co-expression analysis in plant biology. Plant Cell Environ 35(10):1787–1798

Ong Q, Nguyen P, Thao NP, Le L (2016) Curr Bioinformatics Approach in Plant Genomic Research. Genomics 17(4):368–378

Pautasso M (2012) Challenges in the conservation and sustainable use of genetic resources. Biol Lett 8(3):321–323

Popescu GV, Noutsos C, Popescu SC (2016) Big data in plant science: resources and data mining tools for plant genomics and proteomics. Methods Mol Biol 1415:533–547

Price CA, Gilooly JF, Allen AP, Weitz JS, Niklas KJ (2010) The metabolic theory of ecology: prospects and challenges for plant biology. New Phytol 188(3):696–710

Rhee SY, Birnbaum KD, Ehrhardt DW (2019) Towards building a plant cell atlas. Trends Plant Sci 24(4):303–310

Ricroch A, Harwood W, Svobodová Z et al (2016) Challenges facing European agriculture and possible biotechnological solutions. Crit Rev Biotechnol 36(5):875–883

Robinson AJ, Tamiru M, Salby R et al (2018) AgriSeqDB: an online RNA-Seq database for functional studies of agriculturally relevant plant species. BMC Plant Biol 18(1):200. Published 2018 Sep 19. https://doi.org/10.1186/s12870-018-1406-2

Ryan SF, Adamson NL, Aktipis A et al (1891) The role of citizen science in addressing grand challenges in food and agriculture research. Proc Biol Sci 2018(285):20181977

Satheesh V, Zhang H, Wang X, Lei M (2019) Precise editing of plant genomes - prospects and challenges. Semin Cell Dev Biol 96:115–123

Sedeek KEM, Mahas A, Mahfouz M (2019) Plant genome engineering for targeted improvement of crop traits. Front Plant Sci 10:114

Zhang Y, Malzahn AA, Sretenovic S, Qi Y (2019) The emerging and uncultivated potential of CRISPR technology in plant science. Nat Plants 5(8):778–794

Machine Learning and Plant Sciences 32

Abstract

Plant morphology, growth, produce, and immunity to infection bear relation to the agricultural production and economy of the country. Machine learning is having major applications in plant research. We outline in this chapter some of the major areas of application of machine learning in plant research in the domain of (1) plant infection and immunity; (2) plant genomics; (3) stress biology; (4) plant growth; (5) genome selection, plant phenotyping, and morphology; and (6) study of release of airborne pollen.

Keywords

Immunity · Stress · Genome · Pollen · Phenotyping · PTI · ETI

Abbreviations

A thaliana	*Arabidopsis thaliana*
ANN	Artificial neural network
CEBP	Cellulose-binding elicitor proteins
CNN	Convolution neural network
CP	Capsid protein
DL	Deep learning
Ef-Tu	Elongation factors Tu
ETI	Effector triggered immunity
ETL	Exact, transform, load
GS	Genomic selection
HR	Hypersensitive cell death response
IDC	Iron deficiency chlorosis
LPS	Lipopolysaccharide

S. Ghosh, R. Dasgupta, *Machine Learning in Biological Sciences*,
https://doi.org/10.1007/978-981-16-8881-2_32

LRR	Leucine-rich repeat
MAMPs	Microbe-associated molecular patterns
MSV	Maize streak virus
NB	Nucleotide binding
NBS	Nucleotide-binding site
NNI	Nitrogen nutrition index
PAMPs	Pathogen-associated molecular patterns
PRRs	Pattern recognition receptors
PTI	PAMP triggered immunity
R	Resistance
RF	Random forest
RH	Relative humidity
RLK	Receptor-like kinase
RPs	Ribosomal proteins
SVR	Support vector regression
TEs	Transposable elements

32.1 Introduction

Researchers across the globe are trying to extract maximum information from the vast genomic data generated and predict plant behavior in response to altered nutrition, toxins, exposure to stress and pathogens. This in turn finds importance in management of crops, their improvement, disease susceptibility, diagnosis and prognosis and better yield.

Machine learning, multidisciplinary study with properties of predictions, and feature extraction from data have a newly developing application in the domain of biological sciences. Incorporating knowledge and applications from diverse domains of artificial intelligence (AI), computer science, statistics and information science machine learning is enabling analysis of complex datasets from biological systems. This science is being applied in the domain of different branches of biology including research in plant sciences.

Machine learning is finding application in genomics in handling and analysis of huge data effectively including application in functional genomics in plant–microbial interactions, plant–microbe biology, monitoring of plant disease, study of gene regulatory networks (grns) involved in plant biology and microbial interactions, selection of disease resistant strains, and deciphering the roles of pathogen effector genes in plant immunity. It includes development of novel models, biological databases, software packages, and web servers targeted to specific problems in plant biology. This finds particular importance in the study of plant–pathogen interaction in crop plants of agronomic importance.

Techniques of imaging are enabling study of plant morphology, growth, affected physiology when plants are exposed to several stress factors. Machine learning has

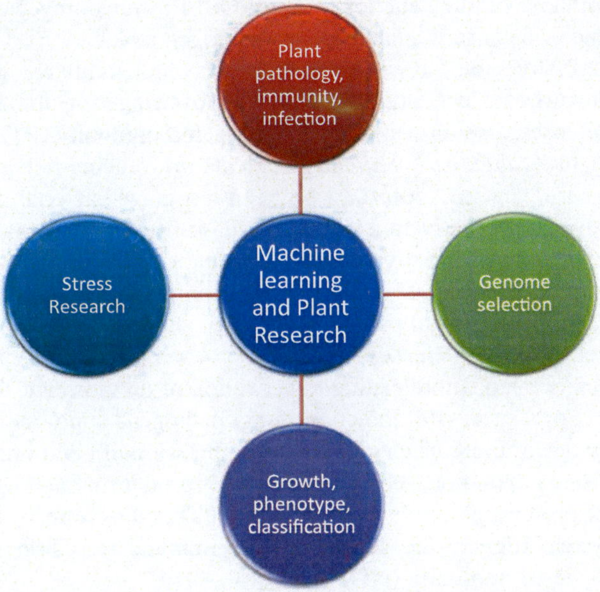

Fig. 32.1 Major applications of Machine Learning in Plant Research

enabled data extraction, prediction of models, and access to public source databases and is a step towards in-depth study of plant molecular biology (Mishra et al. 2019). However, accurate prediction of complex processes in agriculture still remains a major challenge (Fig. 32.1).

32.2 Application of Machine Learning

32.2.1 Plant Immunity and Infection

Pathogens including virus, bacteria, and phytoplasmas being obligate bacterial parasites of plant phloem tissue transmitted by insect vectors invade the plant tissue and establish a parasitic relationship between itself and the plant. Plant disease outbreak is a major concern that can affect agricultural produce causing economic loss of farmers and affect at different levels in the supply chain and can even bear a negative impact on the growth of a country's economies (Varun et al. 2019).

Plant–microbe interaction impairs plant growth and reproduction. The plant lacks adaptive immune system (AIS) that is functional and well developed in vertebrates. Plant immunity encompasses innate immune system. It includes two branches comprising of recognition of pathogens by transmembrane localized pattern recognition receptors (PRRs) and their responses to microbial associated molecular patterns (MAMPS) or pathogen-associated molecular patterns (PAMPs), including flagellin, lipopolysaccharides (LPS) elongation factors Tu (Ef-Tu, Selin et al. 2016), from bacterial pathogens and polysaccharides and chitin from fungal sources cellulose-binding elicitor proteins (CBEP) from oomycete that contributes to the first line of defense in plants against pathogen attack.

PAMPs located within the apoplasts of plants are recognized by PRRs that lead to downstream activation of pathways by triggering the first line of plant immune defenses, also termed as PAMP triggered immunity (PTI).

In *Arabidopsis thaliana (A. thaliana)*, leucine-rich repeat (LRR) receptor-like kinase, flagellin receptor, FLS2 can recognize bacterial flagellin (flg22) and lead to downstream activation of MAP kinase pathway leading to defense response gene activation. PRR activation can lead to altered hormone biosynthesis and deposition of callose comprising of β-1,3-glucan polysaccharide with few β-1,6-branches. Chitin from fungal pathogen has been studied to be recognized by the LysM-RLK (receptor-like kinase) AtCERK1 *A. thaliana* receptor leading to CERK1 homodimerization leading to activation of downstream signaling.

For a successful infection, fungi pathogens apply mechanisms to overcome PTI by one or more mechanisms like suppression of plant immune responses, plant host cell manipulation, secretion of effector proteins by pathogens. These secreted effector proteins play role in suppressing PTI and once recognized by the plant immune system triggers the second line of immune response also called as the effector triggered immunity (ETI, Selin et al. 2016).

These effectors elicit an ETI response and are recognized by plant resistance proteins (R proteins) encoded by the R genes. R genes encode for intracellular proteins containing nucleotide-binding site (NBS) and leucine-rich repeats (LRRs) receptors that function like NLR receptors. Plant NBS LRR proteins comprise of TIR and non-TIR groups based on the sequences preceding the NBS domain wherein the TIR group of proteins includes an amino terminal domain bearing homology to the Toll receptors (TR) and interleukin 1 receptors(IL-1R) and the non-TIR group of proteins in its amino-terminal end contains an α-helical coiled coil like sequence (DeYoung and Innes 2006).

Effector proteins can be recognized both by means of receptor binding and by accessory protein (Selin et al. 2016).

ETI activation has been observed to be associated with pathophysiology of hypersensitive cell death response (HR) at the infection site. The NBS-LRR proteins have been known to play role in recognition of pathogen effectors or avirulence (Avr) proteins leading to a strong HR response.

Although a few effector proteins and those coded by Avr genes have been studied so far from infection in plants caused by fungi and oomycetes, the exact role played by Avr effectors in virulence and plant immune defenses is far from being completely understood. Research in this domain of biology focuses on understanding of their role in plant–microbe interactions (Selin et al. 2016).

Plant immunity is a major domain of importance and machine learning finds important application (Silva et al. 2019). Although the conventional methods to identify R genes and proteins are based on extracting information from sequence similarity and pattern matching sequence annotations of proteins, recently application of machine learning tools is finding importance in their early prediction and detection (Varun et al. 2019) and is an effective and better approach as compared to

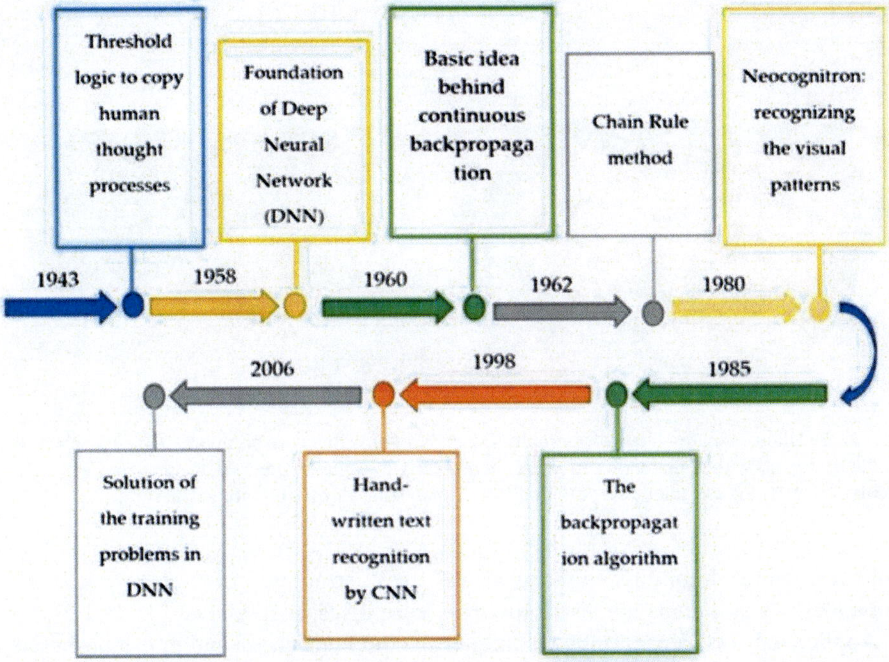

Fig. 32.2 Summary of the evolution of deep learning from 1943 to 2006. Image adapted from (Saleem et al. 2019) under CC-BY License

conventional bioinformatics approaches. A support vector machine (SVM)-based tool has been designed to predict disease resistance proteins in plants (Pal et al. 2016).

The deep learning (DL) branch of machine learning (ML) was introduced in 1943. The application in diverse domains further expanded and improved between 1943 and 2006 including applications of back propagation, chain rule, Neocognitron, handwritten text recognition (LeNET architecture) in the first phase. In the second phase the training problem was resolved and from 2012 onwards deep learning found applications in diverse fields ranging from self-driving cars, applications in healthcare sector, applications in text recognition, predictions of earthquake, marketing, finance, and image recognition tools were further developed. AlexNet is a breakthrough architecture in the field of DL in object recognition ((Saleem et al. 2019), Fig. 32.2). Deep learning found initial application in biological sciences in the understanding of human biological pathways. The application of deep learning in the understanding of plant biology, immune system, plant–pathogen interaction is only a recent one and although at its infancy, the field is rapidly developing.

Infections in plants are caused by virus, bacteria, viroids, prions, fungi, oomycetes, mollicutes, higher plants, invasive climbing plants, blue-green algae, protozoa, nematodes. Adverse environmental factors including temperature,

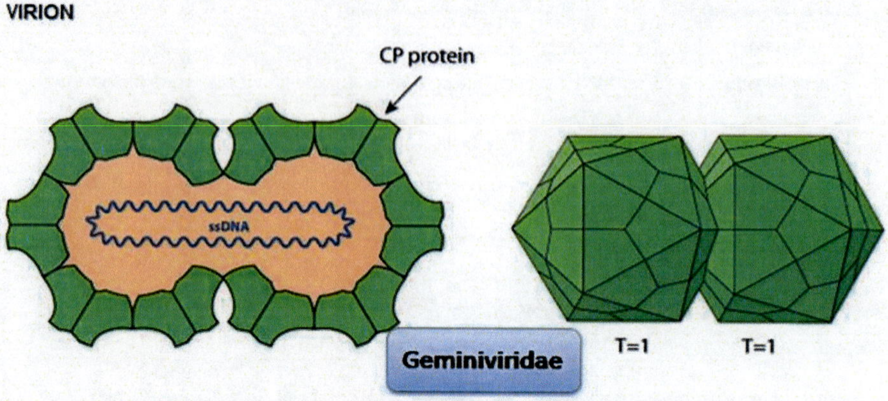

Fig. 32.3 Diagrammatic view of Geminiviridae: Non-enveloped, incomplete T=1 icosahedral symmetry capsid with pentameric capsomers comprising of capsid proteins (CP), presence of single circular ssDNA. Figure adapted from https://viralzone.expasy.org/109 with permission

moisture, wind, humidity, light, soil pH, soil structure, host-plant nutrition, herbicides, air pollutants are also known to cause diseases in plants.

Applications of deep learning techniques in computer vision and threshold-based segmentation together with a series of deep convolutional neural networks (CNNs) have been used to diagnose the severity of the disease in plants and therefore in immensely important from the point of view of application in the domain of modern agriculture (Wang et al. 2017).

Viruses can cause diseases. Family Geminiviridae, or Class II virus including twinned, non-enveloped single-stranded (ss) circular DNA (2.5 to 3.0 kb) is known to infect both dicotyledonous (dicot) and monocotyledonous (monocot) plants causing different diseases affecting crops and is a major cause of concern predominantly in the tropics and subtropics (Fig. 32.3).

Machine leaning tools have enabled high precision and effective gene prediction. Fangorn Forest (Silva et al. 2017a) has enabled the classification of Geminiviridae including nine genera, including *Mastrevirus, Capulavirus, Curtovirus, Grablovirus, Topocuvirus, Eragrovirus, Turncurtovirus*, *Becurtovirus* with monopartite genomes and *Begomovirus* with mono- or bipartite genomes according to their insect vector, host range, genome organization.

Geminiviruses are transmitted to host plants by insects belonging to four families including homopterans including aphids, leafhoppers, treehoppers, and whiteflies. While *Begomovirus* are predominantly transmitted by whiteflies, those of the genera *Becurtovirus, Curtovirus*, *Mastrevirus*, and *Turncurtovirus* are transmitted by leafhoppers, *Capulavirus* is transmitted by an aphid, and *Topocuvirus* are transmitted by treehoppers. The vectors of *Eragrovirus* and *Grablovirus* remain unknown (https://talk.ictvonline.org/ictv-reports/ictv_online_report/ssdna-viruses/w/geminiviridae).

The reconstruction of their phylogeny has been possible through applications of data mining approaches of ETL (extract, transform, load), and application of machine learning (ML) techniques in analysis of data obtained from genomics based studies (Silva et al. 2017b). Subsequence natural vector (SNV) and support vector machine (SVM) tools have been applied in classification of begomoviruses genome and prediction of genes by representing the genome sequence as a vector numerically by SNV method followed by SVM application to datasets, thereby building a classification model followed by application of recursive feature elimination (RFE) to select essential features of the subsequence based on its importance (Silva et al. 2017b).

Eukaryotic filamentous plant pathogens like fungus have been known to secrete effector proteins that can modulate the host cell, thereby facilitating infection, play role in host cell regulation thereby obtaining nutrients and play role in suppressing plant host immune defenses. Understanding the biology of the secreted effector proteins in plant–fungal pathogen interaction finds importance in understanding the role of effectors in triggering plant defense responses, generation of plant resistance, relation of pathogen effectors, plant targets, and plant immune system components.

Mummy berry, caused by fungal infection *Monilinia vaccinii-corymbosi*, infecting the floral and vegetative buds of highbush blueberry in the Pacific Northwest (PNW) of USA causes agricultural loss and acts as a burden to the economy.

Fungal infection poses a threat to production and yield affecting adversely the agricultural crops with an estimated 15% loss of crops per annum due to plant diseases. Farmers use diverse practices of use of resistant varieties, use of fungicide, rotation of crops in controlling plant infections.

However, disease control using the fungicide which is an environmental pollutant poses threat to the environment.

Release of ascospore, based on parameters including temperature of air and soil, precipitation, moisture content of the soil, wetness of leaf, relative humidity (RH), and solar radiation was used to derive a prediction model using logistic regression (LR), multivariate adaptive regression spines, artificial neural networks (ANNs), and random forest (RF, Harteveld et al. 2017) and finds importance in prediction and control of fungal infection.

EffectorP application tool using machine learning finds application in prediction of effector proteins against fungal infection in plants from secreted proteins (Sperschneider et al. 2016).

32.2.2 Plant Genomics

Transposable elements (TEs) with natural repetitive numbers and high structural diversity are also termed as jumping genes were regarded as "junk DNA," but have been known to play profound role in regulating chromosomal structures, expression of gene, adaptation, and evolution. Several bioinformatics tools find applications study of TE. However they suffer from the point of view of reliability, precision, and analysis speed. Machine learning and deep learning find importance in identification of TEs (Orozco-Arias et al. 2019). Genomic selection (GS) enables the rapid selection of better genotypes of plants of better breeds. In the recent times applications of hyperspectral imaging technology are being applied for better and effective selection of genomes (Crossa et al. 2017).

32.2.3 Stress Biology

Adverse abiotic environmental factors including high rainfall, moisture, relative humidity in the air, temperature, wind, light, soil pH, soil structure, role played by minerals in plant nutrition can increase the susceptibility to infection, rendering the plants weak, slower growth, and aging faster in plants and spread the infection in already infected plants.

Automated and high-throughput imaging technologies have been applied to phenotyping of plants affected by biotic and abiotic stress. High-resolution images of stress induced effects in plants and sensor data of plants have enabled their identification, classification, quantification, and prediction (ICQP, Pal et al. 2016; Saleem et al. 2019). Deep learning (DL) finds application in digital image-based plant stress phenotyping (Singh et al. 2018). It has been possible to design an image-phenotyping pipeline for the genetic studies of abiotic stress factors causing iron deficiency chlorosis (IDC) in soybean plants (Zhang et al. 2017). Machine learning algorithms have been instrumental in detecting the roles of long non-protein coding RNAs and their roles in stress responses (Simopoulos et al. 2018). Machine learning (ML) has been applied to predict stress responsive genes in plants (Torregrossa et al. 2018).

32.2.4 Plant Growth, Genome Selection, and Phenotyping

Plant phenotyping involves understanding of genome and its interaction with the environment that affects plant phenotypes by physical and biochemical processes. Conventional studies in this domain are by manually observing these altered traits. Manual observation of traits suffers from the limitation of being subjective and varies with the person concerned in monitoring, prone to human errors, low-throughput in nature, labor-intensive method, may entail cost and time and also difficult in application to larger number of samples. Therefore, image-based plant phenotyping by automated digital cameras provides a promising tool in terms of objectivity, noninvasiveness, being faster, reliable, accurate, and low cost method for analysis. Different superpixel-based machine learning algorithm of Random Forest (RF) has been used to study segmented images and by automated analysis infer at the plant growth (Lee et al. 2018). The plant phenotyping system includes automated robotic features and hardware for image-capturing module, use of sensors for data encompassing environmental parameters, irrigation and light controllers and suitable software with feature extraction, generation of classifiers, and removal of noise for accurate analysis of data.

Machine learning algorithms including neural network (NN), random forest (RF), support vector regression (SVR), and together with advances in imaging technology are instrumental in monitoring plant nitrogen (N). This could be done by detection of nitrogen nutrition index (NNI). The application finds advantages in its low cost, efficiency, timely and accurate method of analysis and finds importance in precision fertilization in studies conducted in pakchoi (*Brassica campestris* ssp. *chinensis* L., (Xiong et al. 2019).

Image-based analysis of plant phenotypes applied in Pheno-Deep Counter could predict leaf count in two-dimensional (2D) plant images of different species with a rosette-shaped appearance (Giuffrida and Doerner 2018). Plant image identification of ornamental plant species in Beijing Forestry University campus has been done from image data collected in mobile phones using a 26 layer DL model with 8 residual building blocks (Sun et al. 2017). An algorithm was proposed for automatic tomato detection (Liu et al. 2019) using image and color analysis of tomatoes.

Leaf reflectance spectroscopy and machine learning to analyze the data have been used to predict parameters of photosynthetic capacity in C3 crop *Brassica oleracea* and C4 crop *Zea mays* and find application in crop improvement (Heckmann et al. 2017). Plant-derived natural products applied in disease prevention and treatment are called as meridians in traditional Chinese medicine (TCM) and its classification has been studied (Wang et al. 2019) by machine learning methods.

Ortho-mosaic and digital elevation models (DEM) were constructed from the raw-multispectral images from cotton plant based on characters of canopy cover, height of the plant height, flower and vegetation index(Xu and Li 2019) using the structure from motion (SfM) algorithm.

Ribosomal proteins (RPs) play important role within all types of cells, as they are major components of ribosomes, playing role in translation of mRNAs. Genes coding for RPs, RPL12, and RPL19 are known to be involved in non-host disease resistance in plant–pathogen interaction in *Nicotiana benthamiana* (*N benthamiana)* plants. Rama, a computational method has been used for the prediction of RPs (Carvalho et al. 2017).

Feature extraction algorithms and deep convolution neural networks (CNN) architecture has enabled processing of images from photos and videos from fruit trees to detect fruits in orchard (Bresilla et al. 2019) even from branches in plants where they remain obscure to the naked eye. Fruit occlusion remains a major bottleneck in harvesting in tree canopy. Therefore accurate estimation of fruit number based on machine learning based model finds importance. This model, trained with images of fruit trees with occluded fruits has enabled detection of even partially occluded fruits which are difficult to be detected visually (Zewdie et al. 2019) (Fig. 32.4).

Fig. 32.4 Pears detected from the updated model. Image adapted from open-access article (Bresilla et al. 2019)

32.2.5 Airborne Pollen Release

It has been possible to detect the concentration of airborne *Ambrosia* pollen through approaches of machine learning and deep learning (Zewdie et al. 2019).

32.3 Discussion

Plant pathogen interaction is a very intriguing and developing branch of Science. A major interest lies in the study of the plant immune system and understanding the different key components and the roles played by them (Jones and Dangl 2006). The interaction of host and pathogens and release of pathogen that trigger the plant molecular immunity in turn activating the downstream signaling processes is in itself a complex process. While we are learning more on the different components of effectors, R genes, receptors involved in plant pathogen interactions, based on data and information from omics based studies, the whole of plant immune system and its response to pathogens and role played by these effectors in establishing successful infection process is immensely complex and far from being completely understood. Application of different tools and techniques of machine learning, ANN and DL has immense application in diverse domains of biological sciences. The application in plant sciences in the domain of plant–pathogen interaction is just started and finds importance in the agriculture and horticulture based industry. It is helping in our understanding of pathophysiology of disease (Saleem et al. 2019) and its early detection (AsefpourVakilian 2020), detection of nitrates (Massah and AsefpourVakilian 2019; Massah et al. 2019), analysis of plant phenotypes from images by feature extraction methods (Lee and Chang 2018).

Yet the science is as its infancy and there is scope for further research in the area of plant biology.

References

AsefpourVakilian K (2020) Machine learning improves our knowledge about miRNA functions towards plant abiotic stresses. Sci Rep 10:3041. https://doi.org/10.1038/s41598-020-59981-6

Bresilla K, Perulli GD, Boini A, Morandi B, Corelli Grappadelli L, Manfrini L (2019) Single-shot convolution neural networks for real-time fruit detection within the tree. Front Plant Sci 10:611. https://doi.org/10.3389/fpls.2019.00611

Carvalho TFM, Silva JCF, Calil IP, Fontes EPB, Cerqueira FR (2017) Rama: a machine learning approach for ribosomal protein prediction in plants. Sci Rep 7(1):16273

Crossa J, Pérez-Rodríguez P, Cuevas J, Montesinos-López O, Jarquín D, de Los CG, Burgueño J, González-Camacho JM, Pérez-Elizalde S, Beyene Y, Dreisigacker S, Singh R, Zhang X, Gowda M, Roorkiwal M, Rutkoski J, Varshney RK (2017) Genomic selection in plant breeding: methods, models, and perspectives. Trends Plant Sci 22(11):961–975

DeYoung B, Innes R (2006) Plant NBS-LRR proteins in pathogen sensing and host defense. Nat Immunol 7:1243–1249

Giuffrida MV, Doerner P, Tsaftaris SA (2018) 4Pheno-deep counter: a unified and versatile deep learning architecture for leaf counting. Plant J 96(4):880–890

Harteveld DOC, Grant MR, Pscheidt JW, Peever TL (2017) Predicting ascospore release of Monilinia vaccinii-corymbosi of blueberry with machine learning. Phytopathology 107(11): 1364–1371

Heckmann D, Schlüter U, Weber APM (2017) Machine learning techniques for predicting crop photosynthetic capacity from leaf reflectance spectra. Mol Plant 10(6):878–890

Jones JDG, Dangl JL (2006) The plant immune system. Nature 444:323–329

Lee U, Chang S, Putra GA, Kim H, Kim DH (2018) An automated, high-throughput plant phenotyping system using machine learning-based plant segmentation and image analysis. PLoS One 13(4):e0196615

Liu G, Mao S, Kim JH (2019) A mature-tomato detection algorithm using machine learning and color analysis. Sensors (Basel) 19(9):E2023

Massah J, AsefpourVakilian K (2019) An intelligent portable biosensor for fast and accurate nitrate determination using cyclic voltammetry. Biosyst Eng 177:49–58

Massah J, AsefpourVakilian K, Torktaz S (2019) Supervised machine learning algorithms can predict penetration resistance in mineral-fertilized soils. Commun Soil Sci Plant 50(17): 2169–2177

Mishra B, Kumar N, Mukhtar MS (2019) Systems biology and machine learning in plant-pathogen interactions. Mol Plant-Microbe Interact 32(1):45–55

Orozco-Arias S, Isaza G, Guyot R (2019) Retrotransposons in plant genomes: structure, identification, and classification through bioinformatics and machine learning. Int J Mol Sci 20(15): E3837

Pal T, Jaiswal V, Chauhan RS (2016) DRPPP: a machine learning based tool for prediction of disease resistance proteins in plants. Comput Biol Med 1(78):42–48

Saleem MH, Potgieter J, Arif KM (2019) Plant disease detection and classification by deep learning. Plan Theory 8:468

Selin C, de Kievit TR, Belmonte MF, Fernando WG (2016) Elucidating the role of effectors in plant-fungal interactions: Progress and challenges. Front Microbiol 7:600. Published 2016 Apr 27. https://doi.org/10.3389/fmicb.2016.00600

Silva JCF, Carvalho TFM, Fontes EPB, Cerqueira FR (2017a) Fangorn Forest (F2): a machine learning approach to classify genes and genera in the family Geminiviridae. BMC Bioinform 18(1):431

Silva JCF, Carvalho TFM, Basso MF, Deguchi M, Pereira WA, Sobrinho RR, Vidigal PMP, Brustolini OJB, Silva FF, Dal-Bianco M, Fontes RLF, Santos AA, Zerbini FM, Cerqueira FR, Fontes EPB (2017b) Geminivirus data warehouse: a database enriched with machine learning approaches. BMC Bioinform 18(1):240

ilva JCF, Teixeira RM, Silva FF, Brommonschenkel SH, Fontes EPB (2019) Machine learning approaches and their current application in plant molecular biology: a systematic review. Plant Sci 284:37–47

Simopoulos CMA, Weretilnyk EA, Golding GB (2018) Prediction of plant lncRNA by ensemble machine learning classifiers. BMC Genomics 19(1):316

Singh AK, Ganapathysubramanian B, Sarkar S, Singh A (2018) Deep learning for plant stress phenotyping: trends and future perspectives. Trends Plant Sci 23(10):883–898

Sperschneider J, Gardiner DM, Dodds PN, Tini F, Covarelli L, Singh KB, Manners JM, Taylor JM (2016) EffectorP: predicting fungal effector proteins from secretomes using machine learning. New Phytol 210(2):743–761

Sun Y, Liu Y, Wang G, Zhang H (2017) Deep learning for plant identification in natural environment. Comput Intell Neurosci 2017:7361042

Torregrossa D, Leopold U, Hernández-Sancho F, Hansen J (2018) Machine learning for energy cost modelling in wastewater treatment plants. J Environ Manag 223:1061–1067

Varun TP, Jaiswal V, Chauhan S (2019) DRPPP: a machine learning based tool for prediction of disease resistance proteins in plants. Mol Plant-Microbe Interact 32(1):45–55

Wang G, Sun Y, Wang J (2017) Automatic image-based plant disease severity estimation using deep learning. Comput Intell Neurosci 2017:2917536

Wang Y, Jafari M, Tang Y, Tang J (2019) Predicting Meridian in Chinese traditional medicine using machine learning approaches. PLoS Comput Biol 15(11):e1007249

Xiong X, Zhang J, Guo D, Chang L, Huang D (2019) Non-invasive sensing of nitrogen in plant using digital images and machine learning for brassica campestris ssp. Chinensis L. Sensors (Basel) 19(11):E2448

Xu R, Li C, Paterson AH (2019) Multispectral imaging and unmanned aerial systems for cotton plant phenotyping. PLoS One 14(2):e0205083

Zewdie GK, Lary DJ, Levetin E, Garuma GF (2019) Applying deep neural networks and ensemble machine learning methods to forecast airborne ambrosia pollen. Int J Environ Res Public Health 16(11):E1992

Zhang J, Naik HS, Assefa T, Sarkar S, Reddy RV, Singh A, Ganapathysubramanian B, Singh AK (2017) Computer vision and machine learning for robust phenotyping in genome-wide studies. Sci Rep 7:44048

Machine Learning in Understanding of Plant–Pathogen Interactions

33

Abstract

Plant pathogens including virus, viroids, bacteria, mollicutes, higher plants and green algae, nematodes, and protozoa can cause infectious diseases in plants and can be a major problem for the agriculture and economy of the country. Environmental factors also pose additional threats to the plant health. Therefore it is important to understand the pathogens causing infection and plant–pathogen interaction in order to control plant diseases. Machine learning is enabling better understanding of plant–pathogen interaction. We highlight in this chapter a brief history of the advances in the study of plant–pathogen interaction, tracing the history of scientific developments in the field of plant pathology and recent applications of machine learning in solving some of the major challenges in the domain of plant pathology.

Keywords

Virus · Viroids · Bacteria · Mollicutes

Abbreviations

A. thaliana	*Arabidopsis thaliana*
A. tumefaciens	*Agrobacterium tumefaciens*
	P. megasperma Phytophthora megasperma
ApEn	Approximate multi-scale entropy
B. distachyon	*Brachypodium distachyon*
BN	Bayesian network
$CuSO_4$	Copper sulfate
DDT	Dichlorodiphenyltrichloroethane
F. graminearum	*Fusarium graminearum*

S. Ghosh, R. Dasgupta, *Machine Learning in Biological Sciences*,
https://doi.org/10.1007/978-981-16-8881-2_33

GRN	*Gene regulatory network*
H. annosum	*Heterobasidion annosum*
HBD	Head blight disease
HC	Hydrocarbons
Hrp	Hypersensitive protein
lncRNA	Long noncoding RNA
miRNA	MicroRNA
MMP	Maximal Mappable Prefix
NaCl	Sodium chloride
P. carotovorum	*Pectobacterium carotovorum*
P. gigantea	*Phlebiopsis gigantea*
Pol	Polymerase
PPIs	Protein–protein interactions
S. typhimurium	*Salmonella typhimurium*
SAR	Systemic acquired resistance
	P. infestans Phytophthora infestans
ss	Single stranded
SVM	Support vector machine
Z. mays	*Zea mays*

33.1 Introduction

Plant pathology involves the study of plant diseases caused by unfavorable environmental conditions and infections by pathogens that affect the cultivation and growth of crops, cause destruction of plant parts, and cause reduction in crop yields. It also encompasses the study of mechanisms involved in induction of disease, plant–pathogen interactions, and the methods involved in disease control. Diseases, insects, and weeds can act on the plants to destroy them contributing to a significant loss of produce. The loss is lower in more developed countries (Agrios 2010).

The disease-causing agents in plants include pathogenic microorganisms including bacteria, mollicutes including a type of bacteria lacking cell wall, fungi, viruses, protozoa, nematodes, and unfavorable environmental conditions including lack or excess nutrients, moisture, and light, exposure to toxic chemicals in air or soil, bad practices in agriculture, and exposure to herbicides. Plants also face competition with other plants, weeds, and damages inflicted by insects, humans, and animals. Research in this domain of biology entails exploration of possibilities for plant survival by the control of infectious agents (Agrios 2010).

33.2 Plant Disease

More than thousands of diseases and adverse environmental conditions affect plant growth and produce. Plant diseases are grouped by the causative symptoms like root rots, wilts, leaf spots, blights, rusts, smut, affecting plant organs including root, stem, and foliages. Plant diseases are also grouped by causative agents including diseases caused by infectious biotic agents including fungi, prokaryotes including bacteria and mollicutes, higher plants and green algae, virus and viroids, which are short strand of circular, single-stranded (ss) RNA lacking protein coat, nematodes, and protozoa and those caused by abiotic factors including very high or low temperature, lack or excessive soil moisture, lack of oxygen, caused by air pollution, nutrient deficiencies, mineral toxicities, soil acidity or alkalinity, toxic pesticides, improper culture.

33.3 A Brief History and the Advancement in the Study in Plant Pathology

Plant diseases have been identified and mentioned in the oldest books dating as early as around 1000 BC to 750 BC. In those earlier times, bacteria, fungi, and nematodes associated with plant diseases could be detected by visualizing through primitive devices as magnifying glasses, lenses, microscopes. Development of pure cultures of fungi and bacteria by Brefeld, Koch, and Petri during the period of time spanning from 1875 to 1912 was a step towards better understanding and study of plant pathogens. In 1887, Koch's postulates enabled the identification of particular micro-organism as pathogen.

Later with the development of compound microscope and plant tissue staining techniques and application of cytological and histological techniques enabled better study of pathogens including bacteria, fungi, and nematodes. After 1940, electron microscope enabled visualization of virus and post-1970, it was possible to detect mollicutes and viroids.

Around 1200 AD it was noticed that a mistletoe infection could be cured by pruning the branch of an infected tree carrying the mistletoe infection.

Chemicals were used in control of infection in the late 1600 when some farmers in south England observed planted wheat seed salvaged from a ship wreck with smut/bunt infection. They then treated the wheat seeds with sodium chloride (NaCl) solution to control bunt infection. Later in mid-1700 copper sulfate ($CuSO_4$) replaced the application of NaCl to improve the control of bunt infection in plants.

Around the 1870s the application of chemicals in disease control of aggressive downy mildew disease in grapes was tested. Millardet, in 1885, observed that spray of Bordeaux mixture comprising of a mixture of copper sulfate and lime on vines could prevent infection as compared to the untreated vines. The fungicide Bordeaux mixture was also observed to be effective against diseases like late blight of potato, other downy mildews, leaf spots and blight infection, and different other plant diseases and revealed that plant diseases could be chemically controlled.

In 1913, organic mercury compounds were being used in seed treatments, but in 1960s treatment and use of mercury based pesticides were banned. 1928 records the discovery of antibiotic penicillin by Alexander Flemming that was effective against bacterial infection. However its large-scale application suffered from its limited availability and cost.

In 1934, studies and experiments with fungicide dithiocarbamate or thiram led to the development of fungicides including compounds like ferbam, zineb, maneb. 1965 marked the discovery of systemic fungicide, carboxin, followed by other such fungicides including benomyl.

Streptomycin antibiotics was used to control bacterial infection in 1950, followed by antibiotic cycloheximide, and tetracycline in 1967, in the control of mollicutes mediated infection and infectious diseases caused by fastidious bacteria surviving in the xylem of host plants.

The major limitation of the application of bactericides and fungicides was due to generation of plant resistant to bactericides and fungicides observed as early as in 1954–1970s. This led to the need for better and effective strategies to control infection like using mixture of fungicides, use of alternating compounds, spray of systemic compounds in early stages of disease, and use of broadspectrum compound in later stages of the disease.

Use of pesticides which are toxics gained importance in killing bacteria, fungi, insects, weeds, rodents, and also humans, plants, or animals affecting the plant and was termed as bacteriocide, fungicide, nematicide, herbicide, etc. Some of these pesticides could be broken down to less or nontoxic forms, but chlorinated hydrocarbons (HC) including dichlorodiphenyltrichloroethane (DDT) were persistent to further breakdown and remained toxic for years, thus acting as environmental contaminants causing harm to the ecosystem, causing unwanted deaths of birds and fishes and leading to bioaccumulation in food chain and thus remained the major limitation of their applicability. This leads to increased alarms and protests voiced as early as in 1950s against their use. The pesticides with suspected carcinogenic and mutagenic effects were banned. This led to the banning of the use of mercury containing pesticide by the U.S Govt in 1960s followed by banning of application of DDT and chlorinated hydrocarbons.

This led the scientists to re-evaluate the practices in protection of plant health. Application of better cultural practices like removal of plant debris, use of pathogen free seeds, rotation of crops, control of weeds, proper irrigation, proper time and rate of sowing of crops, date of harvest, monitoring the development of plants, detection of plant disease epidemics was emphasized for effective management of plant diseases (Agrios 2010; Keen 2000; Raychaudhuri et al. 1972; Galloway 1928; Ristaino et al. 2021).

The observation that some soil microorganisms could inhibit the soil-borne pathogen mediated infection and disease and nonpathogenic microbes like fungi, bacteria that antagonized the growth of plant-pathogenic fungi, bacteria and nematodes marked the beginning of biological control of disease. Infection by fungus *Heterobasidion annosum (H annosum)* caused root and butt rot disease in pines. In 1963, it was observed that inoculation of surfaces of cut pines with spores of a nonpathogenic fungus *Phleviopsis gigantean (P gigantean)* conferred protection to the pines against infection by *H annosum*. Although promising results were observed in experimental studies, application in real plants proved ineffective. Hence alternative approaches were being searched for in the control and prevention of pathogenesis infection process in plants.

Around 1980s with the advancement of genetic engineering and biotechnological approaches introduction of one or more desired viral genes into host plant was possible. This enabled prevention and delay of infection process in the plant.

Another strategy used in prevent infection was by boosting up the plant immune defense system. Pathogenic microbes and chemical compounds were used to cause necrotic lesions in host plants. This led to the activation of the innate immune responses of the plant against subsequent infections by same or different pathogens, called as systemic acquired resistance.

The next period in plant pathogen research focused on research on mechanism of disease, post pathogen infection. In 1886, deBary observed that while invading fungal hyphae could kill host cells and fluid from rotten tissue could cause damage to healthy host tissues, the boiled fluid could not cause infection or cause any damage to the healthy tissues indicating that fungal enzymes and toxins could kill healthy cells. In 1905, L.R. Jones detected the role of cytolytic enzymes in causing rot disease. The wildfire toxin was isolated in 1950.

Around 1970–1980, the mechanism of crown gall disease caused by *Agrobacterium tumefaciens (A tumefaciens)* infection was discovered revealing insights into incorporation of T-DNA of Ti plasmid and its transcription in the plant cell. T-DNA was found to contain genes including those producing indole-3-acetic acid (IAA) as a cytokinin which lead to uncontrolled growth of the plant causing gall disease.

In 1972, further research highlighted common features in the plant and animal defense mechanisms. Later in 1946, studies by Flor on the rust disease of flax highlighted the mechanism of disease resistance and susceptibility, revealing gene for gene relationship wherein it was understood for the first time that a gene for resistance in host has a corresponding gene for avirulence in the pathogen and a gene for virulence in the pathogen has a corresponding gene for susceptibility in the host.

In 1946, studies by E. Gaumann suggested the role of hypersensitivity in host–pathogen interactions by which some sensitive cells and their adjacent cells die immediately, preventing the pathogen from spreading further infection to other healthy cells.

In 1963, studies by Vanderplank revealed two types of resistance, expressed by plant host against infection by one or more species of pathogens as vertical or major resistance and horizontal resistance or minor resistance. Antimicrobial plant substances including phytoalexin have been studied to over accumulate in response to pathogen attack. These studies led to the identification of genes in pathogens involved in pathogenesis. Vanderplank in 1963 in his book *Plant Diseases: Epidemics and Control* documented studies on epidemiology and its importance in plant pathology, highlighting the factors contributing to plant epidemics, discussing on mechanism of disease development, control of monocyclic pathogens that cause

Table 33.1 Major landmarks in plant pathology studies

Year	Landmark
1935	Studies on isolation of tobacco mosaic virus (TMV)
1941	Beadle and Tatum revealed one gene one enzyme hypothesis
1942	Flor in 1942 revealed one gene involved in pathogenicity from flax rust infectious fungal disease
1953	Double helix structure of DNA was identified by Watson and Crick
1960–1970	Marked initiation of computer based approaches in study of plant–pathogen interaction
1970	Structure of Ti plasmid in *A tumefaciens* was deciphered
1984	P Albersheim revealed molecules in oomycete cell wall can destroy plants *Phytophthora megasperma* (*P megasperma*) can elicit defense response in soybean host
1986	Discovery of bacterial hypersensitive protein (hrp)
	Discovery of tobacco plants resistant to TMV infection
1990	Systemic acquired resistance (SAR) was discovered Discovery of the first fungal avirulence gene (avr9)
1993	Plant resistance genes revealed presence of leucine-rich repeats (LRRs)
1998	VM Williamson revealed that a disease resistant gene from tomato can confer resistance to both nematode pathogen and insect pest

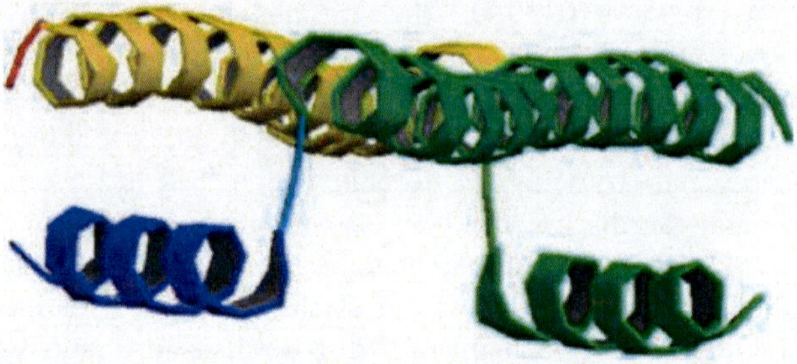

Fig. 33.1 6GU1: 3D view of SFI3 effector protein from the oomycete plant pathogen Phytophthora infestans (P infestans, Hughes and Banfield 2019) from Protein Data Bank (PDB, Berman et al. 2000)

one infection cycle per crop cycle and polycyclic pathogens that cause more than one infection cycle per cycle of the crop.

Late in the1960s and early 1970s scientists made attempts to design models of plant diseases from the information on host and pathogen and their interaction (Table 33.1), based on various time points, different environmental condition, and life cycle of the pathogen (Fig. 33.1). The first computer simulation of plant epidemics was designed on studies from fungal infections causing early blight disease in tomato and potato plants. Since then attempts were being made to predict and forecast disease so that farmers could take appropriate steps prior to disease outbreak.

The advances in the field of genomics enabled better understanding of the genomes of plants and pathogens. However, challenges were posed by plant pathogens including fungus, virus, viroid, bacteria, phytoplasma, and nematode that caused plant diseases affecting plant yield and production and can cause huge loss to the economy. Challenges in the domain of plant pathology lies in the detection of the causative agent of the disease including the pathogen, environmental factor, plant–pathogen interaction, plant immune system.

In response to infection by pathogen, plant PRRs on identification of PAMP triggered the PAMP-triggered immunity (PTI). Activation of PRR, flagellin-sensing 2 (FLS2), and EF-Tu receptor (EFR) by PAMPs led to downstream activation of MAPK signaling pathway further leading to downstream activation and synthesis of plant antimicrobial molecules. Upregulated cytosolic Ca^{2+} levels lead to increased production of reactive oxygen species (ROS) and programmed cell death (PCD) leading to a hypersensitive response (HR).

The secondary response in plants is known as the effector triggered immunity (ETI) leading to resist pathogen growth, leading to generation of resistant varieties. Pathogens can release effector molecules and cause suppression of the PTI, manipulation of the hormone mediated signaling pathways in plants. R proteins in plants can monitor virulence proteins of pathogens (Kanehisa and Sato 2020).

33.4 Machine Learning and Plant Pathogen Interaction

Since the 1960s and 1970s with the application of computational sciences in the understanding of plant pathogen interaction, computational methods have been used to understand the pathways in infection (Fig. 33.2). X-ray crystallography studies (Fig. 33.1) could decipher structures of pathogens and toxins involved in plant pathogenesis.

With the advent of the omics based studies including knowledge from the proteomics, genomics, metabolomics based studies have enabled further deeper understanding molecules and genes involved in plant–pathogen interaction.

Blast disease in rice plants (Fig. 33.3) caused by fungus with rapid adaptability to the host coupled together with durability of genetic resistance contributed by the *R* gene to a specific pathogen and is difficult to control and is a major concern globally. An integrated approach involved in the understanding of the genotypes by next-generation sequencing (NGS), phenotyping, omics based studies, machine learning finds importance in tackling blast disease and prevent the crop loss globally.

Omics data, network-based analyses, and machine learning technologies enable decoding of biological networks, including co-expression, protein–protein interaction (PPI), and gene regulatory networks (GRNs). Advanced applications of machine learning based studies have provided new insights into plant–microbe interactions and plant immunity (Simmi et al. 2020; Kshirsagar et al. 2015; Schleker et al. 2015; Wheeler et al. 2018; Yang et al. 2019; Mahlein et al. 2019; Wekesa et al. 2019; Mishra et al. 2019; Kurubanjerdjit et al. 2013; Kuska et al. 2018; Bourgeois et al. 2018; Guo et al. 2016; Ma et al. 2014; Dobin et al. 2013; Windram et al. 2014).

Applications of time-series techniques and approximate multi-scale entropy (ApEn) patterns on the electrophysiological systemic responses of tomato plants inoculated with a pathogenic fungus under conditions of controlled light and temperature revealed insights into plant pathogen interactions (Simmi et al. 2020).

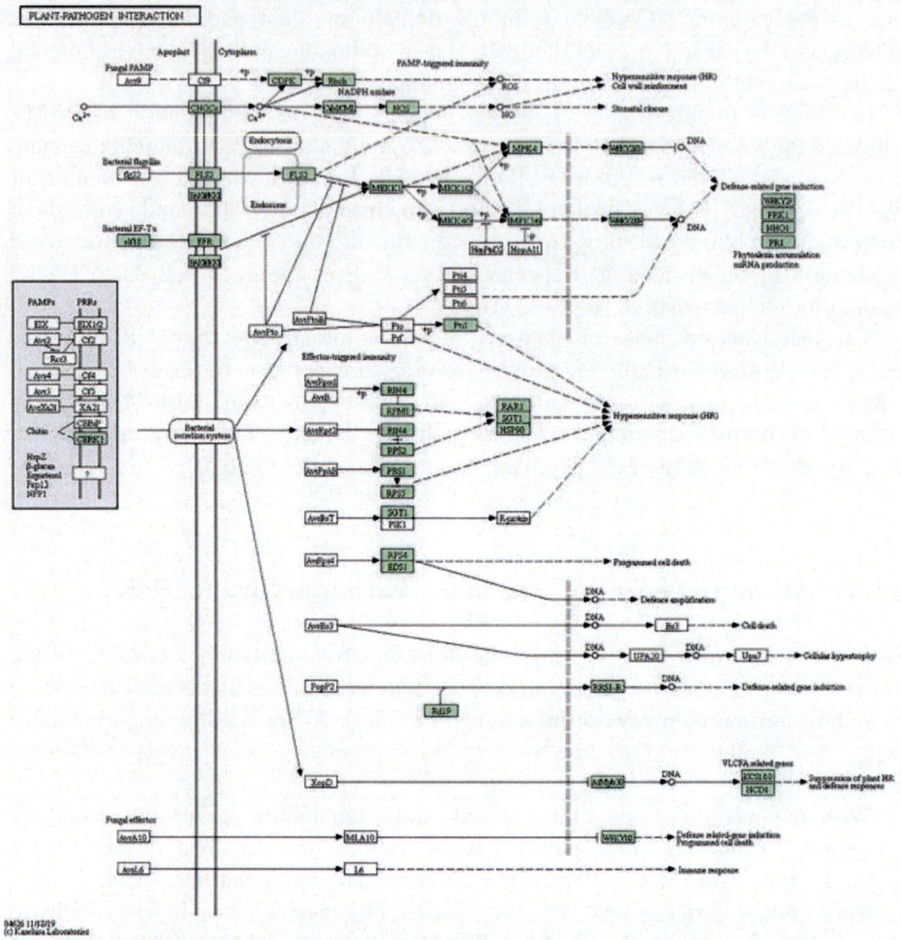

Fig. 33.2 Plant–pathogen interaction in *A. thaliana* (https://www.genome.jp/kegg-bin/show_pathway?ath04626) reproduced with permission (Kanehisa and Sato 2020)

PPI interaction between the bacteria *Salmonella typhimurium (S typhimurium)* and the host *Arabidopsis thaliana (A thaliana)* and their involvement in transcriptional regulation, hormone mediated signaling and defense response (Kshirsagar et al. 2015; Schleker et al. 2015) could be studied by application of machine learning models.

Machine learning also finds importance in study of *Salmonella* and its interaction with the human host in understanding mechanisms of Salmonellosis, a foodborne disease worldwide (Wheeler et al. 2018). Computational methods of interolog, domain–domain interaction, and domain–motif interaction have enabled prediction of plant–pathogen interactions (PPI). Random forest (RF) has been applied to build an inter-species PPI prediction model based on multiple sequence encodings and

Fig. 33.3 Rice blast disease across the globe. Image adapted with permission from (Wang et al. 2014) under Creative Commons Attribution 3.0 License

novel network attributes and the Web server InterSPPI, http://systbio.cau.edu.cn/intersppi/index.php that enabled a better prediction of plant–pathogen interaction (Yang et al. 2019).

Understanding the resistance of crops against fungal pathogens finds importance in selecting disease resistant genotypes. Digital phenotyping and machine learning have enabled identification of susceptible or resistant plants by noninvasive and objective means (Mahlein et al. 2019).

In plants long non-protein-coding RNAs (lncRNAs), >200 nt in length transcribed by RNA Polymerase (Pol) II, IV, and V from the sense or antisense strands are known to play the role in biological processes. lncRNA–protein interaction in plants is known to play role in pathogen resistance which also remains a relatively new and unexplored domain in plant pathogen interaction. PLRPIM, a model designed by application of deep learning and shallow machine learning methods using *k*-mer has been used to detect the lncRNA–protein interaction on *Arabidopsis thaliana (A thaliana)* and *Zea mays* (*Z mays*) datasets (Wekesa et al. 2019). Network modeling using machine learning has been applied to understanding of plant immunity and plant–pathogen interactions (Mishra et al. 2019).

MicroRNAs (miRNAs) are small, endogenous RNAs that are known to play role in major biological processes in plants, including metabolism and development of leaf and flower. Machine learning algorithm has enabled study of miRNAs and target gene interactions in *A thaliana* by integrating prediction scores from PITA, miRanda, and RNAHybrid algorithms using SVM, RF tree, and neural network. The designed database http://ppi.bioinfo.asia.edu.tw/At_miRNA/ plays important roles in understanding of plant–pathogen interaction networks (Kurubanjerdjit et al. 2013).

An integrated approach involving study of the plant physiological process, metabolic phenotyping involving detection of levels of invertase and study of the phenomic profiles enabled efficient quantitative detection of genotypes from near-isogenic lines of barley plants *Hordeum vulgare (H. vulgare,* (Kuska et al. 2018)*)* .A support vector machine (SVM) with a supervised machine learning approach enabled detection of pixel wise quantitative detection of powdery mildew from diseased barley tissue and hypersensitive response (HR) spots and identification of the barley–powdery mildew interaction (Kuska et al. 2018).

Selection pressures on grass *Brachypodium distachyon* (*B distachyon)* was detected by analysis of whole-genome sequencing (WGS) data (Bourgeois et al. 2018). *Fusarium graminearum (F. graminearum)* infection in wheat causes head blight disease (HBD) with production of mycotoxin as contaminant affecting adversely the production of wheat. A global *F. graminearum* gene regulatory network (GRN) generated from transcriptomic data using Bayesian network (BN) inference could infer connections between key regulators and their target genes enabling our understanding of mechanism and pathways involved in pathogenesis with prospects for designing of novel disease control strategies by targeting of master regulators in pathogens (Guo et al. 2016).

Despite advances in the study of genes involved in plant pathogenesis, the role of a large number of hypothetical genes in phytopathogens remains yet to be deciphered. Computational approaches have enabled identification of virulence (vir) genes by searching for sequence similarity to known host–microbe interaction factors from other organisms and identifying islands of genes that discriminate between pathogens of one type and closely related non-pathogens or pathogens of a different type. A supervised machine learning has enabled recognition of candidate host–microbe interaction factors using the soft rot Enterobacteriaceae strains *Dickeya dadantii* 3937*(D dadantii)* and *Pectobacterium carotovorum (P carotovorum)* WPP14 (Ma et al. 2014).

33.5 STAR RNA Sequencing

STAR is a freely available code at Github.com(https://github.com/alexdobin/STAR/releases). STAR is used to generate, analyze, cluster RNA sequences. STAR is being widely used to identify and classify the virus. STAR outperforms other aligners by a factor of >50 in mapping speed, while at the same time improving alignment sensitivity and precision. In addition to unbiased de novo detection of canonical junctions, STAR has been able to discover non-canonical splices and chimeric or fusion transcripts, and has been capable of mapping full-length RNA sequences. The latest STAR version is 2.7.3a available at Github.com. The central idea of the STAR (Dobin et al. 2013) is the sequential search for a Maximal Mappable Prefix (MMP) similar to the Maximal Exact (Unique) Match concept used by the large-scale genome alignment tools.

33.6 Discussions

Machine learning and deep learning tools are finding application in detection and classification of plant diseases (Windram et al. 2014).

Understanding the plant–pathogen interaction is the key to the understanding the pathophysiology of disease and design steps to prevent it so that damages to plants from infection by pathogens can be minimized or prevented. Machine learning has offered tools for the insights into in-depth study of plant–pathogen interaction. Although the applications at this juncture is rather at its infancy, the future is extremely bright. Research in this domain is focused towards development of better and appropriate machine learning tools and algorithms in this domain for better and effective application.

References

Agrios GN (2010) Plant pathology, 5th edn. Elsevier, Amsterdam

Berman HM, Westbrook J, Feng Z, Gilliland G, Bhat TN, Weissig H, Shindyalov IN, Bourne PE (2000) The protein data bank. Nucl Acids Res 28:235–242

Bourgeois Y, Stritt C, Walser JC, Gordon SP, Vogel JP, Roulin AC (2018) Genome-wide scans of selection highlight the impact of biotic and abiotic constraints in natural populations of the model grass Brachypodium distachyon. Plant J 96(2):438–451

Dobin A, Davis CA, Schlesinger F, Drenkow J, Zaleski C, Jha S, Batut P, Chaisson M, Gingeras TR (2013) STAR: ultrafast universal RNA-seq aligner. Bioinformatics 29(1):15–21

Galloway BT (1928) A review of the development of the science in the United States agricultural history. Plant Pathol 2(2):49–60

Guo L, Zhao G, Xu JR, Kistler HC, Gao L, Ma LJ (2016) Compartmentalized gene regulatory network of the pathogenic fungus fusarium graminearum. New Phytol 211(2):527–541

Hughes RK, Banfield MJ (2019) SFI3 effector protein from the oomycete plant pathogen Phytophthora infestans. New Phytol 222:438–454

Kanehisa M, Sato Y (2020) KEGG mapper for inferring cellular functions from protein sequences. Protein Sci 29:28–35

Keen NT (2000) A century of plant pathology: a retrospective view on understanding host-parasite interactions. Annu Rev Phytopathol 38:31–48

Kshirsagar M, Schleker S, Carbonell J, Klein-Seetharaman J (2015) Techniques for transferring host-pathogen protein interactions knowledge to new tasks. Front Microbiol 6:36

Kurubanjerdjit N, Huang CH, Lee YL, Tsai JJ, Ng KL (2013) Prediction of microRNA-regulated protein interaction pathways in arabidopsis using machine learning algorithms. Comput Biol Med 43(11):1645–1652

Kuska MT, Behmann J, Großkinsky DK, Roitsch T, Mahlein AK (2018) Screening of barley resistance against powdery mildew by simultaneous high-throughput enzyme activity signature profiling and multispectral imaging. Front Plant Sci 23(9):1074

Ma B, Charkowski AO, Glasner JD, Perna NT (2014) Identification of host-microbe interaction factors in the genomes of soft rot-associated pathogens Dickeya dadantii 3937 and Pectobacterium carotovorum WPP14 with supervised machine learning. BMC Genomics 15: 508

Mahlein AK, Kuska MT, Thomas S, Wahabzada M, Behmann J, Rascher U, Kersting K (2019) Quantitative and qualitative phenotyping of disease resistance of crops by hyperspectral sensors: seamless interlocking of phytopathology, sensors, and machine learning is needed! Curr Opin Plant Biol 50:156–162

Mishra B, Kumar N, Mukhtar MS (2019) Systems biology and machine learning in plant-pathogen interactions. Mol Plant-Microbe Interact 32(1):45–55

Raychaudhuri SP, JP JPV, Nariani TK, Sen B (1972) The history of plant pathology in India. Annu Rev Phytopathol 10(1):21–36

Ristaino JB, Anderson PK, Bebber DP, Brauman KA, Cunniffe NJ, Fedoroff NV, Finegold C, Garrett KA, Gilligan CA, Jones CM, Martin MD, MacDonald GK, Neenan P, Records A, Schmale DG, Tateosian L, Wei Q (2021) The persistent threat of emerging plant disease pandemics to global food security. Proc Natl Acad Sci U S A 118(23):e2022239118

Schleker S, Kshirsagar M, Klein-Seetharaman J (2015) Comparing human-Salmonella with plant-Salmonella protein-protein interaction predictions. Front Microbiol 6:45

Simmi FZ, Dallagnol LJ, Ferreira AS, Pereira DR, Souza GM (2020) Electrome alterations in a plant-pathogen system: toward early diagnosis. Bioelectrochemistry 21(133):107493

Wang X, Lee S, Wang J, Ma J, Bianco T, Jia Y (2014) Current advances on genetic resistance to rice blast disease. Rice Germpl Genet Imp. https://doi.org/10.5772/56824

Wekesa JS, Luan Y, Chen M, Meng J (2019) A hybrid Prediction method for plant lncRNA-protein interaction. Cell 8(6):521

Wheeler NE, Gardner PP, Barquist L (2018) Machine learning identifies signatures of host adaptation in the bacterial pathogen salmonella enterica. PLoS Genet 14(5):e1007333

Windram O, Penfold CA, Denby KJ (2014) Network modeling to understand plant immunity. Annu Rev Phytopathol 52:93–111

Yang S, Li H, He H, Zhou Y, Zhang Z (2019) Critical assessment and performance improvement of plant-pathogen protein-protein interaction prediction methods. Brief Bioinform 20(1):274–287

Machine Learning in Plant Disease Research 34

Abstract

Biotic agents including virus, viroids, fungi, bacteria, mollicutes, protozoa, and nematodes and adverse environmental conditions can affect the plant and its growth and produce, impacting largely the plant health and economy. In this chapter we highlight the different plant diseases and symptoms caused by the infectious pathogens and adverse environmental conditions. Computational biology and recently machine learning have enabled better understanding of the causative agents, pathophysiology of plant disease, identification and prediction of the disease, understanding the function and role of gene regulatory networks and biological pathways in plant diseases.

Keywords

Virus · Viroids · Fungi · Bacteria · Mollicutes · Protozoa

Abbreviations

2–4-D	2–4-Dicholorophenoxy acetic acid
ANN	Artificial neural network
CMV	Cauliflower mosaic virus
CNN	Convolution neural network
DL	Deep learning
EM	Electron microscope
GRN	Gene regulatory networks
HC	Hydrocarbons
IOT	Internet of things
ML	Machine learning
MR	Multiple regressions

S. Ghosh, R. Dasgupta, *Machine Learning in Biological Sciences*,
https://doi.org/10.1007/978-981-16-8881-2_34

NLB	Northern leaf blight
O_3	Ozone
PAN	Peroxyacyl nitrate
PVX	Potato virus X
QPLS	Quantitative partial least squares
RF	Random forest
SNB	*Stagonospora nodorum blotch*
SVM	Support vector machine
SVR	Support vector regression
T. semipenetrans	*Tylenchulus semipenetrans*
TMV	Tobacco mosaic virus
V.dahliae	*Verticillium dahliae*

34.1 Introduction

Plant diseases could lead to huge loss to the economy of the country. The disease causing agents in plants include infections caused by virus, viroids, fungi, bacteria, mollicutes, protozoa, stress adverse environmental condition (Table 34.1).

Fungi are small submicroscopic spore-bearing organism lacking cell wall with cell wall composed of chitin and glycerol. The fungal filamentous vegetative body is called mycelium, composed of individual branches or hyphae. They reproduce by spores produced in the sporangium, zoospores, and conidiophores by asexual reproduction. Oomycetes are fungi-like organisms with cell wall composed of glucans and cellulose but lack chitin and are included under kingdom Chromista. Myxomycetes are saprophytic fungi-like organisms under kingdom Protozoa and infect plants causing diseases. Oomycete-mediated infectious diseases can be controlled by soil sterilization and pasteurization by steam.

Prokaryotic pathogens including bacteria and mollicutes include DNA and 70S ribosomes causing diseases in plants. Bacterial diseases are difficult to control as they cause harm to crops. Sanitation practices like burning or removing infected plants, decontamination, adjusting fertilizing and watering, controlling the succulence of plants, crop rotation, use of resistant varieties of crops, steam sterilization of soil, application of heat and chemicals like formaldehyde, disinfection of seeds by sodium hypochlorite or HCl solutions, heat treatment of infected seeds at 52° C for 20 min, to disinfect seeds, use of foliar sprays, copper compounds, antibiotics, Bordeaux mixture, cupric hydroxide to control bacterial infection like leaf spots and blights.

Mollicutes including phytoplasmas and spiroplasmas are microbes lacking cell wall, can be visualized under electron microscope (EM) that infect the phloem tissues of plants and can cause more than 200 types of diseases in plants.

Parasitic higher plants, green alga, invasive climbing plants that can penetrate the plant vascular tissues can cause diseases in plants.

Table 34.1 Causative agents for Plants Diseases

1. Biotic factors	Effects
Fungal infection	• Causes leaf spot lesions • Causes blight disease • Canker or localized necrosis • Dieback or necrosis of twigs • Root rot or decay of roots • Damping off with death of young seedlings • Basal stem rot affecting the stem • Anthracnose or ulcer like lesions on stem, fruits, leaves • Scab or localized lesions on host fruits • Decline or loss of vigor • Galls • Clubroot • Warts • Leaf curls • Wilt • Rust • Smut • Mildew • Mitosporic fungi including ascomycetes and deuteromycetes can cause sooty molds • Taphrina can cause leaf curl • Defoliation in peach, leaf, flower, fruit deformation, obligate parasitic fungi can cause powdery mildews
Oomycetes	• Seed-rot • Seedling damping off • Root rot • Soft rot of fleshy fruits • Late blight of potatoes, root, fruit, stem rot • Cankers • Diebacks • Mildews in grape, tobacco, cucurbits. • Downy mildew in corn, sorghum, sugarcane • White rust
Bacteria and mollicutes	• Leaf spots • Blights • Soft rots of fruits, roots, wilts, scabs, cankers, galls by *agrobacterium* sp • Nodules by *rhizobium, Azorhizobium, Bradyrhizobium* species • Bacterial brown spot in beans • Bacterial angular leaf spot in cucumber • *Pseudomonas* and *Xanthomonas* cause bacterial spots and blights • *Pseudomonas* sp. can cause wildfire of tobacco, soybean • Angular leaf spot in cotton caused by *Xanthomonas campestris pv. Malvacearum* • *Xanthomonas campestris pv. Vesicatoria* causes bacterial spot in tomato and pepper affecting leaves,

(continued)

Table 34.1 (continued)

1. Biotic factors	Effects
	stem fruits, blossom drop in flower • *Pseudomonas syringae pv. Tomato* causes bacterial speck in tomato. • Bacterial fruit blotch infection caused in watermelon by *Acidovorax avenue citrulli* • *Xanthomonas axonopodis pv. Manihotis* causes cassava bacterial blight disease • *Xanthomonas arboricola* causes bacterial spot disease in stone fruits • Bacteria causes vascular wilts in cucurbits, solanaceous plants • Causes ring rot of potato • Causes moko disease in banana • Causes bacterial canker and wilt of tomato. • Causes black rot in crucifers • Causes Stewart's wilt of corn • Causes bacterial soft rots of vegetables • Causes pierce disease of grape • Causes ratoon stunting of sugarcane • Causes yellow wine disease of cucurbits
Mollicutes	• Diseases in vines, pears, yellowing of grapes, lethal in coconuts • X disease in peach • Aster yellow in ornamental, vegetables, stolbur in tomato • Spiroplasma can cause corn stunt lethal yellowing of coconut palms • Affects apple proliferation adversely • European stone fruit yellows • Yellowing of ash plants • Peach X disease • Decline in pear production
Parasiting higher plants, invasive climbing plants, parasitic green alga	• Dodder affects growth and yield of infected plants like yellowing of vine strands • Witchweed attacks monocots, corn, millet. • Broomrape affects pods • Dwarf mistletoes of conifers • Leafy mistletoes • Invasive climbing plants • Photosynthetic blue-green cyanobacteria cause the black band diseases leading to bleaching and death of coral symbionts.
Virus	• Oak-leaf pattern in leaves • Affects leaves, stems, roots, and fruits. • Necrotic local lesions • Infected tobacco stalks, caused by TMV • Tobacco rattle disease caused by Tobravirus • Furovirus causes beet necrotic yellow vein virus, rhizomania disease of sugar beets • Hordeivirus affects the grain crops, wild grass

(continued)

Table 34.1 (continued)

1. Biotic factors	Effects
	• Pecluvirus is known to infect peanuts • Pomovirus can infect potatoes causing loss to crop • Benyvirus infects beets • Potexvirus or potato virus X (PVX), affects different crops including orchids • Carlavirus can cause fleck disease • Allexivirus can infect grapes and wines
Nematode	• Nematodes infect together with fungus increasing the severity of the disease • Fusarium wilt reveals severity and incidence during infection by root-knot, lesion • Viruses including grapewine fanleaf virus, tomato ringspot virus are transmitted from soil by nematode vectors to plants causing infection • *Trichodorus* and *Paratrichodorus* transmit tobacco rattle virus (TRV) and pea early browning virus (PEBV) • Root-knot nematodes including *Meloidogyne* sp. can attack more than 2000 plants, bushy roots, cause necrosis, rotting and galls • Underground parts like carrots, peanuts, potatoes, yam develop swellings, distortion, cracking • *Globoderarostochiensis (G rostochiensis)* infects tomato, potato, and eggplant. • *Heterodera sp* infects cereals, soybeans, sugarbeet, crucifers, spinach, clover revealing cysts in roots, with bushy roots, yellowing of soybean leaves, loss in sugarbeet yield by almost 50% • The citrus nematode *Tylenchulus semipenetrans (T semipenetrans)* infects grapewines, olive, lilac causing poor growth, yellowish leaves, leaf drop, reduced fruit formation • Lesion nematodes attack the roots, of cereal plants, vegetables, fruit plants, ornamentals • Infected plants reveal stunted growth, reveal mineral deficiencies, patches of yellowish-green plants, infections on peanuts • *Pratylenchus* sp. can infect roots • *Radopholus sp* infects banana, causing its decreased growth and yield with fewer leaves • *Belonolaimus longicaudatus* can infect the plant sucking the sap, infect the roots • Stubby roots nematodes including *Paratrichodorus* and *Trichodorus* are known to infect roots, causes stunting and chlorosis of plants • *Anguina* sp. causes seed-gall • *Aphelenchoides* sp. infect the above ground plant parts

(continued)

Table 34.1 (continued)

1. Biotic factors	Effects
Protozoa	• *Phytomonas* sp. causes severe loss to economically important plants • Flagellates cause phloem necrosis disease of coffee, hartrot disease of coconut palm • Causes coffee phloem disease
2. Abiotic/environmental factors	
High and low temperature	
Effects of high temperature	• High temperature together with excessive light, drought, lack of oxygen, high winds, low relative humidity causing sunscald injuries • Very high temperature can kill leaves, young seedlings • Cause cankers at crown on stems of older plants, together with reduced oxygen cause blackheart of potatoes • Protein coagulation and denaturation • Disrupted cytoplasmic membranes • Release of toxic products into cells
Effects of low temperature	• Low temperatures, freezing can cause damage to plants that usually grows in warm weather like corns, beans • Excessive sweetening of potatoes • Frost killing of peach, cherry buds, flowers, fruits • Frost bands on apples, pears • Kills roots • Bark splitting • Blackheart condition in woods • Necrosis and affected vascular tissues in potatoes • Damaged tubers • Blotch-type necrosis. • Yellowing of plants and leaf, bud, flower drop in indoor plants • Ice formation between/within cells can damage cells and tissues • Plasmolysis and dehydration of the protoplasm can occur • Ice crystals within cell can disrupt plasma membrane • Ice formation enables susceptibility of infection by epiphytic bacteria like *pseudomonas syringae*, infection by saprophytic bacteria in stomatal leaves
Moisture effects	
Low soil moisture	• Lack of moisture in soil leads to stunted growth in plants, pale green or yellow leaves, spare flower and fruits • Prolonged drought leads to wilting and death
High soil moisture	• Poor drainage, water clogging lead to damage and death of plants • Causes wilt in plants • Damage of roots due to reduced oxygen supply

(continued)

Table 34.1 (continued)

1. Biotic factors	Effects
	• Causes brown or black patches in roots • Edema or swelling on leaves and stem • Leaves can get corky texture • Irregular moisture supply can cause bitter pit in apples
Low relative humidity	• Low atmospheric moisture, high wind, high temperature cause loss of water from foliage • Causes scorching or burning of leaves
Inadequate oxygen	• High moisture and high temperature are associated with inadequate oxygen supply leading to root collapse and death • Low oxygen levels in centers of fruits/vegetables • Causes blackheart of potatoes • Causes dysregulated enzymatic reactions in potato tuber
Light	• Lack of sufficient light slows down chlorophyll formation • Causes pale green leaves, leaf, flower drop called etiolation • Excess light can cause injuries, sunscald in pods, beans, small pods that turns brown when immersed in water, silver like spots on leaves
Air pollution	• Gases, particulate matter, dust can cause injuries in plants • Ethylene (C_2H_4), ammonia (NH_3), ozone (O_3), sulfur dioxide (SO_2), hydrogen fluoride (HF), Nitrogen dioxide (NO_2), Peroxyacyl nitrates (PAN) can cause serious damage to plants • Ozone (O_3) and PAN produced in the atmosphere are secondary pollutants produced as photochemical reaction involving NO_2, O_2, Hydrocarbons (HC), and sunlight • Automobile exhaust which is the source of unburnt HC, NO_2 may cause serious damages to plants • NO_2 can react with atmospheric O_2 in presence of sunlight forming ozone, (O_3) and nitric oxide (NO). • O_3, NO, and HC can lead to the formation of increased O_3 and PAN • All these are harmful and toxic to plants. • The age, plant part, and concentration of each pollutant vary the damage inflicted upon the plant • Acid rain can cause damage to plants.
Nutritional deficiencies toxic to plants	• The major elements include Nitrogen (N_2), Phosphorous (P), Potassium (K), Calcium (Ca^{2+}), Magnesium (Mg), Sulfur (S), while the minor elements include Iron (Fe), Boron (B), Manganese (Mn), Zinc (Zn), Copper (cu), Molybdenum (Mo), and Chlorine (Cl) • Deficiencies or less concentrations of the elements may lead to diseases

(continued)

Table 34.1 (continued)

1. Biotic factors	Effects
Injury by herbicides	• Use of herbicides, weed killers affects leaves causing their distortion, yellowing, browning, shedding
Weather	• Lightning, hail storms can affect the plants
Improper agricultural practice	• Spraying of insecticides, nematocides, fertilizer can lead to injury in leaves and plant parts

Virus containing nucleoprotein either DNA or RNA as genetic material, of different shapes and sizes, including Tobacco mosaic virus (TMV), potato virus X (PVX), cucumber mosaic virus, cauliflower mosaic virus (CMV) can infect plant and cause diseases. They may be transmitted by seed, pollen, insect, mites, nematodes, fungus, and Dodder.

Nematodes belong to kingdom Animalia, feed on plants and animals, and infect plants causing diseases.

Protozoa are single-celled microscopic organisms that occur in single forms or exist in colonies. They may be free living, symbiotic, parasitic, saprophytic. Flagellated forms like Trypanosomatid flagellates can infect plants and affect the plant yield adversely.

Therefore steps to identify the disease and characterize the disease find importance in plant biology. One of the major goals in improved disease management of the plant infection lies in the early diagnosis of the plant disease and taking appropriate steps early in disease management. Fungal diseases are controlled by application of sprays or fungicides. Machine learning is finding importance in the early diagnosis of disease and is enabling better management of plant disease and infection.

34.2 Plant Diseases

Plant diseases affect the growth of plants and their early identification finds importance. Machine learning (ML) models find application in the detection and classification of plant diseases and deep learning (DL) has been used to detect and classify plant diseases (Saleem et al. 2019).

Using hyperspectral imaging and application of machine learning it has been possible to predict presymptomatic detection of tobacco mosaic virus (TMV, Fig. 34.1) infection much earlier than the manifestations of the actual symptoms of the infection (Zhu et al. 2017).

Machine Learning is enabling better understanding of disease etiology and host–pathogen interaction in complex diseases like those caused by soil-borne microorganisms. Supervised machine learning methods were applied to understand the etiology and host–parasite interaction in Verticillium wilt of mint, an infectious disease caused by the fungus *Verticillium dahlia* (*V. dahlia*) and enabled prediction

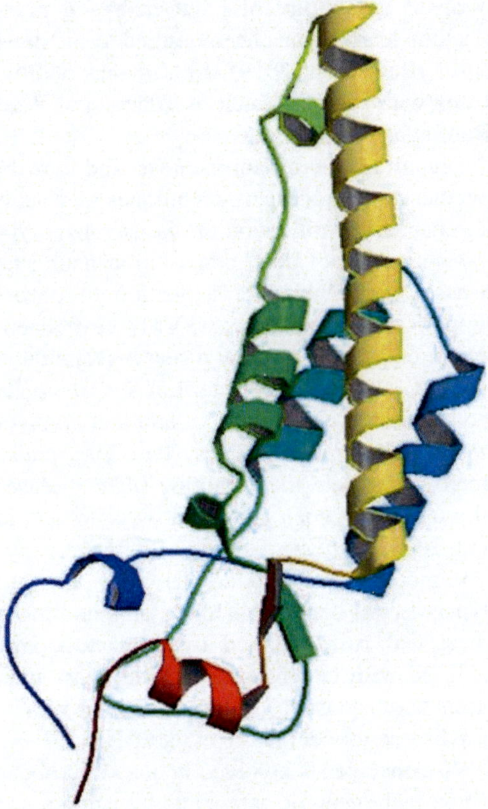

Fig. 34.1 3D view of Tobacco mosaic virus (TMV) 6I5A,pdb from Protein data Bank (PDB) (Song, B., Lenhart, J., Flegler, V.J., Makbul, C., Rasmussen, T., Bottcher, B. Capabilities of the Falcon III detector for single-particle structure determination. (2019) Ultramicroscopy 203: 145–154)

of wilt. Parameters associated with wilt like Wilt foci, V. dahliae, and plant-parasitic nematodes when quantified, using RF(random forest), and artificial neural network (ANN) the results revealed nematodes of genus Pratylenchus were important predictor of wilt, followed by age of the crop and *V. dahlia*, thus revealing their importance as strong indicators of nematode infection (Wheeler et al. 2019).

Machine learning is enabling detection and characterization of fungus-resistant crop genotypes (Mahlein et al. 2019).

Pantoea agglomerans, a bacterium, is known to cause galls on beet (*Beta vulgaris*) and gypsophila (*Gypsophila paniculata*). Application of draft genome sequences and analysis using machine learning tools and techniques has enabled identification of factors contributing to pathogenicity on beet and gypsophila (Nissan et al. 2018). Machine learning algorithm has found application in detection of plant disease from extraction of information from images taken of plants subjected to infection (Wiesner-Hanks et al. 2018). Such studies have been conducted in maize (*Zea mays L.*) leaves' diseases including lesions of northern leaf blight (NLB, Wiesner-Hanks et al. 2018). Hyperspectral sensors and imaging technologies with applications of machine learning have enabled understanding of mechanisms involved in biology of plant–pathogen interactions (PPI, Mahlein et al. 2018). Machine learning has been applied to the detection and classification of diseases in Citrus (Rauf et al. 2019) from images of fruits and leaves of healthy and infected plants with diseases such as Black spot, Canker, Scab, Greening, and Melanose (Rauf et al. 2019).

Digital image of anthracnose and powdery mildew diseases of sandalwood together with processing techniques and analysis by machine learning algorithms has enabled identification of disease phenotypes (Wu and Wang 2017).

Application of deep convolution neural network (CNN) to potato plants with diseases has enabled classification of potato disease (Oppenheim et al. 2019). Support vector machines (SVM) have been used in developing weather-based prediction models of plant diseases (Kaundal et al. 2006).

The Gram-negative bacterium *Xanthomonas euvesicatoria* (Xcv) causes bacterial spot disease in pepper and tomato that involves delivery of effector proteins including type III secretion (T3S) system. They are known to suppress plant immunity and thereby play role in promoting plant disease. Xcv effectors comprising of about 30 proteins has been identified by genome-wide machine learning approach (Teper et al. 2016).

Remotely sensed data has been used for disease monitoring and control. Hyperspectral data of healthy plants and those infected with wheat stripe rust and wheat leaf rust obtained by a black-paper-based measuring method, with the analyzed with techniques of partial least squares (DPLS), SVM, and quantitative partial least squares (QPLS) and support vector regression (SVR) has enabled design of a disease model (Wang et al. 2016).

Virulence genes involved in host–microbe interaction have enabled identification of non-pathogens or pathogens by using a supervised machine learning in studies from the soft rot Enterobacteriaceae strains *Dickeyada dantii 3937* and *Pectobacterium carotovorum WPP14 (*Ma et al. 2014*)*.

Citrus canker and Huanglongbing (HLB), diseases of citrus affect the yield and produce of the plants and can cause economic losses. Images obtained by fluorescence imaging spectroscopy (FIS) and a machine learning technique have enabled characterization of diseases(Wetterich 2016).

Northern leaf blight (NLB) in maize affects the plant growth and production adversely. Images of the health and infected plants when analyzed using convolutional neural networks (CNNs) have enabled identification of NLB from images (DeChant et al. 2017).

Pectobacterium and *Dickeya* spp. are soft rot Pectobacteriaceae infecting agricultural crops, adversely affecting the yield and produce, causing economic losses. Application of infrared (IR) spectroscopy with machine learning methods has been able to identify the bacterial pathogens (Abu-Aqil et al. 2020).

Biomarkers find importance in detecting specific disease resistance. Multispectral imaging studies of barley plants of powdery mildew susceptible and resistant varieties could reveal signatures of host–pathogen interaction which in this case was barley powdery mildew interaction. Support vector machine (SVM) was used to identify and classify the images and thereby enabled quantification of mildew diseased barley tissue and enabled automatic identification of hypersensitive response (HR) spots during establishment of infection, thereby enabling screening of disease resistant strains(Kuska et al. 2018).

Multiple regression (MR) and ML algorithms of artificial neural networks (ANN), categorical regression trees (CRT), and random forests (RF) applied to *Stagonospora nodorum blotch* (SNB) fungal infection caused by *Parastagonospora nodorum*, in winter wheat (*Triticum aestivum*) have enabled in prediction of the risk of incidence of SNB infection in wheat even prior to plantation (Mehra et al. 2016).

Fusarium graminearum (*F. graminearum*) gene regulatory network (GRN) designed from transcriptomic data using Bayesian network inference (BNI) has enabled the connection of key regulators and their target genes (Guo et al. 2016).

34.3 Detection of Plant Diseases Using Deep Learning Algorithm

Detection of disease for plants is an important process in the science of modern agricultural science. Most of the crop growing and vegetable growing firms are using a number of IOT (internet of things) devices and controllers to monitor and control of use of water and resources. In the process the firms collect huge data to detect the health signatures of the plants.

While plant growth rate, size, color, and shape can indicate the health of the plant, it is necessary to track the disease signatures early to take preventive actions. Agricultural scientists with their experience can detect it to some extent. Given the current growth of computational techniques, computer vision, image recognition techniques, and machine learning algorithm, it is wiser to use them to identify the disease early. There are deep learning algorithms such as convolutional neural network (CNN) which can effectively identify the signatures from the image of the plant and cluster the healthy ones and different diseased ones separately.

Mohanty, Hughes, and Salathe (Mohanty et al. 2016) used deep learning algorithm to identify the plant disease. They used a public dataset of 54,306 healthy and diseased plant images to train a CNN to identify 14 crops and 26 diseases with 99.35% accuracy (He et al. 2015; Huang 2007).

34.3.1 An Approach in Creating a Deep Learning Code to Identify Plant Disease

Interested readers may try writing a CNN code on Python on Jupyter Note Book. It is a Python IDE. One can download and install Jupyter Note Book from: https://jupyter.org/install.html.

Once the Jupyter Note Book is installed, one should look for the data. The data may be available on github. Then one may define the dataset structure in Python and invoke the libraries. Here is the list of libraries one needs:

NumPy: It is a library for the Python programming language, adding support for large, multidimensional arrays and matrices, along with a large collection of high-level mathematical functions to operate on these arrays.

SciPy: It is a library for Python Program language adding support for scientific and mathematical functions.

Keras: Keras is an open source neural network library written in Python. It is capable of running on top of TensorFlow, Microsoft Cognitive Toolkit, or Theano. Designed to enable fast experimentation with deep neural networks, it focuses on being user-friendly, modular, and extensible.

Sklearn: It is a machine learning library for Python. It features various classification, regression, and clustering algorithms including support vector machines, random forests, gradient boosting, k-means, and DBSCAN and is designed to interoperate with the Python numerical and scientific libraries NumPy and SciPy.

Matplotib: It is a plotting library for Python Programming language.

The next step is to load the database and convert the images to an array and then use Scikit's Label Binarizer to covert the images to binaries. One can use any Regularization technique before building model using Keras Adam Optimizer (https://keras.io/optimizers/#adam).

To plot the graph one may use Matplotlib for finding the test and validation accuracy. There are many such codes available at Github.com.

34.4 Discussion

In this chapter we have highlighted the application of machine learning in the study of plant disease. Although the science and its application in this domain of biology is at its infancy, nevertheless it has promising scope and application in this domain remains the future scope of study and research. This has immense application in the plant agriculture and horticulture industry and their ease in early detection of disease could save the industry from economic losses.

References

Abu-Aqil G, Tsror L, Shufan E, Adawi S, Mordechai S, Huleihel M, Salman A (2020) Differentiation of Pectobacterium and Dickeya spp. phytopathogens using infrared spectroscopy and machine learning analysis. J Biophotonics 13:e201960156

DeChant C, Wiesner-Hanks T, Chen S, Stewart EL, Yosinski J, Gore MA, Nelson RJ, Lipson H (2017) Automated identification of northern leaf blight-infected maize plants from field imagery using deep learning. Phytopathology 107(11):1426–1432

Guo L, Zhao G, Xu JR, Kistler HC, Gao L, Ma LJ (2016) Compartmentalized gene regulatory network of the pathogenic fungus Fusarium graminearum. New Phytol 211(2):527–541

He, K., Zhang, X., Ren, S., and Sun, J. (2015). Deep residual learning for image recognition. arXiv:1512.03385

Huang KY (2007) Applications of artificial neural network for detecting phalaenopsis seedling disease using color and texture features. Comput Electron Agric 57:3–11

Kaundal R, Kapoor AS, Raghava GP (2006) Machine learning techniques in disease forecasting: a case study on rice blast prediction. BMC Bioinform 3(7):485

Kuska MT, Behmann J, Großkinsky DK, Roitsch T, Mahlein AK (2018) Screening of barley resistance against powdery mildew by simultaneous high-throughput enzyme activity signature profiling and multispectral imaging. Front Plant Sci 9:1074

Ma B, Charkowski AO, Glasner JD, Perna NT (2014) Identification of host-microbe interaction factors in the genomes of soft rot-associated pathogens Dickeya dadantii 3937 and Pectobacterium carotovorum WPP14 with supervised machine learning. BMC Genomics 15: 508

Mahlein AK, Kuska MT, Behmann J, Polder G, Walter A (2018) Hyperspectral sensors and imaging technologies in phytopathology: state of the art. Annu Rev Phytopathol 25(56): 535–558

Mahlein AK, Kuska MT, Thomas S, Wahabzada M, Behmann J, Rascher U, Kersting K (2019) Quantitative and qualitative phenotyping of disease resistance of crops by hyperspectral sensors: seamless interlocking of phytopathology, sensors, and machine learning is needed! Curr Opin Plant Biol 50:156–162

Mehra LK, Cowger C, Gross K, OjiamboPS. (2016) Predicting pre-planting risk of Stagonospora nodorum blotch in winter wheat using machine learning models. Front Plant Sci 30(7):390

Mohanty SP, Hughes DP, Salathe M (2016) Using deep learning for image based plant disease detection. Front Plant Sci 22:1419

Nissan G, Gershovits M, Morozov M, Chalupowicz L, Sessa G, Manulis-Sasson S, Barash I, Pupko T (2018) Revealing the inventory of type III effectors in Pantoea agglomerans gall-forming pathovars using draft genome sequences and a machine-learning approach. Mol Plant Pathol 19(2):381–392

Oppenheim D, Shani G, Erlich O, Tsror L (2019) Using deep learning for image-based potato tuber disease detection. Phytopathology 109(6):1083–1087

Rauf HT, Saleem BA, Lali MIU, Khan MA, Sharif M, Bukhari SAC (2019) A citrus fruits and leaves dataset for detection and classification of citrus diseases through machine learning. Data Brief 26:104340

Saleem MH, Potgieter J, Mahmood Arif K (2019) Plant disease detection and classification by deep learning. Plants (Basel) 8(11):E468

Teper D, Burstein D, Salomon D, Gershovitz M, Pupko T, Sessa G (2016) Identification of novel Xanthomonas euvesicatoria type III effector proteins by a machine-learning approach. Mol Plant Pathol 17(3):398–411

Wang H, Qin F, Ruan L, Wang R, Liu Q, Ma Z, Li X, Cheng P, Wang H (2016) Identification and severity determination of wheat stripe rust and wheat leaf rust based on hyperspectral data acquired using a black-paper-based measuring method. PLoS One 11(4):e0154648

Wetterich CB, Felipe de Oliveira Neves R, Belasque J, Marcassa LG (2016) Detection of citrus canker and Huanglongbing using fluorescence imaging spectroscopy and support vector machine technique. Appl Opt 55(2):400–407

Wheeler DL, Scott J, Dung JKS, Johnson DA (2019) Evidence of a trans-kingdom plant disease complex between a fungus and plant-parasitic nematodes. PLoS One 14(2):e0211508

Wiesner-Hanks T, Stewart EL, Kaczmar N, DeChant C, Wu H, Nelson RJ, Lipson H, Gore MA (2018) Image set for deep learning: field images of maize annotated with disease symptoms. BMC Res Notes 11(1):440

Wu C, Wang X (2017) Preliminary research on the identification system for anthracnose and powdery mildew of sandalwood leaf based on image processing. PLoS One 12(7):e0181537

Zhu H, Chu B, Zhang C, Liu F, Jiang L, He Y (2017) Hyperspectral imaging for Presymptomatic detection of tobacco disease with successive projections algorithm and machine-learning classifiers. Sci Rep 7(1):4125

Biorobots 35

Abstract

A machine performs single or multiple actions in a repetitive manner with or without human intervention. Robots are machines which are created to be far more intelligent and can function with minimal human interference. Recent developments in robotics include intuitive, designed robots with biomechanical functions with biomedical applications. A revolution in the domain of robotics including its advanced design and function has been developing in recent times. In this chapter we discuss the nature of biorobots and molecular robots and their applications in science and technology focusing on recent developments in biological sciences, highlighting their application in (1) detection systems and infectious disorders, (2) the study of immune responses to infection, (3) brain and neurological functions, and (4) locomotion. We also highlight the recent applications of machine learning in the domain of biorobots.

Keywords

Biorobots · Molecular robots · Locomotion surgery · CNN · ANN · Biomechanics

Abbreviations

2D	Two-dimensional
3D	Three-dimensional
AF	Amniotic fluid
APR	Autocycling proximity recording
B. licheniformis	*Bacillus licheniformis*
BioMEMS	Biomedical microelectromechanical systems
BMI	Brain–machine interface
CNS	Central nervous system

S. Ghosh, R. Dasgupta, *Machine Learning in Biological Sciences*,
https://doi.org/10.1007/978-981-16-8881-2_35

DNA	Deoxyribonucleic acid
E. coli	*Escherichia coli*
HAV	Hepatitis A virus
HBV	Hepatitis B virus
HCV	Hepatitis C virus
HIV	*Human immunodeficiency virus*
Mm	Millimeter
NA	Nucleic acid
PCR	Polymerase chain reaction
PER	Primer exchange reactions
RL	Reinforcement learning
RNA	Ribonucleic acid
T. gondii	*Toxoplasma gondii*

35.1 Introduction

Nanorobotics is an emerging field of nanoelectromechanical functioning at molecular level with bio-medical applications. Its size confers its potential advantage over conventional tools with bio-medical application. They are finding bio-medical and clinical applications in diagnosis of different deadly diseases including HIV, cancer, lost tissue restoration, monitoring disease progression and monitoring of neuroelectric signals. Nanorobots are microscopic in size and can perform both microscopic and macroscopic tasks.

They include components like carbon nanotube, motor, biosensors, and DNA joints. Biochips find applications in the design of surgical instrumentation, disease diagnosis, and drug delivery.

Nanorobots integrated as nanoelectronics devices enable tele-operation and advanced capabilities for medical instrumentation. Nubots or nucleic acid robots is described as the several DNA walkers. Positionally controlled nanoassemblies of diamond mechano-synthesis are being used in building medical nanorobots. Models are being designed based on *Escherichia coli (E coli)* wherein the model uses a flagellum like body for propulsion and electromagnetic fields control the motion of the biological integrated device.

35.1.1 Biorobotics

Biorobotics (Fig. 35.1) is an interdisciplinary science to develop new technologies to create machines that imitate biological systems. It is a science developed by integrating process of biology and mechanical systems encompassing concepts from biomedical engineering, cybernetics, and robotics. Biorobots supersede the

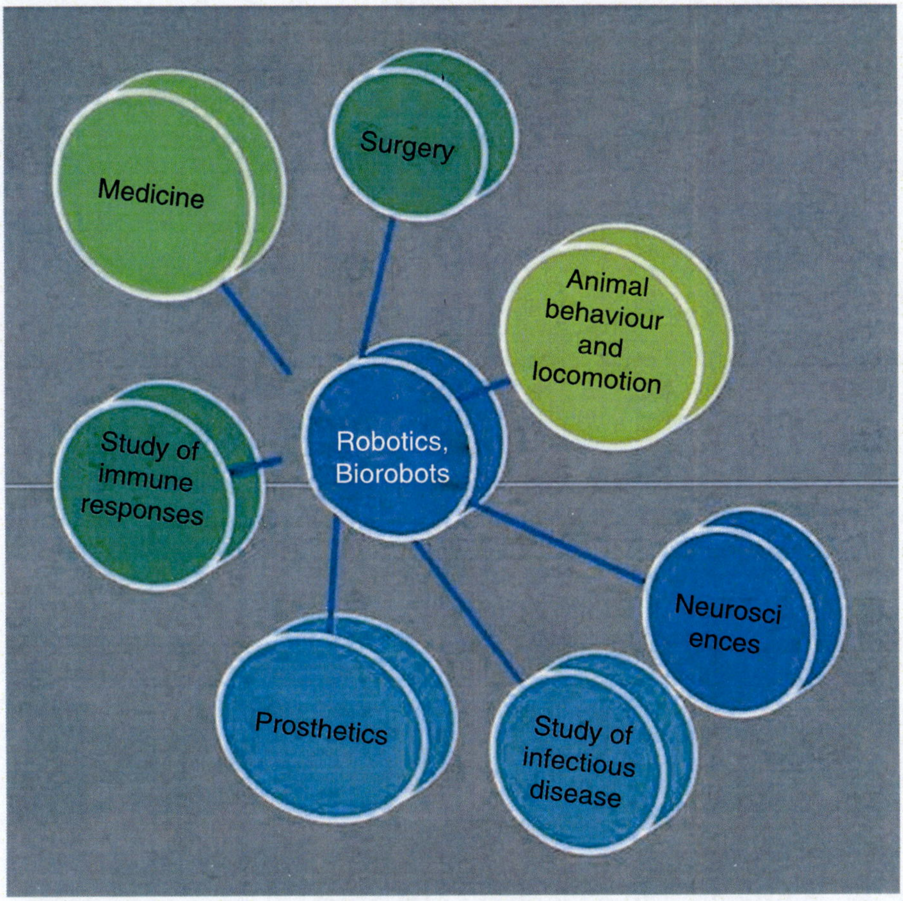

Fig. 35.1 Applications of robotics in biological sciences

traditional forms of robots in possessing greater mobility, flexibility, and sensory abilities.

Biorobotics technology based devices with highly advanced prosthetics have been applied in diverse applications including neural engineering, detection and monitoring of disease, monitoring of a patients recovery from an injury or stroke, etc.

There are several forms and applications of biorobotics technologies. It has been applied to the design and development of biomimetics and BioMEMS, or biomedical microelectromechanical systems, with applications in biomedical devices, like artificial skin designed to sense and detect pressure on contact, which finds importance in the restoration of touch sensation in diabetic patients with peripheral neuropathy by mimicking the sensations felt by finger pads and feet.

Contact sensors are being designed for early diagnosis of autism.

Biorobotics based strategies are being applied in training of surgeons and dentists using virtual environments and finds applications towards offering assistance to surgeons and medical practitioners while performing real life surgeries.

35.1.2 Biomechanics

Biomechanics includes application of engineering mechanics to biology, physiology, pathophysiology, medical diagnostics, and treatment. It has been applied in understanding mechanical influences on bone and joint health, mechanics of growth and form and in the design and manufacture of medical implants. Biomechanical models designs of soft organs and hard tissues enable study of their physiology and complex mechanical behavior.

Biomechanics research has enabled study of growth and development in understanding regulation of growth, tissue formation, organism morphogenesis, tissue regeneration and has been applied in study of injury, post-injury survival, and repair of central nervous system (CNS) in patients suffering from head-injury, pediatric head trauma, and spinal cord injury (sci).

35.1.3 Molecular Robots

Molecular robots are the ultimate biorobots, created out of the programmable biomaterials such DNA. They can also simple biomolecules capable of getting manipulated for a particular process. These biomolecules can help industrial processes as nanoscale robots.

35.2 Biorobots and Biological Application

Biorobots have been designed to study animal locomotion and adaptive behavior, with potential applications in the domain of hydrodynamics, biomechanics, prosthetics, movement, and neuroscience (Karakasiliotis et al. 2016).

35.2.1 Application in Biology

We highlight in this chapter the different applications of biorobots in diverse domains of biology. It is known that biomolecules including DNA, RNA, and proteins reveal the property of self-assembly, showing responses to the changes in the environment without any help from any external energy source. Such biological molecules like DNA can be programmed to make molecular robots as nanoscale devices and systems with industrial applications. Molecular robotics has been applied to manipulate DNA, RNA, exploiting their property of self-assembly as building block for molecular sensors, structures, computers, and actuators and to create new nanoscale structures and devices with diverse functions.

The self-assembly nature of DNA has been exploited into formation of DNA bricks with specific shape and create novel devices for electronics, photonics, and nanoscale machines.

Molecular robotics has enabled synthesis of new DNA sequences from a mixture of short fragments in vitro by primer exchange reactions (PER) which together with the procedure of autocycling proximity recording (APR) has better applications in DNA sequencing technology. Toehold switches involve RNA-based hairpins which are synthetic circuits that can mimic the AND,OR, and NOT system and can be designed to generate a synthetic toxin to target spreading of an infection or spreading of cancer. Super-resolution microscopy can be done by application of DNA-PAINT (https://wyss.harvard.edu/news/molecular-robotics-at-the-wyss-institute/) (Fig. 35.1).

35.2.2 Application in Detection Systems and in Detection of Infectious Disorders

Biorobots have been applied to development of a sensitive, automated polymerase chain reaction (PCR) in molecular detection of *Toxoplasma gondii (T. gondii)* infection. It has been applied in diagnosis of prenatal and neonatal congenital toxoplasmosis (CT) from Toxoplasma DNA isolated from amniotic fluid (AF, Yera et al. 2018) by automated methods.

The major challenges in design of assay systems for detection of viral infections lie in increased sensitivity and specificity with minimal or negligible effects of cross contamination.

Biorobots have been applied in automated sample preparation systems for routine diagnosis for virus mediated infectious diseases like human immunodeficiency virus (HIV-1), hepatitis C virus (HCV), hepatitis B virus (HBV) with greater precision, sensitivity, and minimal effects of cross contamination (Schneider et al. 2009). Biorobots have been used in the automated nucleic acid (NA) extraction systems from nasopharyngeal aspirates of patients suffering from infection by respiratory viruses and detection of their RNA (ribonucleic acid) and DNA (deoxyribonucleic acid) (Chan et al. 2008). Hepatitis A virus (HAV) infection is the leading cause of acute viral hepatitis (Perelle et al. 2009) and its effective control lies in rapid detection of HAV in drinking water contaminated with feces.

Biorobots have been designed for effective viral RNA purification to be used in PCR assays for detection of disease (Perelle et al. 2009).

35.2.3 Study of Immune Responses to Infection

Gram-negative intracellular bacterium *Francisella tularensis* (*F tularensis)* is known to cause tularaemia. The infection is known to initiate T cell responses but the study of the entire cellular immune response remained largely unknown due to the lack of sufficient tools. However studies from genomics with solid-phase antigen delivery system together with T cell functional assay and application of biorobotics have enabled identification of T cell epitopes (Valentino et al. 2011) that contribute to the immune responses in the disease.

35.2.4 Brain and Neurological Function

Robots find importance in study of human brain functions (Chen et al. 2013). Brain–machine interface (BMI) enables communication of artificial signaling between brain and machine. Designed in rat animal model it has been instrumental in infering of neural activities and generating sensations and detecting navigation (Webb 2001).Rat-robots or Ratbot have been designed that has been applied to the study and function of neurons (Chen et al. 2013; Sun et al. 2011; Sun et al. 2012). Automatic navigation method designed in Ratbots, utilizing rat's habit of thigmotaxis and its reward-seeking behavior together with rat's intrinsic intelligence has enabled rat-bots to navigate in unknown environments avoiding obstacles (Gao et al. 2013).

Dragonfly hybrid biorobots act as a living sensor, in which visual information is performed in the brain of the dragonfly and recording and processing neural signals are done in software or hardware (Pham and Higgins 2014).

35.2.5 In the Study of Animal Locomotion

Movement in animals involves the coordinated interaction of nervous system, muscles and skeletal system together with the complex interplay of the environment which is extremely complex to understand.

The mechanism of locomotion in snakes that lack limbs and are subject to diverse conditions has been enriched by evolutionary forces and has been intriguing to the human mind. The mechanism underlying of the complex locomotion pattern of snakes including the scaffold based concertina movement with anchored head, tail pulled forward followed by the extension of the head with the tail part anchored, three-dimensional (3D) locomotion patterns with ground contact alteration in a spatiotemporal manner, like side-winding and sinus-lifting, through decentralized control mechanism, remains yet to be completely unknown. Application of mathematical models and two-dimensional (2D) and 3D snake-like robot, Tegotae-based decentralized control mechanism has found application in study of scaffold-based, concertina locomotion, 3D-locomotion observed in snakes (Kano and Ishiguro 2019). Biorobots has been applied to understand the movements in snakes including movements like lateral oscillation and cantilevering movements (Fu et al. 2020). Biohybrid robots with onboard microelectronics have been designed to induce swimming in live jellyfish with less external power per mass and are being applied as tools in monitoring of the ocean (Xu and Dabiri 2020). Neural networks have been used to study sprawling locomotion in salamanders, sensory feedback, and body-limb coordination (Suzuki et al. 2019). Pleurobot has been a designed to study motor skills in salamanders (Karakasiliotis et al. 2016).

Alternatively animals and their biology of movement have greatly influenced the design of mobile robots with diverse application. Robots have been designed with an animal like motion using a numerical simulation (Park and Cha 2019). A bioengineered robot with cockroach like features, termed as the robot-roach hybrid designed at the University of Connecticut, Mansfield, USA has been designed with prospective applications in national defense (Dutta 2019).

35.2.6 Other Major Biological Applications

Biosensors or biorobots could enable detection of cell motility in *Escherichia coli* (*E. coli)* based on signal gradient (Ravichandar et al. 2017). Cellulases are enzymes from microbial sources including bacteria and fungi that have found application in different industries including pharmaceutical, textile, food processing, paper recycling, and detergent synthesis. Due to their large-scale applications in the industry, they also are known as the industrial biorobots. Therefore, different ecological niches and thermophilic bacteria, *Bacillus licheniformis* (*B licheniformis*) strain TLW-3 have been explored for the isolation of cellulolytic microbes (Kiran et al. 2018) from which cellulases can be extracted for large-scale industrial application.

The Biorobot EZ1 has found application in extraction of DNA from blood, saliva, and semen which is effective over conventional methods and finds application in forensic sciences (Montpetit et al. 2005).

35.3 Molecular Robotics and its Application in Science and Technology

Don Ingber Wyss the Founding Director, and Judah Folkman Professor of Vascular Biology at HMS and the Vascular Biology Program at Boston Children's Hospital, Professor of Bioengineering at Harvard SEAS has stated highlighted the importance of molecules that can function at its own, and trouble with robots that lack properties of self-assembly, autonomous movement, responses to the environment and function dependent on motors and external power supply (https://wyss.harvard.edu/news/molecular-robotics-at-the-wyss-institute/). Programmable biological molecules like DNA have wide potential for creating nanoscale devices and systems that could be transformational in a large number of industries and fields.

In September 2017, Nature Magazine (Kassem et al. 2017) reported the design of world's first "*molecular robot*" with advanced properties including building other molecules by Scientists at The University of Manchester. Biorobots with size in the range of millions of a millimeter (mm) are being applied in the design to build molecular messenger, designed to carry out chemical reactions in a controlled and programmed environment when placed in solutions. This feature has enabled their prospective applications in medicine, manufacturing, building molecular factories, and assembly lines.

Professor David Leigh, who led the research at University's School of Chemistry, while explaining about the robots designed by their laboratory, explained that while atoms are the basic building units of molecules the robots designed by their laboratory are similar to that of one like Lego bricks that can respond to a series of simple commands that are programmed with chemical inputs by a scientist. He has explained the functional mechanism of designed molecular robots that can function at the molecular level with diverse potential applications as similar to those observed in a car assembly line.

The design of a microscale machinery offers advantages of drastic reduction on the demands for materials, accelerated and improved features in drug discovery, and reduction in power requirements with significantly increased miniaturization of other products which offers its extremely varied and exciting potential applications.

Eminent scientist Prof Leigh reflected that molecular robotics involves design of miniature form and smallest form of machinery and he anticipated that by next one or two decades robotics may be involved in making of molecules and their assembly (University of Manchester 2017).

35.4 Machine Learning and Biorobots

Machine learning including tools of reinforcement learning (RL) reward generating algorithms together with the knowledge from the learning intelligence of animals, spatial recognition skills in rats have been applied to improvise biorobots in study of animals like rats and their biology during navigation and exploration of unknown areas (Zhang et al. 2013). A real-time video tracking algorithm has been used to study trends in motion, the location, and spread in rat robots by extracting their motion states in complex environment using optical flow, which has been revealed to be more accurate over conventional methods (Zhang et al. 2012). Research in this domain is being carried out in the understanding of the underlying mechanism involved in the movement of snakes by approaches of geometric mechanics and machine learning approaches of Bayesian Optimization and Pareto optimal solutions (http://biorobotics.ri.cmu.edu/projects/modsnake/; Saha et al. 2020). Berlin Robotics (http://berlinbiorobotics.blog) in their "Bees Book" project highlighted barcode like markers on honey bees using convolutional neural network algorithm (http://berlinbiorobotics.blog/2017/04/26/ben-and-leon-iclr-2017/). They used RenderGAN (https://arxiv.org/pdf/1611.01331.pdf) method which learned to generate realistic marker images, underlying structure of the barcode design and angles of bend, etc.

35.5 Discussions

Biorobotics is an interdisciplinary science and is at its infancy but with wide and prospective applications in diverse domains of biology in future.

The major prospective applications lie in exploiting the functions of macromolecules involved in biological processes, signaling cascades, systems, regulatory network to decipher the complexity and diversity in biological systems.

Some of the major applications include chemotactic and biosensors for biomedical applications, medical nanorobot for biohazard defense, designed with a set of integrated circuit block as an ASIC or application-specific integrated circuit.

The nanorobot comprised of carbon-metal nano composites, attached to artificial glycocalyx surface, is being used to minimize fibrinogen and other blood proteins adsorption or bioactivity, with sufficient biocompatibility to avoid immune reaction.

Biorobotics is being used in the design of chemotactic sensors with binding sites with different affinity for different molecule that can also detect obstacles in its environment. They are being applied in biomedical and clinical field to detect protein overexpression for continuous real-time monitoring in prognosis of disease (Cavalcanti et al. 2008). Biosensors can also detect chemicals and different cell types.

Nanorobots have enabled precise detection of α-NAGA or α-N-acetylgalactosaminidase in initial stages of influenza infection. They have been applied to design new treatments and holds promise in several biomedical and clinical applications as in surgery and medical procedures like catheterization, microtechnology in cardiac procedures, probable and future application in chemotherapy to combat cancer by precise dosage administration of chemicals, of drugs and molecules, delivery of anti-HIV drugs, monitoring diabetes and controlling glucose levels, dissolving kidney stones, locate atherosclerotic lesions in stenosed blood vessels, and coronary circulation, and to treat them either mechanical, chemical or pharmacological means (Cavalcanti et al. 2008; Bhardwaj et al. 2014).

The clinical domain of nanodentistry has been witnessing a large and diverse application of nanorobots inducing performance of oral analgesia, straightening of irregular teeth, tooth repair and replacement. Use of nano-structured composite material sapphire in replacing the upper enamel layer which increases tooth durability, hardness and appearance, use of nanofillers of alumino-silicate powder increases tooth strength are few of the clinical applications in nanodentistry. They have also promising applications in the domain of brain aneurysm with applications in tracking vessel endothelial injury before subarachnoid hemorrhage (Bhardwaj et al. 2014).

They find importance in cancer detection and treatment. Nanorobots with embedded chemical biosensors are being applied in detection of tumor cells in early stages of development. This application is most important in solving the major problems associated with cancer biology that lies with early detection of cancer. Recent applications of nanobots in the domain of cancer biology have revealed their importance in applications involving finding intensity of cell adhesion molecule E-cadherin in cancerous cells as loss of the E-cadherin has been associated with development and progression of epithelial cancers or carcinomas. Medical nanorobots find application in gene therapy, and treatment of genetic diseases, chromosomal replacement therapy (Kutty et al. 2020).

However, there are several factors associated with their large-scale application in the clinical medicine industry, including involvement of hugh cost, complicated in design and implementation, for use in the body, concerns on biocompatibility, maintenance of privacy, electrical fields created can affect the normal biology and calls in for more research in this domain prior to their large-scale application. Thus research in the area of biorobotics would focus on improvisation of tools and techniques and applications in making them user-friendly, with availability at low cost, endowed with properties for accurate detection when designed for disease detection with accuracy, short time of analysis, flexibility, robustness and closer resemblance and adaptation to the biological system.

The other major areas of application lie in study of movement, behavior, actuation, feedback regulations, dynamics, sensing, regulation and control of biological, processes. Robotics have been applied to answer intriguing question in the domain of ecology related to the understanding of organism dynamics with the habitat and niche. Survival of animals in aerial, terrestrial, and water habitats, migration, the sensory and neuromuscular systems involved in locomotion and movement of animals through different zones of air, land, and water are all important domains that are a major focus of bio-robotics.

Robots can be designed inspired by real life biological processes including adaptation, learning, and involvement of neurological and brain functions. One of the major area of research in this domain remains in design of completely organic functional robots. Recently functional biohybrid robots are being attempted to be designed. One such attempt has been made by linking a muscle tissue from the mollusk sea slug with swimming potential to a 3D-printed frame (https://phys.org/news/2017-12-biorobotic-future.html). This finds application in study of underwater environment.

Applications of advanced robust sensor based technology, artificial intelligence (AI), neuroscience, robotics, modeling and simulation are the key for design of improved biorobotic application in biology to solve the unsolved mysteries in life.

References

Bhardwaj A, Bhardwaj A, Misuriya A, Maroli S, Manjula S, Singh AK (2014) Nanotechnology in dentistry: present and future. J Int Oral Health 6(1):121–126

Cavalcanti A, Shirinzadeh B, Zhang M, Kretly LC (2008) Nanorobot hardware architecture for medical Defense. Sensors 8:2932–2958

Chan KH, Yam WC, Pang CM, Chan KM, Lam SL, Lo KF, Poon LLM, Peiris JSM (2008) Comparison of the NucliSenseasyMAG and Qiagen BioRobot9604 nucleic acid extraction systems for detection of RNA and DNA respiratory viruses in nasopharyngeal aspirate samples comparative study. J Clin Microbiol 46(7):2195–2199

Chen S, Qu Y, Guo S, Shi Z, Xu K, Zheng X (2013) Encode the "STOP" command by photo-stimulation for precise control of rat-robot. Conf Proc IEEE Eng Med Biol Soc 2013:2172–2175

Dutta A (2019) Cyborg insects could someday save your life. IEEE Pulse 10(3):24–25. https://doi.org/10.1109/MPULS.2019.2911818

Fu Q, Gart SW, Mitchel TW, Kim JS, Chirikjian GS, Li C (2020 Mar) Lateral oscillation and body compliance help snakes and snake robots stably traverse large, smooth obstacles. Integr Comp Biol 26:icaa013. https://doi.org/10.1093/icb/icaa013

Gao L, Sun C, Zhang C, Zheng N, Chen W, Zheng X (2013) Ratbot automatic navigation by electrical reward stimulation based on distance measurement in unknown environments. Conf Proc IEEE Eng Med Biol Soc 2013:5315–5318

Kano T, Ishiguro A (2019) Decoding decentralized control mechanism underlying adaptive and versatile locomotion of snakes. Integr Comp Biol 60(1):232–247

Karakasiliotis K, Thandiackal R, Melo K, Horvat T, Mahabadi NK, Tsitkov S, Cabelguen JM, Ijspeert AJ (2016) From cineradiography to biorobots: an approach for designing robots to emulate and study animal locomotion. J R Soc Interface 13(119):20151089

Kassem S, Lee ATL, Leigh DA, Marcos V, Palmer LI, Pisano S (2017) Stereodivergent synthesis with a programmable molecular machine. Nature 549(7672):374

Kiran T, Asad W, Ajaz M, Hanif M, Rasool SA (2018) Industrially Relevant Cellulase Production by Indigenous Thermophilic Bacillus Licheniformis TLW-3 Strain: isolation-molecular identification and enzyme yield optimization. Pak J Pharm Sci 31(6):2333–2340

Kutty CC, Patil AB, Asha Spandana KM, Gowda DM, Preeti S (2020) Nanorobots in cancer therapy. Int J Res Pharmaceut Sci 11(3):3626–3636. https://doi.org/10.26452/ijrps.v11i3.2523

Montpetit SA, Fitch IT, O'Donnell PT (2005) A simple automated instrument for DNA extraction in forensic casework. J Forensic Sci 50(3):555–563

Park T, Cha Y (2019) Soft mobile robot inspired by animal-like running motion. Sci Rep 9(1): 14700. https://doi.org/10.1038/s41598-019-51308-4

Perelle S, Cavellini L, Blaise-Boisseau C, Hennechart-Collette S, Merle C, Patrick Fach G (2009) Use of a robotic RNA purification protocol based on the NucliSenseasyMAG for Real-Time RT-PCR Detection of Hepatitis A Virus in Bottled Water comparative study. J Virol Methods 157(1):80-3Apr

Pham TT, Higgins CM (2014) A visual motion detecting module for dragonfly-controlled robots. In: EMBC 2014 36th annual international conference of the IEEE engineering in medicine and biology society. IEEE, Piscataway

Ravichandar JD, Bower AG, Julius AA, Collins CH (2017) Transcriptional control of motility enables directional movement of escherichia coli in a signal gradient. Sci Rep 7(1):8959

Saha D, Mehta D, Atlan E et al (2020) Explosive sensing with insect-based biorobots. Biosens Bioelectron X. https://doi.org/10.1016/j.biosx.2020.100050. bioRxiv 2020.02.10.940866; doi: https://doi.org/10.1101/2020.02.10.940866

Schneider GJ, Kuper KG, Abravaya K, Mullen CR, Schmidt M, Bunse-Grassmann A, Sprenger-Haussels M (2009) Performance evaluation of the QIAGEN EZ1 DSP virus kit with abbott real-time HIV-1, HBV and HCV assays. J Clin Virol 44(4):292–296

Sun C, Zhang X, Zheng N, Chen W, Zheng X (2012) Bio-robots automatic navigation with electrical reward stimulation. Conf Proc IEEE Eng Med Biol Soc 2012:348–351

Sun C, Zheng N, Zhang X, Chen W, Zheng X (2011) An automatic control model for rat-robot. Conf Proc IEEE Eng Med Biol Soc 2011:7413–7416

Suzuki S, Kano T, Ijspeert AJ, Ishiguro A (2019) Decentralized control with cross-coupled sensory feedback between body and limbs in sprawling locomotion. Bioinspir Biomim 14(6):066010. https://doi.org/10.1088/1748-3190/ab3ef6

University of Manchester. "World's first 'molecular robot' capable of building molecules". ScienceDaily. ScienceDaily, 20 September 2017. www.sciencedaily.com/releases/2017/09/170920131744.htm

Valentino MD, Maben ZJ, Hensley LL, Woolard MD, Kawula TH, Frelinger JA, Frelinger JG (2011) Identification of T-cell epitopes in Francisella Tularensis using an ordered protein array of serological targets. Immunology 132(3):348–360

Webb B (2001) Robots can be (good) models. Behav Brain Sci 24(6):1081–1087

Xu NW, Dabiri JO (2020) Low-power microelectronics embedded in live jellyfish enhance propulsion. Sci Adv 6(5):eaaz3194. https://doi.org/10.1126/sciadv.aaz3194

Yera H, Ménégaut L, Brenier-Pinchart M-P, Touafek F, Bastien P, Dalle F (2018) Molecular Biology study group of the French National Reference Centre for Toxoplasmosis. Evaluation of Five Automated and One Manual Method for Toxoplasma and Human DNA Extraction From Artificially Spiked Amniotic Fluid. Clin Microbiol Infect 24(10):1100.e7–1100.e11

Zhang C, Sun C, Gao L, Zheng N, Chen W, Zheng X (2013) Bio-robots automatic navigation with graded electric reward stimulation based on reinforcement learning. Conf Proc IEEE Eng Med Biol Soc 2013:6901–6904

Zhang X, Sun C, Zheng N, Chen W, Zheng X (2012) Motion states extraction with optical flow for rat-robot automatic navigation. Conf Proc IEEE Eng Med Biol Soc 2012:976–979

The Challenges to Application of Machine Learning in Biological Sciences

36

Abstract

Despite applications of machine learning in life sciences and its increasing number of applications in medical and different domains of biological sciences, machine learning itself suffers from some limitations at this current time. We highlight in this chapter certain areas that currently pose as limitations towards far more varied and large-scale applications of the machine learning in the domain of biology. At the same time these aspects contribute to the scope of further research in this interdisciplinary domain of science.

Keywords

Machine learning · Challenges · Biological sciences

Abbreviations

ANN	Artificial neural networks
DL	Deep learning
GPU	Graphics processing unit
NLP	Natural language processing
PCA	Principal component analysis
SGD	Stochastic gradient descent

S. Ghosh, R. Dasgupta, *Machine Learning in Biological Sciences*,
https://doi.org/10.1007/978-981-16-8881-2_36

36.1 Introduction

Machine learning algorithm has been a part of the advanced biology (Fig. 36.1) since the evolution of the subject of computational biology and is finding applications in diverse domains of biology. While machine learning is very successful in analyzing complex disease symptoms such as breast cancer, glaucoma, and brain tumor, data analysis often suffers from challenges of false-positive and false-negative results, due to overfitting of data, ignoring significant outliers and wrong weight distribution, and scoring in artificial neural network (ANN) applications.

Supervised learning techniques require accurate data labeling and feature identifications. In complex biological phenomena labeling large set of parameters and corresponding data remains a challenge. Wrong labeling may lead to wrong mapping and therefore remains a major challenge in data analysis.

Analysis of magnetic resonance imaging (MRI) data and images finds importance in identification of brain disease such as brain tumor, signatures of memory loss, accidental damages, and complex situations like frontal hygroma. The success of a computer vision or applications of image based diagnostic or prognostic methods depends on the quality of the image and accurate analysis and remains a major challenge. Moreover the storage requirement and the graphics processing unit or GPU processor requirement further adds to the challenge in accurate data storage and analysis. The problem of memory and processor has been partly resolved by the availability of low cost cloud infrastructure such as Google Cloud, Microsoft Azure, and Amazon Web service (AWS).

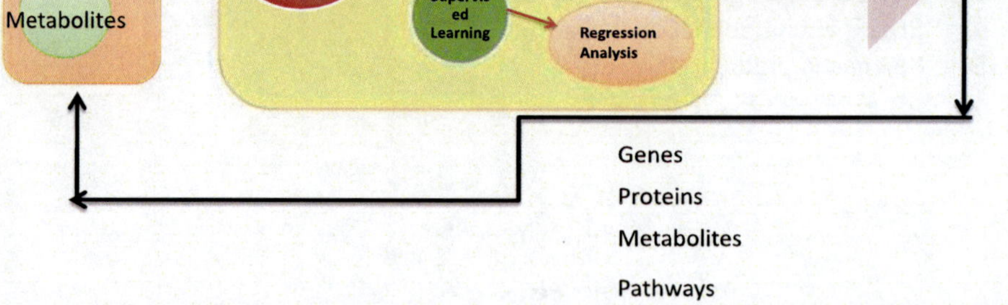

Fig. 36.1 Machine Learning in Biological data analysis application Figure and legend adapted with permission from (Camacho et al. 2018)

In this chapter we discuss the challenges in this domain under the broad headings of (1) the overall challenges faced by machine learning, (2) challenges of machine learning in applications in biology.

36.2 Machine Learning: Challenges in Recent Times

Despite the fact that machine learning has several applications in biological sciences, it suffers from certain challenges.

36.2.1 Search for Better Memory Networks

Memory networks are a relatively new class of models designed to alleviate the problem of learning long-term dependencies in sequential data, by providing an explicit memory representation for each token in the sequence. The original concept of memory networks has been discussed by Sukhabaatar et al. (2015) and Weston et al. (2014).

The input data on memory networks consists of a history of list of inputs so far, a query or current input and, during supervised training, a label or what the model should output based on the given history and query. Each element in the history is embedded as a fixed-size vector and so is the query (commonly, using the same embedding matrix). Sukhabataar et al. (2015) and Weston et al. (2014) implemented an end to end memory network, introduced a neural network with a recurrent attention model over a possibly large external memory.

36.2.2 Improved Natural Language Processing (NLP)

The current status of the NLP includes training the computers to represent languages and simulate reasoning (Fig. 36.2) which needs to be further improvised in its performance. Refined methods are required for NLP and understanding particularly more so for the deep networks (Ching et al. 2018).

36.2.3 Involvement of Attention to Integrate Data

Although attention is used to integrate sets of features by the human visual systems, machine learning largely works focused on one at a time input stimuli, and then their subsequent integration with end results. Thus machine learning lacks the exact nature of working by a human visual system in building neural networks and needs to be fine tuned.

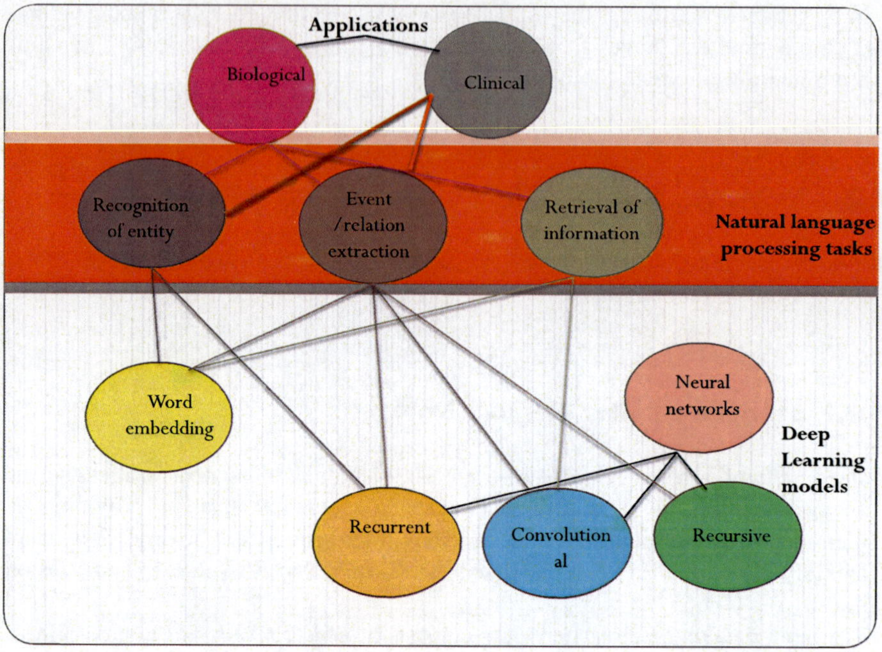

Fig. 36.2 Deep learning, models based on NLP. Image adapted from (Ching et al. 2018) with permission

36.2.4 Understand Deep Nets Training

Deep neural network works on understanding the mapping of input and output. Depending of the level of errors by the model on the training data, the weight factors are assigned. Updates are made to continually reduce this error until either a good enough model is found or the learning process gets stuck and stops. It is a very complex process and one of the most challenging part of the deep network.

Neural network models can be thought to learn by navigating a non-convex error surface.

A model with a specific set of weights can be evaluated on the training dataset and the average error over all training datasets can be thought of as the error of the model. A change to the model weights will result in a change to the model error. Therefore, a set of weights are required that may result in a model with a small error.

This involves repeating the steps of evaluating the model and updating the model parameters in order to step down the error surface. This process is repeated until a set of parameters is found that is good enough or the search process gets stuck.

This is a search or an optimization process and we refer to optimization algorithms that operate in this way as gradient optimization algorithms, as they naively follow along the error gradient. They are computationally expensive, slow, and their empirical behavior means that using them in practice is more art than science. The algorithm that is most commonly used to navigate the error surface is called stochastic gradient descent (SGD). The use of nonlinear activation functions in the neural network refers to optimization of problem to solve in order to find model parameters. Sometimes analyzing data faces challenges of misleading sets of parameters and optimization suffers challenges of local optima.

36.2.5 Approaches of One-Shot Learning

Traditional gradient-based networks requires enormous large data to learn and train and therefore lacks one-shot learning related to computer vision which needs to be improvised.

36.2.6 Deep Reinforcement Learning and Control of Robots

Further research in the application of deep reinforcement learning in control of robots is an area which needs to be researched for.

36.2.7 Semantic Segmentation

Semantic segmentation of images involving classification of each pixel of the image to a definite corresponding label. This finds major important application in medical imaging like in the understanding and accurate study of an organ or lesion from images. This remains a challenge in machine learning.

36.2.8 Video Training Data

Machine learning systems have to be designed to get the training from video data, with better skills of listening and observing rather that the current dependence of static images as inputs.

36.2.9 Detection of Object

Object detection remains a major challenge in machine learning which need to be improvised.

36.2.10 Application of AI to Big Data

Application of AI to big data would enable better understanding of complicated problems in life processes.

36.3 Challenges Faced by Machine Learning in Biology

In the last decade considerable progress has been made in the application of deep learning (DL) algorithms and models capable of combining raw inputs into layers of intermediate features and has application in biology and medical sciences, involving patients, their categorization, processes, treatment, study of disease, processes, and the genomics (Ching et al. 2018; Xiong et al. 2016; Li et al. 2017; Adiyaman and McGuffin 2019; Xu and Jackson 2019). Considerable progress has been made in medical imaging of disorders including diabetic retinopathy, macular edema in diabetic individuals, tuberculosis, skin disorders with the use of accurate classifiers. It has been possible to screen targets from a large number of probable drug candidates and small molecules.

Machine learning approaches have been used to analyze labeled or unlabeled data by unsupervised methods like principal component analysis (PCA) or hierarchical clustering, or by supervised learning method and generate a predictive model for either regression or classification based on the data and based on input labels and thereafter validation of prediction. New data generated or collected can be used to refine the existing model, with better prediction, and study the biological process with greater accuracy.

The application of machine learning to biological data suffers from the disadvantage of immensely complex biological gene, protein regulatory networks, challenges from handling of huge datasets of immensely complex data and requires tailor made algorithm keeping in mind the complexity of data in each separate biological problem. Moreover the diversity of biological process calls in for tailor made algorithms to analyze each data and tackle each unique biological problem. Studies at large are needed to tackle the unique problems in biology in case to case basis and accordingly the algorithms need to be fine tuned to suit the problems of specific biological systems.

Another major limitation in biology is the availability of properly labeled data for training problems and the assessment of false negatives.

Although improvement in speech recognition has been observed with error rates falling from above 20% in 2009 to 5.1% reported by IBM in 2018–2019. Challenges encompass in achieving near human performance (Weston and Chopra 2014; Ching et al. 2018).

Although considerable progress has been made in prediction of protein structure, challenges remain in generating improved 3D structures especially for membrane protein (Xiong et al. 2016).

Challenges arise in attaining perfect accuracy in prediction of gene regulation in disease due to errors in sequence variant calling arising from genome-wide association studies (GWAS).

While attempts are being made to apply deep learning models towards achieving human like performance in biomedicine, algorithms are prone to commit some mistakes not made by human, such as misclassification of adversarial examples and therefore this area needs further research. Challenges lies in proper interpretation of models, understanding of complex biological processes, integration of heterogeneity in different data, dimensionality, and problems related to scarcity, multi-co linearity, and overfitting in omics datasets (Li et al. 2017).

36.4 Discussions

Despite advances in machine learning methods with applications in diverse domains of biology, leading to increased deciphering of biological phenomena, challenges in machine learning in study of biology still suffer from complexity in population size and heterogeneity in samples or complexity in their evolutionary relationships.

Biological datasets should be tackled with removal of confounders or that affects both the dependent and independent variable and that cannot be studied by either associations or correlations. If these parameters are not properly analyzed and included in a biological dataset, it can lead to overfit of models, causing false-positive results.

Associated problems can arise from conditions that are specific for a population, which if left not properly analyzed in datasets can affect model performances.

Although convolutional neural network (CNN) and recurrent neural network (RNN), deep neural network (DNN), random forest (RF), support vector machine (SVM), have been applied to different specific biological problems yet a single machine learning tools to address different biological questions remains a matter of further research.

It remains concepts from computer science, engineering, statistics, physics, and biology to enrich machine learning tools in the years to come and remain the scope of further research.

References

Adiyaman R, McGuffin LJ (2019) Methods for the refinement of protein structure 3D models. Int J Mol Sci 20(9):2301

Camacho DM, Collins KM, Powers RK, Costello JC, Collins JJ (2018) Next-generation machine learning for biological networks. Cell 173(7):1581–1592

Ching T, Himmelstein DS, Beaulieu-Jones BK, Kalinin AA, Do BT, Way GP, Ferrero E, Agapow PM, Zietz M, Hoffman MM, Xie W, Rosen GL, Lengerich BJ, Israeli J, Lanchantin J, Woloszynek S, Carpenter AE, Shrikumar A, Xu J, Cofer EM, Lavender CA, Turaga SC, Alexandari AM, Lu Z, Harris DJ, DeCaprio D, Qi Y, Kundaje A, Peng Y, Wiley LK, Segler MHS, Boca SM, Swamidass SJ, Huang A, Gitter A, Greene CS (2018) Opportunities and obstacles for deep learning in biology and medicine. J R Soc Interface 15(141):20170387. https://doi.org/10.1098/rsif.2017.0387

Li Z, Wang S, Yu Y, Xu J (2017) Predicting membrane protein contacts from non-membrane proteins by deep transfer learning. arXiv: https://arxiv.org/abs/1704.07207v1

Sukhabaatar et al. (2015) End to end memory networks. https://arxiv.org/abs/1503.08895

Weston J, Chopra S, Bordes A (2014) Memory networks. https://arxiv.org/abs/1410.3916

Xiong W, Droppo J, Huang X, Seide F, Seltzer M, Stolcke A, Yu D, Zweig G. (2016) Achieving human parity in conversational speech recognition. arXiv:1610.05256v2 [cs.CL]

Xu C, Jackson SA (2019) Machine learning and complex biological data. Genome Biol 20:76. https://doi.org/10.1186/s13059-019-1689-0

The Future of Machine Learning 37

Abstract

Machine learning has diverse applications in different domains of technology and more so in the different and diverse domains of science of biology and industry. We have discussed in the previous chapters the different applications of machine learning in biology. In this chapter we will discuss the promises and future scope of research areas in machine learning.

Keywords

Robots · Artificial intelligence · Computer vision · Quantum computing

37.1 Introduction

Machine learning (ML) enables inferring of models or relationships by learning from data and is finding major applications in process modeling, computer vision, signal processing, speech recognition and language understanding and processing and life and health sciences. In the domain of biology it has wide scale applications in the domain of plant biology, animal sciences, veterinary sciences, ecology, agriculture Sciences, crop science, cell biology, microbiology, toxicology, health and disease sciences. Its application is gradually increasing day by day. We discuss in this chapter an overview of the machine learning and its applications in future.

Artificial intelligence (AI) has promised application in industry, health, agriculture with promises of a market growth by 2025. Thus development and deployment of AI tools find importance in the tech industry and involvement of industry giants to boost business with Apple, Facebook, Google, Microsoft, and Amazon are the key leaders in this field.

We have discussed in the foregoing chapters the diverse artificial intelligence-based applications integrated Google products including Google Search Engines,

S. Ghosh, R. Dasgupta, *Machine Learning in Biological Sciences*,
https://doi.org/10.1007/978-981-16-8881-2_37

Google Maps, Google Photos, YouTube Safe Content, Gmail Smart Reply, Google Trend technology like TensorFlow and Cloud AutoML are helping developers and helping in research in the field of Biology and environmental sciences. In the preceding chapters, we have highlighted the applications of Google Trends in the domain of predicting epidemics and thus helping in the spread of mass awareness and education and helping information to reach the masses.

Google Brain, Google's RankBrain, Google's Cloud are other such interesting applications in the domain of science of health and biology.

Facebook, Amazon, Microsoft, Microsoft's Azure cloud has been applied to biological sciences with key applications in the domain of health sciences. Social sites like Facebook have a tremendous application in the study of psychology, predicting risks of suicides and understanding the mental health of an individual.

Automation of microscopes, imaging technologies, application of algorithms has enabled high-throughput cell biology studies, microbiology, disease biology and unraveling the immensely complex internal cellular machinery and its function in health and disease. Refining of image analysis by the application of image data processing pipeline is aiding in image-based diagnosis of disease.

Machine learning has helped in the study of animal behavior, its diversity and complexity, species–species interaction in the niche, detection of health and disease from observed behavior in pets and wild animals. It has been possible to study the animal reservoirs that carry infectious pathogens by machine learning techniques. Rapid identification of insects finds importance to understand the vectors of diseases and to understand biodiversity. Recently machine learning applications of feature extraction property of convolutional neural networks (CNNs) together with training sets of pictures of insects and feature transfer are being used in identification of insects as better and faster tools in taxonomy.

Precision animal agriculture is enabling better management, production, welfare, sustainability, health surveillance, early prediction of disease and thereby affords care and treatment based on information collected from animals and farms and monitoring of environmental factors to improve animal agriculture and therefore helping in the growth of the economy.

An important application of machine learning is in the domain of plant sciences, applied to improve plant disease phenotyping, detection of optimal growth conditions, maximize the concentration of secondary metabolites, detecting disease, pathogen attack, increase or decrease attraction by pollinators, growth of disease resistant varieties, sort genes for specialized metabolites in plants.

Nanorobotics is emerging field nanoelectromechanical functioning at molecular level with medical applications in diverse disorders including HIV, cancer, regeneration of lost tissue, disease diagnosis and monitoring disease progression and to monitor neuro-electric signals.

37.2 Machine Learning and the Future

Although many applications of the machine learning focused on the different diverse domain of biology, has been discussed in the preceding chapters, there remains many areas of biology where the applications are yet to be done and also there is need of improved, tailor made machine learning algorithms to problems and challenges in biology that are hitherto unresolved. Improved unsupervised algorithms are required to suit some intricate problems in biology remains one of the major focus in the future of Machine Learning.

Application of quantum computing using quantum sciences phenomena of entanglement and superposition in research and industry seems to be one of the major applications of machine learning, with ability of faster data processing, better analysis of a given dataset, increased performance to generate results is a step above the conventional and classical machine learning that is being researched now and finds importance in shaping future of machine learning. Microsoft and Google may be the pioneers in this field.

Machine learning algorithms in future may be focused with tailor made strategies fulfilling customer or industry or problem demands, which certainly remains the future in machine learning research. This has particular significance in biology, since the problems in diverse domains of biology are extremely unique and different from the others and therefore each problems needs specialized analysis and tackling features of machine learning tools.

Improved and development is necessitated on cognitive applications and applications with intelligent capabilities in apps using machine learning like visual recognition, speech detection and is a future scope of machine learning applications globally.

There is a need to create platforms for increased collaborative research and is one of the main focus of research and further development of applications of machine learning in the near future. This has already started to find its importance in plant biology research, health sciences, taxonomy, agriculture, predicting epidemics. More and more applications in biology are the future of machine learning in the coming years.

Robots are certainly one of the major focus of and future scope of machine learning applications in carrying out tasks with increased intelligence and efficiency. Advancements in the algorithms for language processing or natural language processing (NLP) model are helping to generate better answers for diverse queries, difficult to get by any other conventional method.

37.3 Discussion

With the advancement of application of computer vision which involves giving vision to a computer or machine and its application to recognize and analyze graphics, images and videos has been applied in the domain of diagnosis diabetic retinopathy (DR). This is proving to be better over and above the existing method of extensive eye examination in diagnosis and prognosis in diabetic retinopathy. This has an advantage to patients in interior, rural, and village areas who are poor, residing in areas with poor access and availability to health services and medical facilities, paucity of doctors. A machine learning based computer vision for the disease diagnosis thus can be extremely helpful.

In the healthcare sector, machine learning in future would be a great boon wherein it may be envisioned that practitioners together with machines will enable timely and accurate diagnosis of patient conditions and outcomes, with better disease care and management and monitoring and prognosis in the post-therapy or prognosis period to improve the quality of life of patients.

At this juncture, we hope our on-going discussions of the diverse applications of machine learning in the preceding chapters in the diverse domain of biology reflects the current day state of the art. Its future scope of development and application is full of promises and may inspire the readers to travel the path of future research in resolving the problems in diverse domains of biology through improved tools of machine learning.